ELECTRICAL CONTROL FOR MACHINES

Third Edition

ELECTRICAL CONTROL FOR MACHINES

Third Edition

KENNETH B. REXFORD

 Delmar Publishers Inc.®

NOTICE TO THE READER

COVER PHOTO: Courtesy of HPM Corporation, Mt. Gilead, OH

DEDICATION

I dedicate this book to my wife, Olive, for her patience and understanding during hours of research, writing, and checking.

DELMAR STAFF

Developmental Editor: Marjorie A. Bruce
Production Editor: Carol A. Micheli
Design Coordinator: Susan C. Mathews

COPYRIGHT © 1987
BY DELMAR PUBLISHERS INC.

10 9 8 7 6 5 4 3

Printed in the United States of America
Published simultaneously in Canada
by Nelson Canada
a Division of International Thomson Limited

Library of Congress Cataloging in Publication Data

Rexford, Kenneth B.
  Electrical control for machines.

  Includes index.
  1. Electric controllers. I. Title.
TK2851.R47 1987   629.8'043   87-8907
ISBN 0-8273-2792-7
ISBN 0-8273-2793-5 (instructor's guide)

TABLE OF CONTENTS

PREFACE

The third edition of ELECTRICAL CONTROL FOR MACHINES reflects the continuing technological evolution of the controls industry. The text addresses the diversity of control devices and applications, including the expanding use of solid-state controls and programmable controllers. At the same time, the text thoroughly covers the basic concepts of machine control to enable the learner to build technical competence upon a firm understanding of principles.

The field of electrical control of machines has grown in response to the increasing requirements of production. While much has been published on machine control, a thorough and practical approach to control principles and applications is more important than ever because of the increasing sophistication of the technology. This text is designed to provide the necessary framework for the electrician, maintenance technician, mechanical or fluid power systems designer, and sales personnel.

It is assumed that the reader has a basic knowledge of electrical theory. The text presents a practical approach to circuit design, maintenance and trouble-shooting. The material is general and can be applied to almost all machine control. One of the first problems for many learners is identifying the electrical components and their symbols. The early chapters of this text show the components and their symbols as the component is used in an elementary circuit.

Typical machine control systems are explained and illustrated, with the basic circuit shown for each application. An understanding of the electrical control circuits is further enhanced by the step-by-step description of the sequence of operations for each circuit.

Designing for easy maintenance and troubleshooting are presented with particular emphasis for the electrician, and electrical service and maintenance personnel. Hours of labor can be saved by the proper approach to a problem.

A discussion of codes and standards is presented as a guide for promoting safety and reducing maintenance and troubleshooting problems.

Beginning in the early 1960s, solid-state technology was introduced into machine control. Modern control systems are designed using both discrete and integrated circuit elements. The resulting control systems are particularly useful in adverse environments, for complex control situations, and where high-speed operation is required. Chapter 18 of this text has been revised to give up-to-date information on solid-state machine control.

In the late 1960s and early 1970s, programmable controllers were introduced in industry. The ability to program a given control system and make changes without requiring physical component or wiring changes has further improved the performance and reliability of machine control systems. Completely revised chapter 19 provides an introduction to the principles of programmable control and the application of programmable controllers.

FEATURES OF THE THIRD EDITION

- Content was updated throughout the text, with new photographs showing equipment and components currently used in industry
- Revised to cover the latest electrical codes, with an emphasis on NFPA 79-1985 *Electrical Standard for Industrial Machinery*
- Revised information on equipment protection devices, with added information on protection requirements of solid-state devices and programmable controllers
- Covers proximity switches, pressure transducers, linear position-displacement transducers
- Circuit diagrams use heavy lines to differentiate the conductors carrying load current at line voltage from the rest of the circuit
- Circuit operations are written in a step-by-step sequence to help learners visualize circuit actions
- Questions and circuit problems were added to the Achievement Reviews to reinforce the new concepts presented
- Revised and expanded chapter 19 on programmable controllers provides common industrial applications
- Stresses the use of ladder diagrams for control circuits to promote understanding of the operational sequence
- Troubleshooting chapter provides a logical approach to analyzing circuit problems

To accompany the text, a revised and expanded Student Manual provides more circuit problems for analysis and solution. The Instructor's Guide provides a pretest to evaluate the learner's level of understanding at the beginning of the course, answers to the text review questions, a posttest, and the answers to the questions and circuit problems in the Student Manual.

ABOUT THE AUTHOR

Kenneth B. Rexford received his professional degree from the College of Engineering at the University of Cincinnati. He was registered as a Professional Electrical Engineer in the state of Ohio.

Mr. Rexford was an electrical test engineer for the Dayton Power and Light Company, and was with HPM Corporation in Mount Gilead, Ohio, for thirty-four years as Chief Electrical Engineer and Chief of Electrical Development. He has served on several national electrical standards committees.

Mr. Rexford, who is retired from industry, now devotes his time to consulting, writing and teaching. He is a frequent guest speaker at schools and technical society meetings.

ACKNOWLEDGMENTS

The author wishes to thank two colleagues who reviewed the revised manuscript and offered their professional critiques.

Pete Giuliani
Franklin University
Columbus, OH

Dave Quinzon
Solid Controls Inc.
Minneapolis, MN

Appreciation is expressed to the following instructors who suggested revisions for the third edition and evaluated the final manuscript.

James Owens
Kellogg Community College
Battle Creek, MI 49016

John Webb
NCTI
Wausau, WI 54401

Duane Olson
Granite Falls Area Vocational-Technical School
Granite Falls, MN 56041

Don Arney
Indiana Vocational Technical College
Terre Haute, IN 47802

James B. Chapman
Florence-Darlington Technical College
Florence, SC 29501

Nick Johnson
Scioto County Joint Vocational School
Lucasville, OH 45648

Robert Morin
W.F. Kaynor Regional Vocational Technical School
Waterbury, CT 06708

Technical guidance and illustrations were provided by the following companies. Appreciation is expressed to them for their cooperation and assistance.

Allen-Bradley Company, A Subsidiary of Rockwell International, Milwaukee, WI 53204
Anderson Bolds Inc., Cleveland, OH 44120
Automatic Timing and Controls Co., King of Prussia, PA 19406
Barksdale Controls Division, Transamerica Delaval Inc., Los Angeles, CA 90058-0843
Barber-Colman Company, Loves Park, IL 61132-2940
Eagle Signal Controls, Division of Gulf & Western Manufacturing Co., Davenport, IA 52803
Eaton Corporation, Cutler Hammer Products, Columbus, OH 43221
Eaton Corporation, Cutler Hammer Products, Industrial Control and Power Distribution Division, Milwaukee, WI 53216
Edwin L. Wiegand Division, Emerson Electric Co., Pittsburgh, PA 15212
efector, inc., Exton, PA 19341
EMCORP, Columbus, OH 43215-2593
Fenwal Inc., Division of Kidde, Inc., Ashland, MA 01721
General Electric Company, Automatic Controls Operations, Bloomington, IL 61702-2913
Gould Inc., Circuit Protection Division, Newburyport, MA 01950
Gould Inc., Industrial Control Division, Bedford, OH 44146
Hoffman Engineering Company, Division of Federal Cartridge Corporation, Anoka, MN 55303

HPM Corporation, Mount Gilead, OH 43338-1095

Mercury Displacement Industries, Inc., Edwardsburg, MI 49112

MTS Systems Corporation, Temposonics Sensors Division, Plainview, NY 11803

Noral, Inc., Cleveland, OH 44122

Siemens Energy & Automation, Inc., Programmable Controls Division, Peabody, MA 01960

Solid Controls Inc., Minneapolis, MN 55435

Square D Company, Milwaukee, WI 53201

The Superior Electric Company, Bristol, CT 06010

Westinghouse Electric Corporation, Pittsburgh, PA 15222

FOREWORD

We need to teach the basic fundamentals of machine tool electrification. Those mainly affected are the thousands of small builders and users, and to some extent, the large manufacturer. Most selling organizations, both large and small, include many thousands of manufacturers' agents, distributors, sales engineers and maintenance personnel. These are the people charged with the responsibility for selling machines and keeping them operating. The success or failure of installations may depend on the ability of these people to maintain and troubleshoot equipment properly. The personnel involved need to understand electrical components and their symbols. With this knowledge, they are in a better position to read and understand elementary circuit diagrams.

Three areas where education is needed are in troubleshooting, maintenance and electrical standards.

1. Troubleshooting machine control circuits involves locating and properly identifying the nature and magnitude of a fault or error. This fault may be in the circuit design, physical wiring or in the components and equipment used. The time required and the technique or system used to locate and identify the error is important. Of like importance are the time and expense involved to put the machine back into normal operating condition.

2. Preventative maintenance would eliminate the need for most troubleshooting. Many machines are allowed to operate until they literally "fall apart."

3. A reasonable set of standards should be followed. If the intended result is the improvement of design and application to reduce down-time and promote safety, electrical standards can be extremely helpful. Where should the education start? The answer is at the beginning and keep it simple. Even a basic concept, such as the relation between a component and its symbol, can be of benefit to the user.

Kenneth B. Rexford

INTRODUCTION TO ELECTRICAL CONTROL — THE DEVELOPMENT OF CIRCUITS

To understand electrical control circuits, it is necessary to examine three basic steps.

The FIRST step is to know what work or function is to be performed.

For example, a simple problem may be to light a lamp. This can be done by completing a path for electrical energy from a source such as a battery to a lamp. For convenience, a switch is used to open or close the path. When the switch is opened, electrical energy is removed from the lamp and it is said to be deenergized. When the path of electrical energy is completed to the lamp, it is said to be energized and the lamp performs a function of illumination. See Figures 1 and 2.

In Figures 1 and 2, a battery is used as the source of electrical energy. These drawings are known as *pictorial* drawings as they attempt to show a picture of the components (battery, switch and lamp).

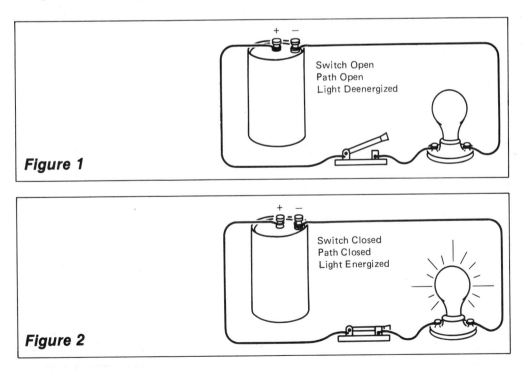

Figure 1

Switch Open
Path Open
Light Deenergized

Figure 2

Switch Closed
Path Closed
Light Energized

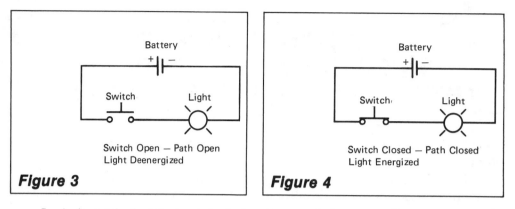

Figure 3

Figure 4

In industrial electrical control circuits, symbols are used to represent the components. Therefore, the first circuits using symbols for the components (battery, switch and lamp) will appear as shown in Figures 3 and 4. These diagrams are referred to as *schematic* or *elementary circuit diagrams.*

Figures 3 and 4 can now be redrawn in a slightly different form. The circuit performs exactly the same as the lamp is energized when the switch is closed. This type of drawing is called a *ladder* diagram. In all future circuit drawings, the ladder-type diagram will be used. In these drawings, the source of electrical energy will always be the two vertical lines (or sides of the ladder). See Figures 5 and 6.

Looking at Figures 5 and 6, an important symbol is introduced. The symbol used is "conductors connected" and is shown in Figure 7. Another similar symbol which is not used here, but appears many times in later diagrams, is "conductors not connected." This is shown in Figure 8.

The SECOND step is to know the operating conditions under which the starting, stopping and controlling of the process is to take place.

Practically all conditions fall into one or more general groups, as affected by:

    A. Position   C. Pressure
    B. Time   D. Temperature

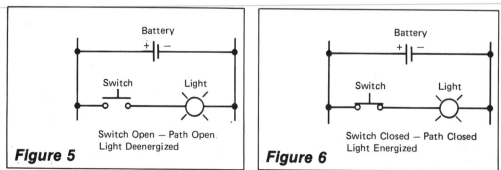

Figure 5

Figure 6

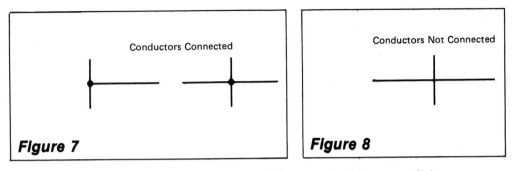

In the chapters on components that follow, each of these conditions can generally be associated with certain components.

In many cases, while the actual initiating of a cycle may be through the manually operated push-button switch, certain conditions must be met before the circuit can be closed. This could be one or any combination of the conditions previously listed.

The THIRD step in the development of a circuit is selecting, dividing and/or isolating.

There are many times that a circuit must be capable of operating under certain sets of conditions to produce the desired results.

For example, a circuit may be required to operate a machine under manual, semiautomatic (single cycle) or fully automatic (continuous cycle) operation.

After a decision is made on which of these types of operation is to be used, a selection is made. For reasons of safety to both the machine and operating personnel, the machine must operate in the selected manner.

Review the three basic steps in developing the elementary diagram.

1. Know what work or function is to be performed.
2. Know the conditions for starting, controlling and stopping the process.
3. Arrange for selecting, dividing or isolating as required.

As the student progresses through the study of components, the symbol for each component will be prominently displayed. It is very important that the symbol becomes closely associated with the component.

At the close of the section on components, all of the symbols used will be shown for review.

In understanding electrical control circuit diagrams, there are fundamental problems that should be recognized. These problems must be removed if progress is to be made.

Some of these problems are:

1. Starting with a circuit that is too large or too complicated.
2. Failing to carry a mental picture of the component through into the electrical circuit.
3. Failing to relate physical action to conditions that are required to perform a job over into electrical signals, through components.
4. Failing to understand that an electrical circuit must perform the correct functions and not perform those actions that will result in damaged components, danger to the operator or machine, or a faulty product.

One of the greatest problems in reading circuits is a clear understanding of a switch or contact condition. The condition must be properly presented in the elementary circuit diagram and the user must properly interpret its use in a process or on a machine. Elementary or schematic circuit diagrams will be shown in each of the component sections. As a new component is introduced, the symbol will be shown in the circuit. These circuits will show methods of obtaining specific actions through the use of electrical components. In many requirements, it may be possible to group some of these small part circuits into a complete circuit for the operation of a machine.

The important point here is to become acquainted with components and their use in the small circuits, and working with only one or a few components at a time.

CHAPTER 1

CONTROL TRANSFORMERS

OBJECTIVES

After studying this chapter, the student will be able to:

- Give two reasons for energizing machine control systems at 120 volts.

- Explain how to obtain 120 volts from a higher line voltage through the use of a transformer.

- Define *turns ratio* in a transformer.

- Identify the symbol for a dual-primary, single-secondary control transformer.

- Explain what causes temperature rise in a transformer.

- Define *regulation* in a transformer.

- Draw a connection diagram for a dual-primary, single-secondary control transformer to a higher voltage line and to a 120-volt control circuit.

- Calculate the size of a transformer for a given load.

1.1 CONTROL TRANSFORMERS

In the electrical control circuits shown in Figures 1 through 6, in the development of circuits a battery is the source of electrical energy. A battery supplies a form of electrical energy known as *direct current* (dc). Most industrial electrical controls use a form of electrical energy called *alternating current* (ac). Ac power is supplied to industry as three phase, 240 volts (V) or 480 V, 60 hertz (Hz). The use of 480 V is most prominent. On circuit diagrams, the three-phase

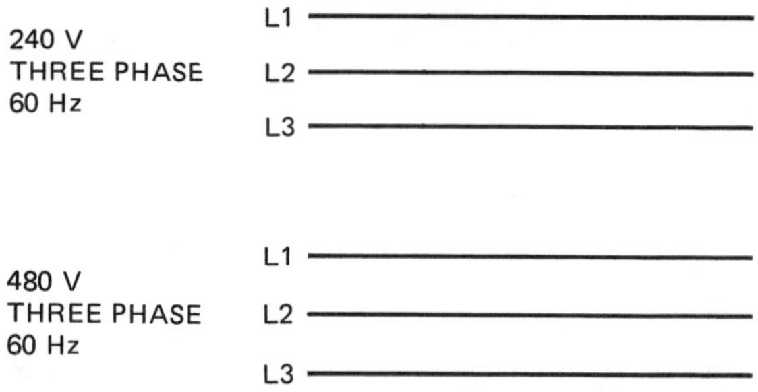

240 V
THREE PHASE
60 Hz

L1
L2
L3

480 V
THREE PHASE
60 Hz

L1
L2
L3

Figure 1-1

power source is normally shown as three parallel horizontal lines marked L1, L2, and L3, Figure 1-1.

Most electrical control systems on machines are energized at 120 V ac. There are at least two good reasons for this: safety, and the use of standard designed components.

Electrical energy at this lower voltage can be obtained by using a *control transformer.* The control transformer is single phase, requiring a connection to any two of the three power lines. The transformer used in industrial machine control consists of at least two separate coils wound on a laminated steel core. The line voltage is connected to one coil, called the *primary.* The control load is connected to the other coil, called the *secondary.* Where the voltage is reduced from the primary to the secondary, the transformer is called a *step-down trans- former.* In a *step-up transformer,* the voltage is increased from the primary to the secondary.

The simplest arrangement uses only two coils. One coil is used for the primary. The other is used for the secondary. However, multiple coils are often used, especially on the primary side. This gives convenience in connecting to different power voltages. Special primary windings with multiple taps for 200- 208-240-480-575 V and a 120-V secondary are available.

Control transformers are available in sizes from 250 volt-amperes (VA) to 10 kilovolt-amperes (kVA). (10 kVA = 10,000 VA) The most widely used control transformers have a dual-voltage primary of 240/480 V. They have an isolated secondary winding to provide 120 V for the load.

The ratio of voltage reduction from 240 V to 120 V is 2 to 1 (2:1). From 480 V to 120 V, the ratio is 4:1. The voltage is directly proportional to the num- ber of turns in each coil. Therefore, this ratio is often referred to as the *turns ratio.* Thus, the control voltage will always be a direct ratio of the line voltage. For example, it is possible for the line voltage in a plant to vary from approximately 460 V to 500 V or from 230 V to 250 V. If the line voltage in a given plant dropped from its normal 480 V to 460 V, a transformer using a 4:1 ratio would deliver

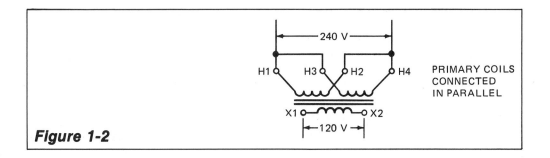

Figure 1-2

115 V to the secondary. Similarly, if a plant's power source voltage dropped from its normal 240 V to 230 V, a transformer using a 2:1 ratio would deliver 115 V to the secondary.

The symbol for the control transformer is shown in Figures 1-2 and 1-3. Two primary coils and one secondary coil are used. When the control transformer primary coils are connected to a 240-V power source, the two coils are connected in *parallel*. When the primary coils are connected to a 480-V power source, the coils are connected in *series*.

When two coils with the same number of turns (same voltage rating) are connected in parallel, the effective number of turns for determining the turns ratio remains the same as if only one coil were used. When two coils with the same number of turns are connected in series, the number of turns on each coil are added together.

When two separate coils are used on the primary side, this is called a *dual primary*. When two separate coils are used on the secondary side, this is called a *dual secondary*.

The transformer can be obtained either in an open type or in its own enclosure. At one time the general practice was to use the open type and panel mount it in the control cabinet. With the present transformer design, however, high allowable temperature rise may create unwanted high temperatures in the control cabinet. With this condition, the transformer in its own enclosure should be used, mounted on the outside of the cabinet.

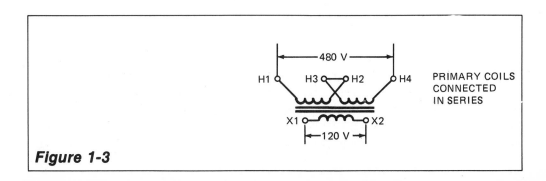

Figure 1-3

The transformer either has screw-type terminals on the coil, or the leads are brought out to a terminal block. In some cases, fuses or circuit breakers are offered as an integral part of the transformer installation.

1.2 SELECTING A TRANSFORMER

The first item to consider in selecting a transformer for a specific control circuit application is to determine the primary and secondary voltages and frequency needed.

Then, consideration should be given to the continuous or sealed and inrush current characteristics of the coils on relays, contactors, motor starters, solenoids, etc. This information is available from the manufacturer of the component.

Using this information, calculate the total maximum continuous or sealed current by adding the continuous or sealed current drawn by all the coils that will be in an energized condition at the same time and multiply this figure by 5/4. Then calculate the total maximum inrush current by adding the inrush current of all the coils that will be energized together at any one time and multiply this figure by 1/4.

Using the larger of the two figures you have just calculated, multiply this current by the control voltage. This product is the volt-amperes (VA) required by the transformer load. In case this resulting figure drops below a commercially available transformer size, it is generally advisable to use the next larger size. For example, you may have calculated that the maximum VA required is 698. Then use a 750 VA transformer which is commercially available.

Another method is used where specific regulation requirements must be met. Regulation is the rise in secondary voltage when the full load is removed. Regulation is expressed as a percent (%) of the full load voltage. A problem can arise with the drop in secondary voltage when a momentary heavy load is applied. If the secondary voltage drops too low, many of the component coils that are in an energized condition are deenergized. This action is called "dropping out." The point of dropping out may vary with the size of component. More on this action is given in Chapter 4.

Regulation curves for transformers are available from the manufacturer. Regulation curves are generally shown for 100% PF (power factor) loads and 20% PF loads. Since a coil presents a low power factor load, it is generally advisable to use the 20% PF curve. (More about power factor in Chapter 17.) Figure 1-4 shows the regulation curves for both 100% and 20% power factor loads. Note that these curves give the relationship between secondary current and secondary voltage.

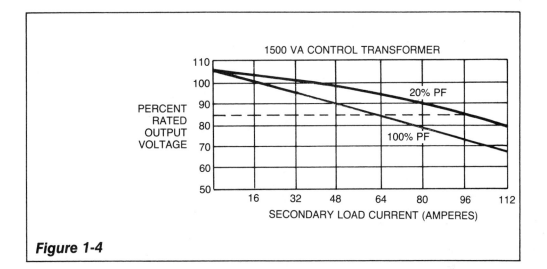

Figure 1-4

Now calculate the continuous or sealed current of all the coils that will be in an energized condition at the same time. Multiply this figure by the secondary voltage. This gives the continuous or sealed VA. Next, calculate the inrush current of all the coils that will be energized together at any one time. Multiply this figure by the secondary voltage. This gives the inrush VA.

The transformer VA required then, is:

$$\sqrt{(VA\ continuous\ or\ sealed)^2 + (VA\ inrush)^2}$$

Go now to the 20% power factor regulation curve for the transformer size you have just determined. The location where the power factor curve intersects at the 85% voltage output point is the maximum secondary amperes that can be allowed for the condition you have calculated. The 85% figure is used as NEMA standards require most magnetic devices to operate at 85% of rated voltage. The maximum amperes is shown in Figure 1-4 for 85% secondary voltage. The curves shown are for a 1500 VA transformer.

Another important factor in selecting a transformer is temperature rise. Temperature rise is caused by losses in both the copper winding and the iron core. The ability to dissipate heat from the surface of the coil and core is important. The transformer is designed to carry full load continuously if the heat is properly dissipated. The continuous load current is thus an important factor in determining the transformer's volt-ampere rating to use for a given job.

A typical transformer is shown in Figure 1-5. A three-phase power supply and control transformer diagram is shown in Figure 1-6. The primary side of the transformer is connected to one phase of a 480-volt, three-phase, 60-HZ power

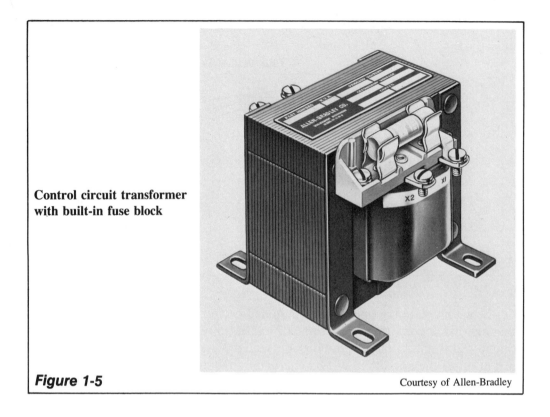

**Control circuit transformer
with built-in fuse block**

Figure 1-5   Courtesy of Allen-Bradley

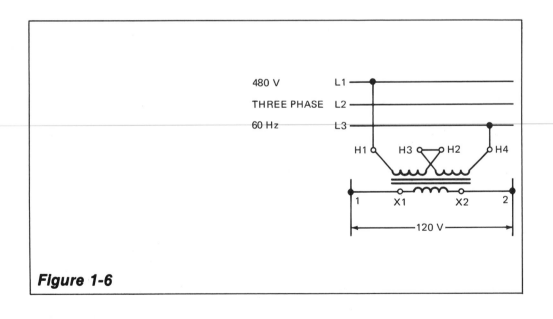

Figure 1-6

source. The secondary is supplying 120 volts AC (to the vertical sides of the ladder-type control circuit diagram).

All machine tool electrical control circuits shown in this text through Chapter 11 use the basic ladder-type diagram. The voltage source between the two vertical sides in these diagrams is 120 volts AC. Therefore, the complete three-phase power circuit and transformer symbol will not always be shown. However, they are usually shown on industrial schematics.

ACHIEVEMENT REVIEW

1. What type of electrical energy is normally used to supply industry?

2. What are the two important reasons for using 120 V in machine control systems?

3. There is 480 V ac available in a given plant. To obtain 120 V, what turns ratio is required between the primary and secondary of the control transformer?

4. You find that under unusually heavy loads, the voltage in your plant drops to 456 V. What will be the resulting secondary control voltage if you use a control transformer with a 4:1 primary-to-secondary turns ratio?

5. Draw the symbol for a dual-primary, single-secondary control transformer. Show all lead designations.

6. What parts of a transformer generally contribute to temperature rise in the transformer?
   a. The enclosure
   b. Copper winding
   c. Iron core

7. What may happen to components that are energized in a control system if, due to poor transformer regulation, the voltage drops to an unusually low level?

8. Draw a complete circuit showing the primary of a dual-primary control transformer connected to a three-phase, 480-V power line, and the single secondary connected to a 120-V control system.

9. In a given transformer the voltage is reduced from the primary to the secondary. This transformer is called a
   a. current transformer.
   b. step-up transformer.
   c. step-down transformer.

10. On a given job, you have calculated that the total inrush current of all coils

energized at any one time is 42 amperes. The calculated continuous or sealed current of all coils energized at the same time is 8 amperes. Using the first method explained in this text, determine the size of transformer you should use. The control voltage is 120. Commercially available transformers are 750, 1000, 1500 and 2000 VA.

CHAPTER 2
FUSES, DISCONNECT SWITCHES, AND CIRCUIT BREAKERS

OBJECTIVES

After studying this chapter, the student will be able to:

- Describe basic fuse construction.
- List three different types of fuses and some of their uses.
- Identify four different types of circuit breakers and uses for each.
- Describe the steps to take when first setting up electrical control and power circuits.
- Explain why time delay fuses are used with motor starter circuits.
- List the voltage and current ratings available for fuses and circuit breakers.
- Discuss the important factors to consider when selecting protective devices.
- Draw the symbols for important protective and disconnecting devices.
- Know what is meant by interrupting capacity.
- Know the use of rejection type fuses.

2.1 PROTECTIVE FACTORS

Once electrical power and control voltage are available, the next important step is to provide for two factors:

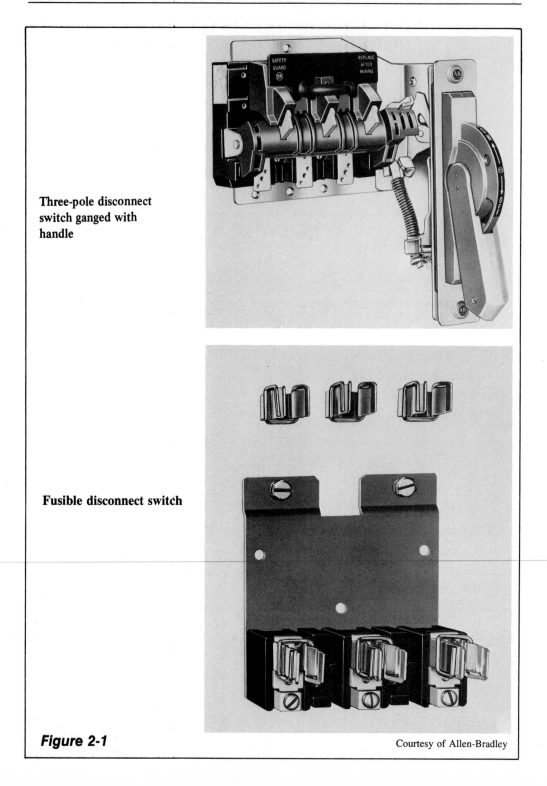

Three-pole disconnect switch ganged with handle

Fusible disconnect switch

Figure 2-1

Courtesy of Allen-Bradley

1. A means of disconnecting electrical energy from the circuits.
2. Protection against sustained overloads and short circuits.

The power circuit can be disconnected by using a disconnect switch or non-automatic circuit breaker (circuit interrupter). Protection is provided by adding adequate fusing to the disconnect switch, and thermal and/or magnetic trip units to the circuit interrupter.

The control circuit is normally protected by a single fuse or circuit breaker. In some cases, two or more fuses may be used. This is covered in more detail in Chapter 15, Troubleshooting.

Figure 2-1 is an example of a commercially available three-pole, fusible-disconnect switch, ganged with the handle.

2.2 FUSE CONSTRUCTION AND OPERATION*

The typical fuse consists of an element which is surrounded by a filler and enclosed by the fuse body. The element is welded or soldered to the fuse contacts (blades or ferrules). See Figure 2-2.

The element is a calibrated conductor. Its configuration, its mass and the materials employed are varied to achieve the desired electrical and thermal characteristics. The element provides the current path through the fuse. It generates heat at a rate that is dependent upon its resistance and the load current.

The heat generated by the element is absorbed by the filler and passed through the fuse body to the surrounding air. A filler such as quartz sand provides effective heat transfer and allows for the small element cross-section typical in modern fuses. The effective heat transfer allows the fuse to carry harmless overloads. The small element cross-section melts quickly under short circuit conditions. The filler also aids fuse performance by absorbing arc energy when the fuse clears an overload or short circuit.

When a sustained overload occurs, the element will generate heat at a faster rate than the heat can be passed to the filler. If the overload persists, the element will reach its melting point and open. Increasing the applied current will heat the element faster and cause the fuse to open sooner. Thus, fuses have an inverse time-current characteristic, i.e., the greater the overcurrent the less time required for the fuse to open the circuit.

*Information supplied through the courtesy of Gould-Shawmut

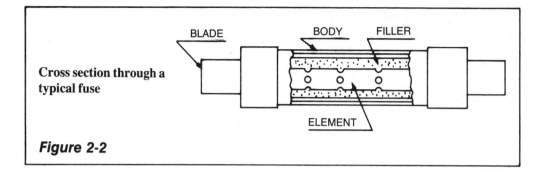

Cross section through a typical fuse

Figure 2-2

2.3 FUSE TYPES

Fuses are available in several types:

1. Standard one-time fuse (Figure 2-3)
2. Time delay fuse (Figure 2-4)
3. Current-limiting fuse (Figure 2-5)

Standard voltage ratings for fuses are 250 V and 600 V. Higher voltage ratings are available. Current ratings range from a fraction of an ampere (A) to 600 amperes (A), in both voltage ratings. This is true of all fuses except the current-limiting type. The current-limiting type may extend up through 6000 amperes (A).

The ability of a protective device (fuses or circuit breakers), to interrupt excessive current in an electrical circuit is important. All protective devices have a published interrupting capacity. This is defined as the highest current at rated voltage that a device can interrupt.

It is therefore important when installing or replacing a protective device that at least three items must be considered:

1. Voltage rating
2. Current rating
3. Interrupting capacity

Correct fuses must be used as replacements for continued safety in the protection of equipment. To help in this area, manufacturers now provide a "rejection" type fuse. It is called Class R (R for rejection). The ferrule sizes (0–60A) have an annular groove in one ferrule. The blade sizes (61–600 A) have a slot in one

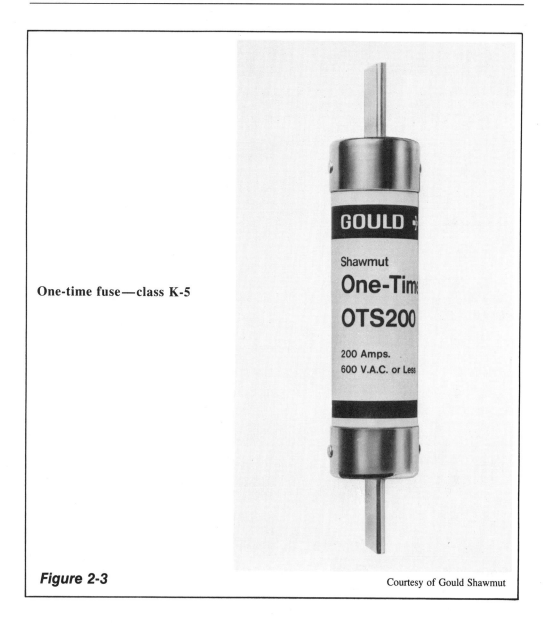

One-time fuse—class K-5

Figure 2-3

Courtesy of Gould Shawmut

blade. Replacement of this fuse with a fuse of lower voltage or lower interrupting rating is not possible provided this fuse is used with rejection fuse blocks. The rejection fuse block is similar to the standard fuse block except physical changes are made in the block to accommodate the annular ring in the ferrule type fuse and the slot in blade type.

The physical configuration of the rejection type fuse is shown in Figure 2-5.

**Time delay fuse—
class RK–5**

Figure 2-4   Courtesy of Gould Shawmut

The rejection type fuse has a 200,000 ampere interrupting rating as contrasted with the standard fuse configuration, ferrule or blade, with an interrupting capacity of 50,000 amperes.

   To further understand the operation of fuses, a melting time-current data curve is shown in Figure 2-6. This curve is for a typical 100-ampere, 250-V, time-delay fuse. It shows an inverse time relationship between current and melting time. That is, the higher the current, the faster the melting time. For example,

**Current-limiting fuse—
class RK-1**

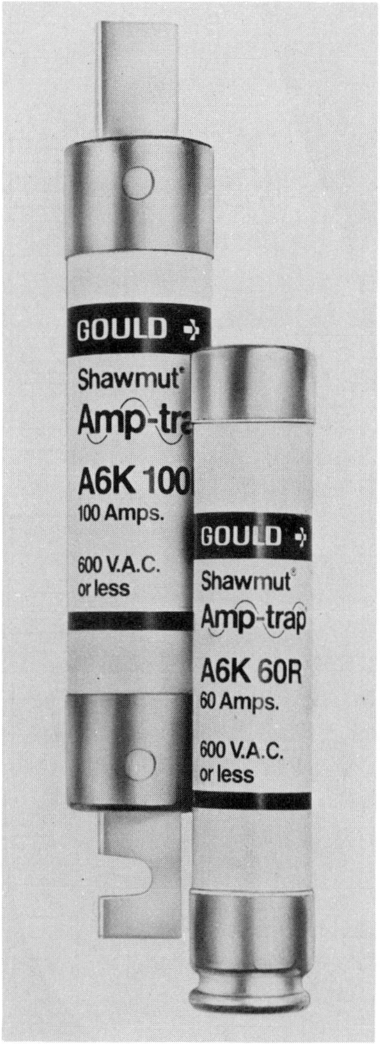

Figure 2-5   Courtesy of Gould Shawmut

referring to this curve, it can be seen that at 1000 amperes, the melting time is 0.2 seconds. At 200 amperes, the melting time is 150 seconds. This characteristic is desirable because it parallels the characteristic of conductors, motors, transformers and other electrical apparatus. This equipment can carry low level overloads for relatively long times without damage. However, under high current conditions damage can occur quickly. Because of the inverse time characteristic, a properly applied fuse can provide effective protection over a broad current range from

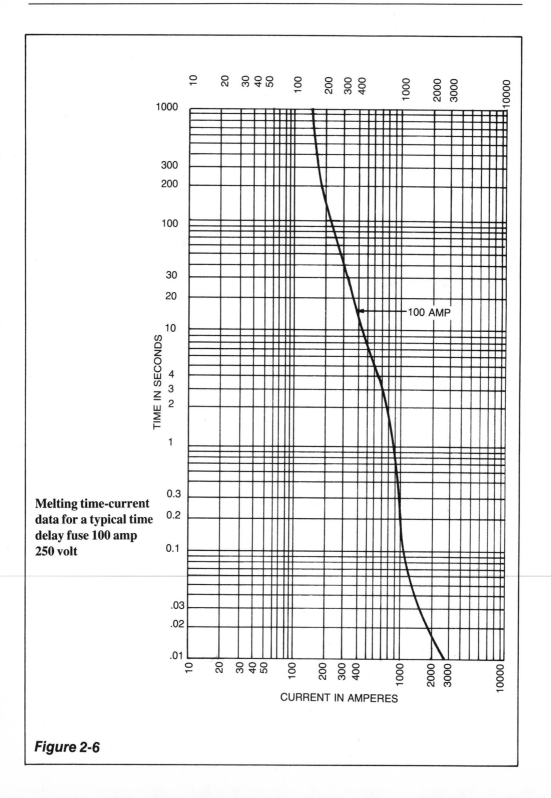

Melting time-current data for a typical time delay fuse 100 amp 250 volt

Figure 2-6

overloads to short circuits.

The standard one-time fuse (Class K-5) consists of a low temperature melting metal strip with one or more reduced area sections. On overloads that exceed the rating of the fuse the narrow section will melt, thus opening the circuit. The heat to melt the strip comes from an excessive current passed through the fuse. In case of short circuits the entire strip may disintegrate. The tube holding the strip (fuse link) must be of sufficient strength to withstand the internal pressure developed during this time.

The Class K-5 fuses are suitable for the protection of mains, feeders and branch circuits serving lighting, heating and other non-motor loads.

The time delay fuses (Class RK-5) were developed for use where a heavy overload might exist for a short time. It is necessary to hold the circuit closed through a normal load and during a short overload period. However, the fuses must open if the overload continues. An example of a short overload is normal motor starting. See Figure 2-4. These fuses are also suitable for general purpose protection of transformers, service entrance equipment, feeder circuits and branch circuits. The time delay characteristic of an efficient fuse allows it to withstand normal surge conditions without compromising short circuit protection.

In cases where extremely high current is available to feed into a fault on a short circuit, the current-limiting fuse is used. In some electrical power systems, a fault can produce currents so great that much damage results. The current-limiting fuse will limit both the magnitude and duration of current flow under short circuit conditions. This fuse will clear available fault currents in less than one half cycle, thus limiting the actual magnitude of current flow. They are generally supplied as a rejection type fuse. The replacement of these fuses with a fuse of another UL class is not possible.

The current-limiting fuse (Class RK-1), is available with interrupting capacity of 200,000 amperes and higher. See Figure 2-5. They are available as a fast acting fuse suitable for the protection of capacitors, circuit breakers, load centers, panel boards, switch boards and bus ducts where high available short circuit currents may exist. They are also available providing time delay for motor loads and a high degree of current limitation for minimizing short circuit current damage.

With the recent advent of solid-state devices, fuse designs have changed to match solid-state protection demands.

Solid-state devices operate at high current densities. Cooling is a prime consideration. Cycling conditions must be considered. The ability of solid-state devices to switch high currents at high speed subjects fuses to thermal and mechanical stresses. Solid-state devices have relatively short thermal time constants. A short circuit current which may not harm an electromechanical device can cause catastrophic failure of a solid-state device.

Most programmable controllers (covered in Chapter 19), use a semiconductor fuse with a blown fuse indicator. There also may be add-on switches that can be used to energize an indicator light. This light can be used for remote indication. The trigger actuator may be a part of the fuse or a field mounted blown fuse indicator. This would be wired in parallel with the fuse being monitored.

2.4 CIRCUIT-BREAKER TYPES

The circuit breaker is available in four types:

- Nonautomatic (circuit interrupter)
- Thermal
- Magnetic
- Thermal magnetic

The *nonautomatic circuit breaker* (circuit interrupter) is used for load switching and isolation. Adding a thermal strip unit to the nonautomatic circuit breaker provides automatic tripping. This is accomplished by using a bimetallic element in each pole of the breaker. This unit then carries the load current. When the conductors carry a current in excess of the normal load, the breaker thermal element increases in temperature, as it carries the same current. This temperature increase deflects the thermal element and trips the breaker. Since the tripping action depends on temperature rise, a time lag is present. Therefore, the tripping action is not affected by momentary overloads.

To clear a circuit in case of a short circuit, a more rapid opening system is required. To obtain this, a *magnetic trip unit* is added either to the nonautomatic breaker or to the breaker with the thermal trip unit. The magnetic trip unit operates through a magnet that trips the breaker instantaneously on short-circuit current.

A combination of thermal and magnetic trip units is desirable. The *thermal element* provides inverse time tripping on overloads. The magnetic trip provides instantaneous trip on short circuits. *Molded case* circuit breakers are now available from 100 A to 2500 A. The voltage ratings are 240 V, 480 V, and 600 V. Recent developments have increased the interrupting capacity in some circuit breakers to 100,000 A. Figure 2-7A is a cutaway view of a molded case circuit breaker.

The fused disconnect switch and the circuit breaker are available in many types of enclosures. They can also be obtained as open types for panel mounting. Remote operators can be used to mount on the door of the control cabinet and

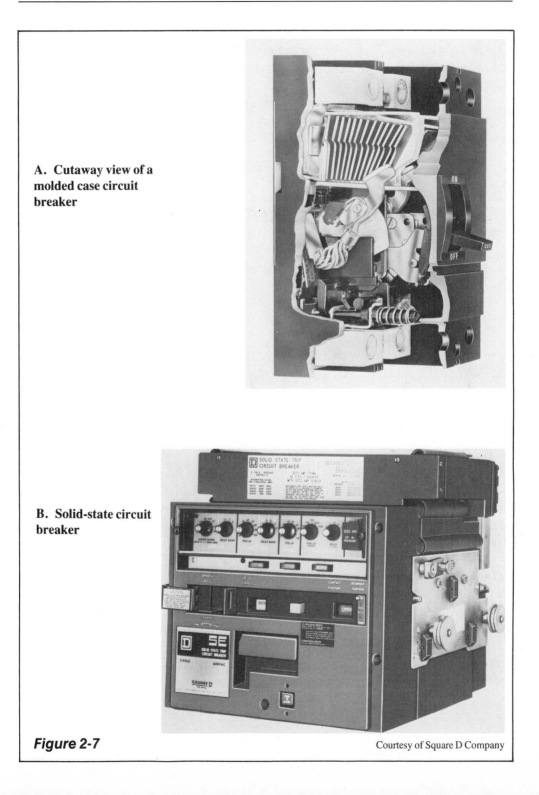

A. Cutaway view of a molded case circuit breaker

B. Solid-state circuit breaker

Figure 2-7   Courtesy of Square D Company

interlock mechanically with the door. The operator may also be mounted in a "dead" or stationary portion of the cabinet. The door is mechanically interlocked. However, this allows the breaker to be locked open or closed with the cabinet door open.

A more recent development in the circuit breaker field is the solid-state circuit breaker. This is shown in Figure 2-7B. The new micrologic trip unit system provides standard and optional features that permit greater versatility, improved selectivity and more reliability in circuit protection.

2.5 SELECTING PROTECTIVE DEVICES

Important factors to consider when selecting protective devices are:
- Size (current rating)
- Is a time lag required?
- Interrupting capacity
- Ambient temperature where the device is to be located
- Voltage rating
- Number of poles
- Mounting requirements
- Type of operator
- Enclosure, if required

If the line voltage and load in a circuit are known, the size and voltage rating of the protective device can be determined. The conductor size to feed the load must be properly determined. Tables are available to supply this information. (Refer to the Appendix.)

The actual sizing of the protective device should come from a source such as the National Fire Protection Association (NFPA), the National Electrical Code* (NEC) and the NFPA-79 Electrical Standard for Industrial Machinery-1985. Local electrical code requirements should also be consulted. The manufacturers of protective devices can also be helpful in obtaining this information.

The problem in providing a time lag element depends on the nature of the load. If inductive devices such as motors, motor starters, solenoids and contactors are involved, some time lag is generally needed. Here again, the manufacturers

*National Electrical Code® is a Registered Trademark of the National Fire Protection Association, Inc., Quincy, MA 02269.

Symbols for protective devices

A. Single-fuse element

B. Three-pole disconnect
switch, ganged with handle

C. Three-pole fused
disconnect switch,
ganged with handle

D. One-pole, thermal-
magnetic circuit breaker

E. Three-pole circuit interrupter, ganged with handle

F. Three-pole, thermal-
magnetic circuit breaker,
ganged with handle

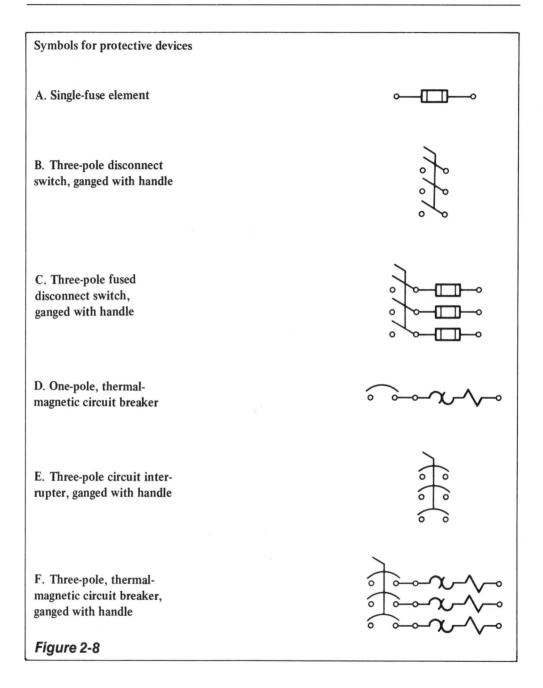

Figure 2-8

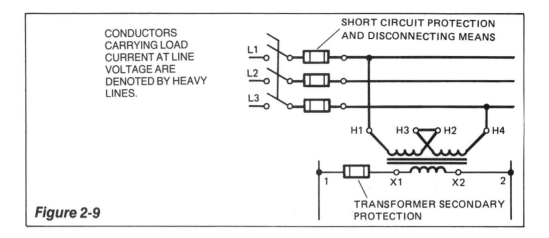

Figure 2-9

of protective devices can help.

In selecting a protective device, the available short circuit current that is available in your plant should be known. Your local power company can advise you how much short circuit current is available in your plant. This figure can be used to determine the required interrupting capacity of your protective devices.

Ambient temperature in the location where the protective device is to be located is important. Overload protection devices can be thermally operated. Therefore, an increase in the ambient temperature can affect the trip setting of the device. In some cases, ambient temperature compensation is provided by the manufacturer in the design of the device.

Standard pole arrangements for most disconnecting devices and protective devices are two pole for single phase lines and three poles for three phase lines. One- and four-pole devices are also available. The multiple-pole (multipole) devices are generally ganged together with a single common operator.

Symbols for various protective and disconnecting devices are shown in Figure 2-8. By using symbols, these components can be added to power source and control source diagrams.

Figure 2-9 illustrates a complete diagram: a three-phase power source, disconnect and protection, and control transformer with protection added in the secondary circuit.

ACHIEVEMENT REVIEW

1. What two factors must be provided for in any electrical power or control system?

2. Under what conditions would you use the time-delay fuse? Give an example.

3. What are the four types of circuit breakers?

4. Under what conditions would you want to use the magnetic trip feature in a circuit breaker?

5. List at least five important factors to consider when selecting a protective device.

6. Are more than two poles available on disconnecting devices? If so, how many?

7. Draw the symbol for each of the following:
    a. Single-fuse element
    b. Three-pole fused disconnect switch, ganged with handle
    c. Three-pole circuit interrupter, ganged with handle
    d. Three-pole, thermal-magnetic circuit breaker, ganged with handle

8. What is meant by the interrupting capacity of a fuse?

9. Quartz sand filler in a fuse provides
    a. increased voltage rating.
    b. effective heat transfer.
    c. an inexpensive filler.

10. The inverse time characteristic of a fuse is important because
    a. the fuse cost is reduced.
    b. it increases the interrupting capacity.
    c. it parallels the characteristics of conductors, motors, transformers and other electrical apparatus.

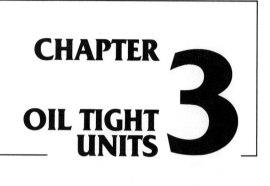

CHAPTER 3

OIL TIGHT UNITS

OBJECTIVES

After studying this chapter, the student will be able to:

- Describe two methods of mounting push-button switch units.
- List four types of operators for the push-button switch.
- Explain why different colors are used for push-button switch operators.
- Draw the symbols for push-button switch units with:
  1. Flush or extended head
  2. Mushroom head
  3. Maintained contact attachment
- List several arrangements available for selector switches.
- Draw the symbol for the selector switch.
- Draw the symbol for the foot switch.
- Discuss the advantages of push-to-test pilot lights.
- Draw the symbol and detailed circuit for the push-to-test pilot light.
- Explain the meaning of the letter in the pilot light.
- Describe the difference between the selector switch and the drum switch.
- Draw simple basic circuits using selector switches or drum switches.

3.1 OIL-TIGHT UNITS

Almost all push-button switches, selector switches, and pilot lights offered to industry today are termed *oil-tight units*. Since this text is concerned with industrial electrical control, only oil-tight types of units are discussed.

3.2 PUSH-BUTTON SWITCHES

Switches generally consist of two parts: the contact unit and the operator. This allows for many combinations that cover almost every application required.

Contact units can be obtained in blocks which contain one normally open (NO) and one normally closed (NC) contact. (*Normal* can be defined as not being

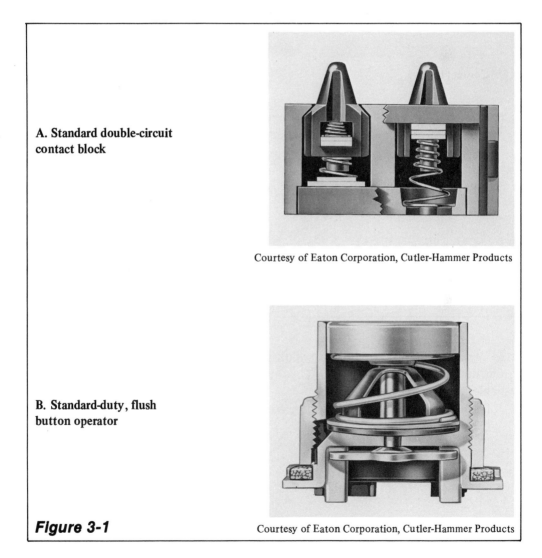

A. Standard double-circuit contact block

Courtesy of Eaton Corporation, Cutler-Hammer Products

B. Standard-duty, flush button operator

Figure 3-1

Courtesy of Eaton Corporation, Cutler-Hammer Products

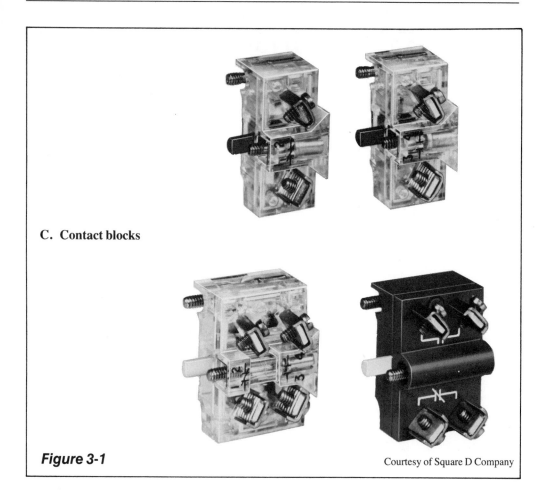

C. Contact blocks

Figure 3-1   Courtesy of Square D Company

acted upon by any external force.) Multiples of these blocks can be assembled to obtain up to four NO and four NC contacts. With some switches that are now available, more units can be added.

The contact rating for a typical heavy-duty, oil-tight unit is as follows:

Ac Volts: 110–125
Amperes Normal: 6.0
Amperes Inrush: 60.0

Figure 3-1A shows a cutaway section of a typical contact block. Figure 3-1B shows a cutaway section of a typical operator. Figure 3-1C also shows typical contact blocks.

In some cases, the contact block is *base mounted* in the push-button enclosure. Thus, the units can be prewired before the cover with the operators is put in place. This method also eliminates cabling conductors to the cover.

The alternate method is *panel mounting*. The base of the operator, with the contact block attached, is mounted through an opening in the panel. It is then secured in place by a threaded ring installed from the front of the panel. The ring is part of the operator assembly. This arrangement has the advantage of providing a space for a terminal block installation in the base of the enclosure. Thus, all connecting circuits can be terminated at an easy checkpoint.

There are slight differences in the way manufacturers machine mounting holes. Therefore, modifications may be required when substituting one unit for another.

Many types of operators are available to suit almost any application. The following are some types of operators:

- Recessed button (Figure 3-2A)
- Mushroom head (Figure 3-2B)
- Time delay (Figure 3-2C)
- Illuminated push-pull (Figure 3-2D)
- Keylock (Figure 3-2E)

Color designation is an important factor in push-button switch operators. Color designation not only provides an attractive panel, but, more important, it lends itself to safety.

Quick identification is important. Certain functions soon become associated with a specific color. Standards have been developed which specify certain colors for particular functions. For example, in the machine tool industry, the colors red, yellow, and black are assigned the following functions:

Red — Stop, Emergency Stop
Yellow — Return, Emergency Return
Black — Start Motors, Cycle

Another operator of a special class is the *maintained contact attachment*. Operation of one unit operator will operate the attached contacts. The contacts remain in an operated condition until the second operator is operated.

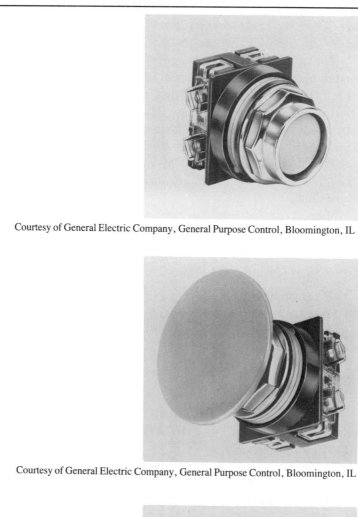

A. Recessed button push-button unit

Courtesy of General Electric Company, General Purpose Control, Bloomington, IL

B. Mushroom head push-button unit

Courtesy of General Electric Company, General Purpose Control, Bloomington, IL

C. Time delay push-button unit

Figure 3-2

Courtesy of General Electric Company, General Purpose Control, Bloomington, IL

D. Illuminated push-pull unit

Courtesy of General Electric Company, General Purpose Control, Bloomington, IL

E. Cylinder lock push-button unit

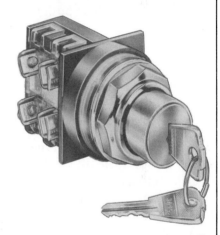

Figure 3-2   Courtesy of General Electric Company, General Purpose Control, Bloomington, IL

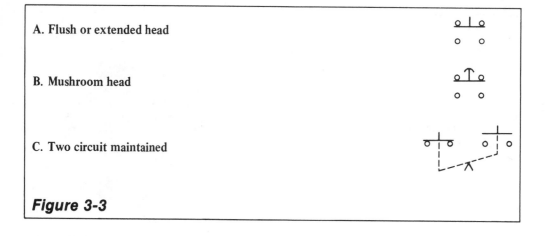

A. Flush or extended head

B. Mushroom head

C. Two circuit maintained

Figure 3-3

Symbols for three push-button units are illustrated in Figure 3-3.

3.3 SELECTOR SWITCHES

Selector switches can be obtained with up to four positions. They can be the maintained contact type. The three-position switch can be arranged for spring return from the right, from the left, or from both right and left. Up to eight contacts are available per device.

The operators are available in standard knob, knob lever, or wing lever types. A cylinder lock can be used, and the switch locked in any one position or all positions. The arrangement for opening or closing contacts in any one position or more depends on a cam in the operator.

Figure 3-4A shows a selector switch with a lever operator. Figure 3-4B shows a selector switch keylock operator.

In the more simple form, a selector switch with two contacts arranged for two positions may be called a *double-pole, double-throw selector switch.* Similarly, a selector switch with two contacts arranged for three positions may be called a *double-pole, double-throw with neutral, selector switch.* As the number of positions increases to four and the number of contacts (poles) increases to eight, the manufacturer's reference to a specific operator is generally coded through a symbol chart or function table. Such charts or tables display which contacts are closed or open in the different positions of the selector switch operator.

Two of the simpler arrangements in selector switches are shown in Figures 3-4C and 3-4D. The symbol is shown at the top. An expansion of the symbol is

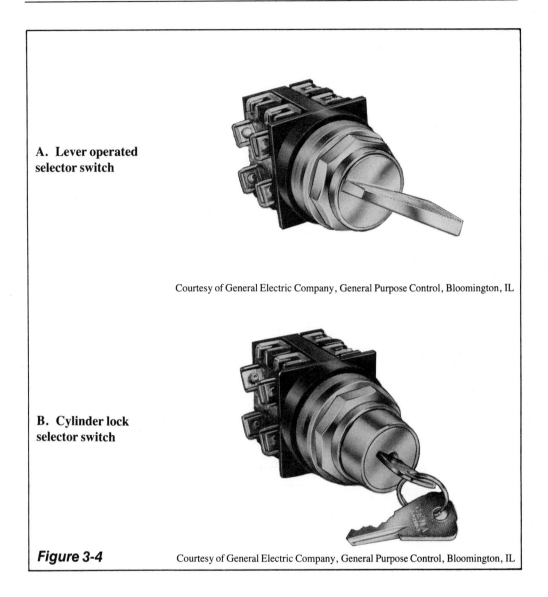

A. Lever operated selector switch

Courtesy of General Electric Company, General Purpose Control, Bloomington, IL

B. Cylinder lock selector switch

Figure 3-4   Courtesy of General Electric Company, General Purpose Control, Bloomington, IL

provided to acquaint the reader with the actual conditions of the contacts in the positions shown in the symbol. In Figure 3-4D, note the use of the X under the position number. This indicates that the contact in line with the X is closed in that position.

Figure 3-4E shows a four-contact (pole), four-position selector switch. Both the symbol and an expansion of the symbol are included.

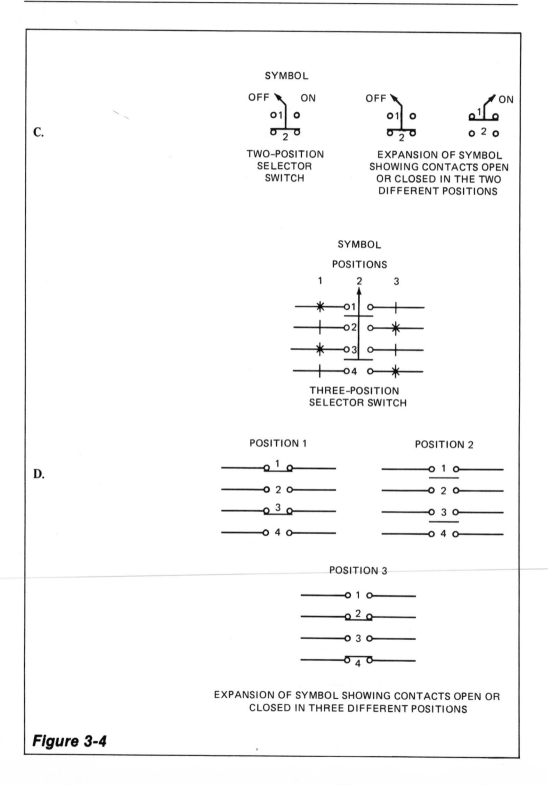

C.

SYMBOL

TWO-POSITION
SELECTOR
SWITCH

EXPANSION OF SYMBOL
SHOWING CONTACTS OPEN
OR CLOSED IN THE TWO
DIFFERENT POSITIONS

SYMBOL

POSITIONS

THREE-POSITION
SELECTOR SWITCH

D.

POSITION 1

POSITION 2

POSITION 3

EXPANSION OF SYMBOL SHOWING CONTACTS OPEN OR
CLOSED IN THREE DIFFERENT POSITIONS

Figure 3-4

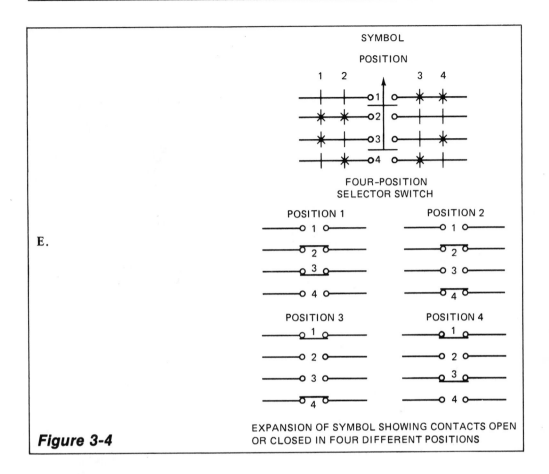

E.

SYMBOL

POSITION

FOUR-POSITION
SELECTOR SWITCH

POSITION 1   POSITION 2

POSITION 3   POSITION 4

EXPANSION OF SYMBOL SHOWING CONTACTS OPEN
OR CLOSED IN FOUR DIFFERENT POSITIONS

Figure 3-4

3.4 HEAVY-DUTY SWITCHES

Heavy-duty switches and special application switches under heavy duty, special service, oil-tight units are those where the switch is made an integral part of the operator and enclosures. A typical example of such a switch is the foot-operated unit, Figure 3-5A. The symbols are shown in Figure 3-5B.

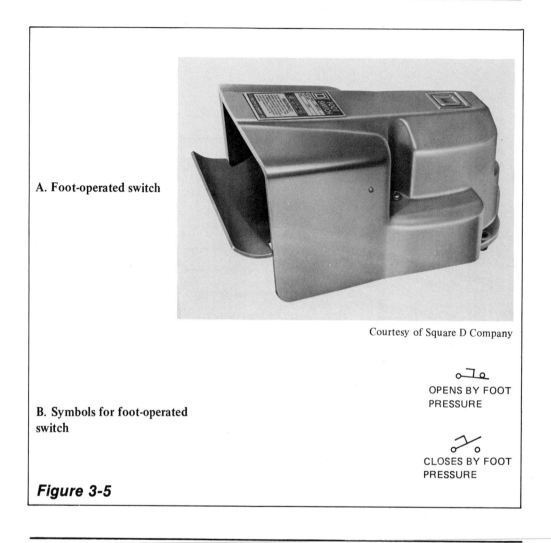

A. Foot-operated switch

Courtesy of Square D Company

OPENS BY FOOT
PRESSURE

B. Symbols for foot-operated switch

CLOSES BY FOOT
PRESSURE

Figure 3-5

3.5 INDICATING LIGHTS

The indicating or pilot light is available in three basic types:

- Full voltage
- Resistor
- Transformer

Due to the vibration normally present in machines, the low-voltage bulb is preferred. The low-voltage bulb operates at 6 V to 8 V obtained through a resistance or transformer unit.

The lens is available either in plastic or glass, and in a variety of colors. Again, as for push-button operators, colors can be used for selection purposes to increase safety of operation. For example:

Red — Danger, Abnormal Conditions
Amber — Attention
Green — Safe Condition
White or Clear — Normal Condition

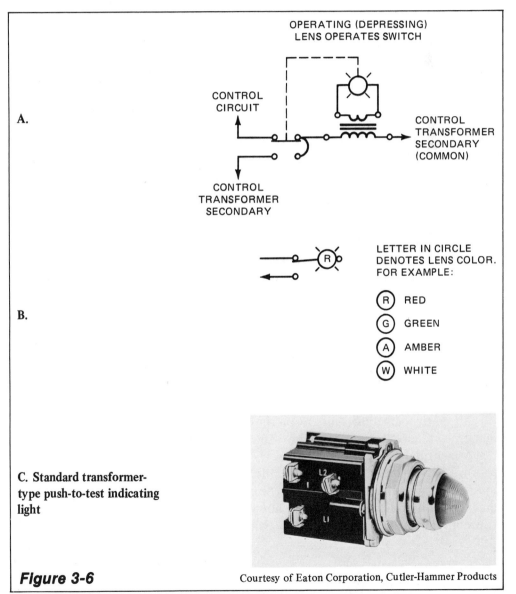

A.

OPERATING (DEPRESSING)
LENS OPERATES SWITCH

CONTROL
CIRCUIT

CONTROL
TRANSFORMER
SECONDARY
(COMMON)

CONTROL
TRANSFORMER
SECONDARY

B.

LETTER IN CIRCLE
DENOTES LENS COLOR.
FOR EXAMPLE:

R RED
G GREEN
A AMBER
W WHITE

C. Standard transformer-type push-to-test indicating light

Figure 3-6

Courtesy of Eaton Corporation, Cutler-Hammer Products

The push-to-test indicating light provides an additional feature. Consider, for example, an indicating lamp that will not illuminate. It may be that the bulb is not energized, or the bulb may be burned out. Depressing the lens unit will connect the bulb directly across the control voltage source. This provides a check on the condition of the bulb.

A detail circuit of the transformer-type, push-to-test indicating light is illustrated in Figure 3-6A. Figure 3-6B shows the symbol for the push-to-test indicating light. Figure 3-6C shows one type of standard transformer-type, push-to-test indicating light.

Note that the push-to-test pilot light consists of a 120–6 volt transformer, a 6-volt bulb covered by a lens and a standard contact block. When the pilot light lens is depressed (operated), it mechanically operates the contact block. The normally closed contacts open and the normally open contacts close. The normally closed contacts are connected to the control circuit. The normally open contacts are connected directly to a source of 120 volts. Thus, when the lens is depressed (operated) the pilot light transformer primary is connected directly across 120 volts. Six volts is now applied to the bulb.

3.5.1 Miniature Oil-tight Units   In addition to the standard line of indicating lights, a line of miniature oil-tight push buttons, selector switches, and pilot lights is now available. This line covers about the same selection as the standard line. These units can often be used to an advantage where space is at a premium. This is particularly true where a great deal of indicating is required. Figure 3-7 shows a group of miniature oil-tight push buttons.

3.6 GENERAL INFORMATION FOR OIL-TIGHT UNITS

All oil-tight operators mount on the panel of an enclosure with some type of sealing ring to give them an oil-tight feature.

A nameplate is provided to properly identify the unit. Nameplates are available in different sizes. The size generally depends on the amount of information required on the nameplate to properly describe the function of the unit. Typical nameplates for oil-tight units are shown in Figure 3-8A. A group of several different types of oil-tight units is shown in Figure 3-8B.

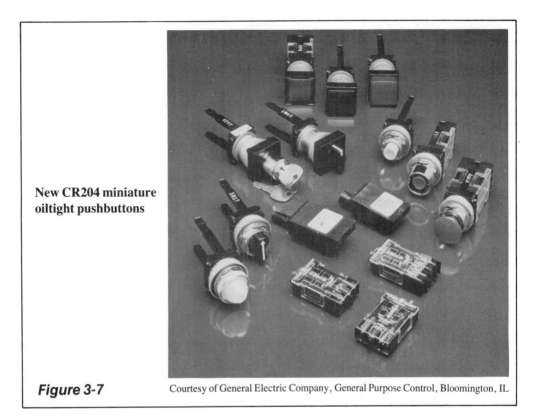

New CR204 miniature oiltight pushbuttons

Figure 3-7

Courtesy of General Electric Company, General Purpose Control, Bloomington, IL

3.7 DRUM SWITCHES

The *drum switch* is a component which is usually not directly included in the oil-tight line. However, the drum switch could be considered the "big brother" of the selector switch. Basically, the drum switch performs the same functions: selecting, dividing, and isolating. The differences between the two components are the size, the number of contacts, and positions.

The selector switch is generally limited to about four positions with four stages (eight contacts), and is rated at 10 A–15 A. The drum switch can be obtained in banks or contacts up to 36, 12 positions, and contact ratings up through 200 A. Generally, larger contact ratings are associated with a smaller number of contacts.

The contacts may be arranged in banks and gear driven from a centrally located operator. The operation can be either manual, solenoid operated, or motor driven.

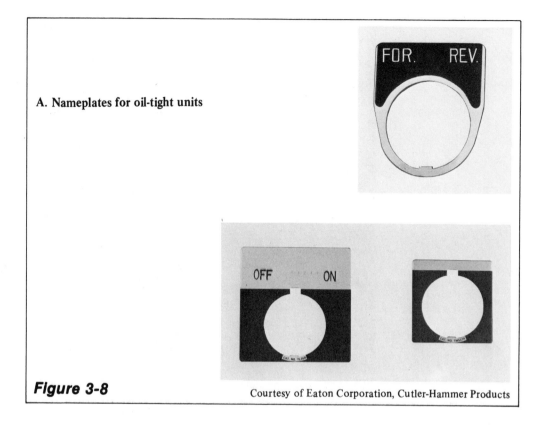

A. Nameplates for oil-tight units

Figure 3-8   Courtesy of Eaton Corporation, Cutler-Hammer Products

Drum switches can be obtained in open types for surface, panel, switchboard, or cavity mounting. They also are available in enclosures suitable for almost any application.

These units are available in several different action types. They can be arranged for maintained contact, self-centering, or spring return from any position to another position. Many arrangements for open or closed contacts in various positions of the operating handle are possible.

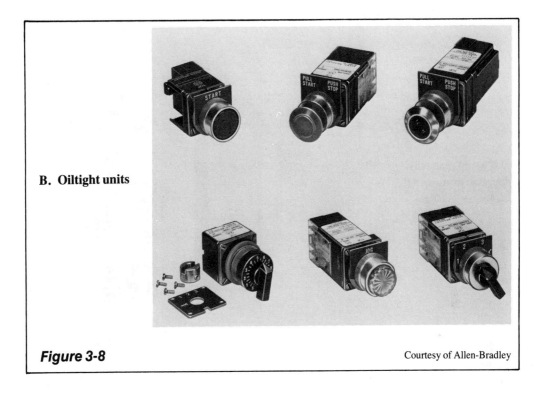

B. Oiltight units

Figure 3-8   Courtesy of Allen-Bradley

Drum switches may also be referred to as rotary pilot or transfer switches. Figure 3-9A shows a typical manually operated drum switch. Figure 3-9B illustrates the symbol for the drum switch. Note that it is similar to the selector switch.

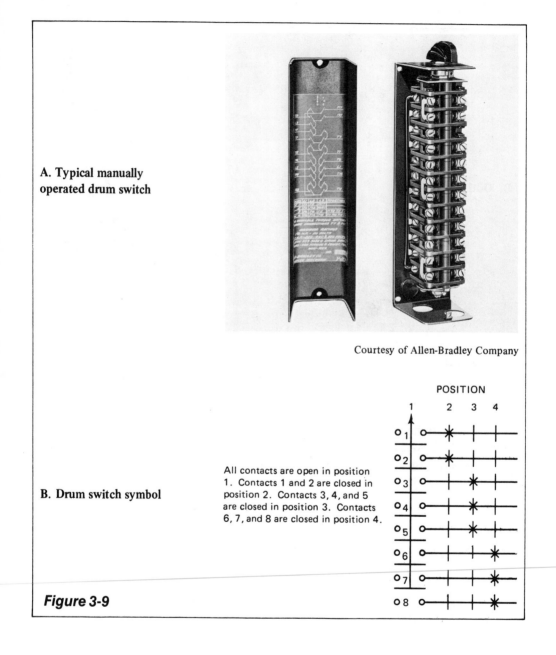

A. Typical manually operated drum switch

Courtesy of Allen-Bradley Company

B. Drum switch symbol

All contacts are open in position 1. Contacts 1 and 2 are closed in position 2. Contacts 3, 4, and 5 are closed in position 3. Contacts 6, 7, and 8 are closed in position 4.

Figure 3-9

3.8 CIRCUIT APPLICATIONS

We can now show a complete circuit, using the symbols illustrated. The electrical power source, electrical control source disconnecting device, protection, and an output or load that can be controlled are shown in Figures 3-10A, 3-10B, 3-10C, and 3-10D. The complete circuit is shown in Figure 3-10A. After this

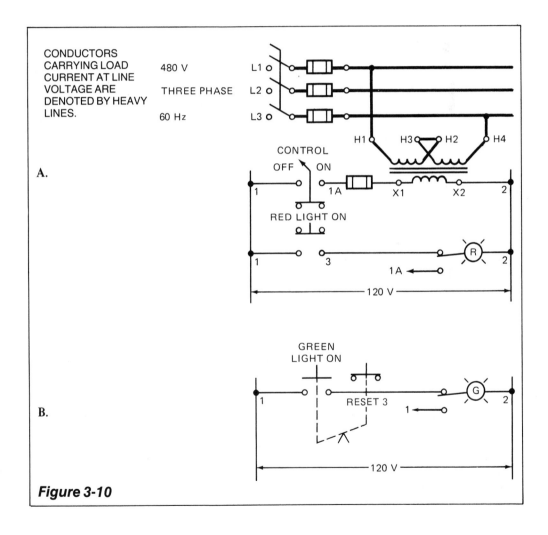

Figure 3-10

C.

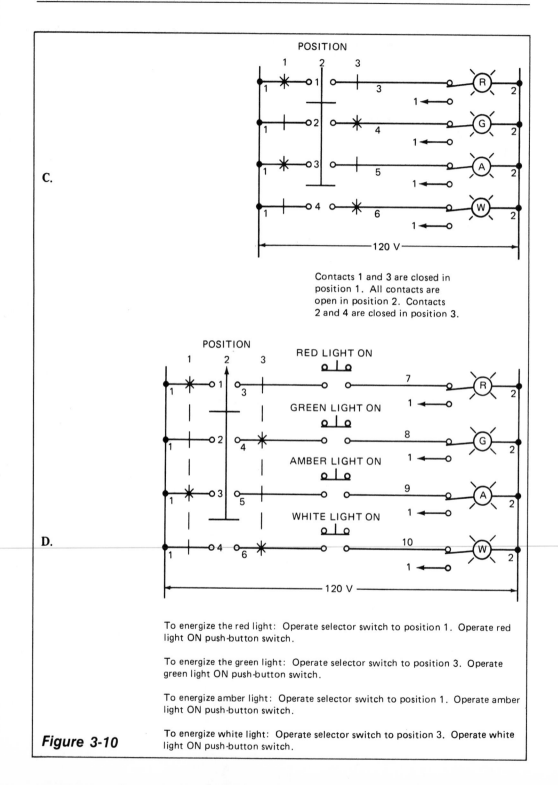

Contacts 1 and 3 are closed in position 1. All contacts are open in position 2. Contacts 2 and 4 are closed in position 3.

D.

To energize the red light: Operate selector switch to position 1. Operate red light ON push-button switch.

To energize the green light: Operate selector switch to position 3. Operate green light ON push-button switch.

To energize amber light: Operate selector switch to position 1. Operate amber light ON push-button switch.

To energize white light: Operate selector switch to position 3. Operate white light ON push-button switch.

Figure 3-10

diagram, the primary power source and control power source are not always shown in circuit diagrams, as the control voltage available between the two vertical lines (sides of the ladder) is generally 120 V ac.

All control circuits are assigned numbers for convenience in checking the circuit. The number changes each time the circuit passes through a component. This practice is discussed in more detail in Chapter 16, Designing Control Systems for Easy Maintenance.

In Figure 3-10A, the selector switch is operated to the ON position, closing the contact between 1 and X1. This puts 120 V ac on the two vertical lines 1 and 2.

The red pilot light can now be energized by operating the red light push-button switch. The red light remains energized as long as the push-button switch is operated.

Figure 3-10B shows the use of a maintained contact push-button switch. Operation of the green light push-button switch energizes the green pilot light. The green light push-button switch will remain in the operating position until the RESET push button is operated.

In the circuit in Figure 3-10C, a three-position, four-contact selector switch is used. The symbol shows that contacts 1 and 3 are closed in position 1. All contacts are open in position 2. Contacts 2 and 4 are closed in position 3.

The pilot light load is connected so that the red and amber lights are energized in position 1 of the selector switch, and the green and white lights are energized in position 3. None of the pilot lights is energized in position 2, as all of the selector switch contacts are open.

Figure 3-10D shows additional control added to the circuit in Figure 3-10C. Momentary contact push-button switches are added to each pilot light circuit. Now, in addition to selecting either position 1 or position 3 of the selector switch, a push-button switch must be operated to energize a pilot light.

ACHIEVEMENT REVIEW

1. Describe how you would panel mount an oil-tight unit.

2. The low voltage bulb in the pilot light is preferred because it
   a. Is much smaller in size.
   b. Will withstand vibration.
   c. Will not consume as much electrical energy.

3. Draw the symbol for a mushroom head push-button switch (one contact block-two contacts; one normally open, one normally closed).

4. Are lenses for pilot lights available in more than one color? If so, why?

5. The advantage of panel mounting a contact block is that the mounting
   a. Provides space for terminal block installation.
   b. Allows for more space on the panel.
   c. Is required by electrical standards.

6. Draw the symbol for a red push-to-test pilot light.

7. What arrangements can be obtained with different types of drum switches?

8. Draw the symbol for a three-position, four-contact selector switch. Contacts 1 and 3 are closed in position 1, all contacts are open in position 2 and contacts 2 and 4 are closed in position 3.

9. Explain how the push-to-test pilot light functions. Draw the complete circuit diagram.

10. Draw a circuit showing a two-position selector switch and a green push-to-test pilot light. The pilot light is to be energized when the selector switch is operated to the on position.

CHAPTER 4

CONTROL RELAYS TIME DELAY RELAYS, AND CONTACTORS

OBJECTIVES

After studying this chapter, the student will be able to:

- Identify two main uses for the control relay.
- Show how the control relay is constructed mechanically.
- List the three published ratings for relays.
- Explain why silver is used in relay contacts.
- Discuss several factors involved with relay operation, such as contact bounce, overlap contacts, contact wipe, and split or bifurcated contacts.
- Describe the interlock circuit and explain why it is used.
- Draw the symbols for the relay coil, relay contacts, and time-delay relay contacts.
- Explain the difference between inrush and holding current in a relay coil.
- Describe how the latching relay operates.
- List the basic uses for the contactor.

4.1 RELAY USES

The *relay* is basically a communication carrier. Relays are used in the control of fluid power valves and in many machine sequence controls such as drilling, boring, milling, and grinding operations. Relays are also used as power amplifiers.

The relay is an electromechanical device. The major parts of the relay are shown in Figures 4-1A and 4-1B. There are both stationary and moving contacts. The moving contacts are attached to the plunger. Contacts are referred to as normally open (NO) and normally closed (NC). When the coil of the relay is energized, the plunger moves through the coil, closing the normally open contacts and opening the normally closed contacts. Figure 4-1A shows the contacts in the normal or deenergized condition of the coil.

When the coil of the relay is connected to a source of electrical energy, the coil is energized. When the coil is energized, the plunger moves up, as shown in Figure 4-1B, due to a force produced by a magnetic field. The distance that the plunger moves is generally short — about one-fourth inch (in) or less.

There is a difference in the current in the relay coil from the time the coil is first energized and when the contacts are completely operated. When the coil is

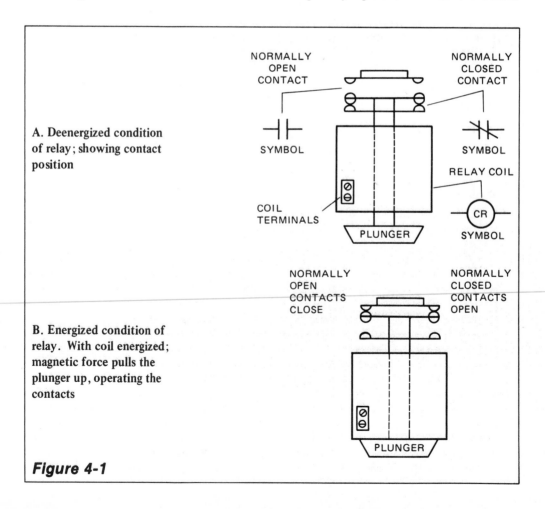

A. Deenergized condition of relay; showing contact position

NORMALLY OPEN CONTACT

SYMBOL

COIL TERMINALS

NORMALLY CLOSED CONTACT

SYMBOL

RELAY COIL

CR SYMBOL

PLUNGER

B. Energized condition of relay. With coil energized; magnetic force pulls the plunger up, operating the contacts

NORMALLY OPEN CONTACTS CLOSE

NORMALLY CLOSED CONTACTS OPEN

PLUNGER

Figure 4-1

energized, the plunger is in an out position. Due to the open gap in the magnetic path (circuit), the initial current in the coil is high. The current level at this time is known as *inrush current.* As the plunger moves into the coil, closing the gap, the current level drops to a lower value. This lower value is called *sealed current.* The inrush current approximates 6 to 8 times the sealed current.

Figure 4-1A also shows symbols for the relay coil and contact.

One of the latest designs of control relays has electrical clearances up to 600 V. The contacts are convertible from open to closed, or from closed to open. Coils are available to cover most standard voltages from 24 V to 600 V. (The standard voltage used in machine control is 120 V.) Relay coils are now being made of a molded construction. This aids in reducing moisture absorption, and increases mechanical strength.

The level of voltage at which the relay coil is energized, resulting in the contacts moving from their normal, unoperated position to their operated position, is called *pick-up voltage.* After the relay is energized, the level of voltage on the relay coil at which the contacts return to their unoperated condition is called *drop-out voltage.*

Coils on electromechanical devices such as relays, contactors, and motor starters are designed so as not to drop out (deenergize) until the voltage drops to a minimum of 85% of the rated voltage. The relay coils also will not pick up (energize) until the voltage rises to 85% of the rated voltage. This voltage level is set by the National Electrical Manufacturers Association (NEMA). Generally, 85% is found to be a conservative figure. Most electromechanical devices will not drop out until a lower voltage level is reached. Also, most electromechanical devices will pick up at a rising lower voltage level. Generally, coils on electro-mechanical devices will operate continuously at 110% of the rated voltage without damage to the coil.

The two important factors in a relay are the coil and contacts. Of these, the contacts generally require greater consideration in practical circuit design.

There are some single-break contacts used in industrial relays. However, most of the relays used in machine tool control have double-break contacts. The rating of contacts can be misleading. The three ratings generally published are:

1. Inrush or "make contact" capacity
2. Normal or continuous carrying capacity
3. The opening or break capacity

For example, a typical industrial relay has the following contact ratings:

- 10 A noninductive continuous load (ac)
- 6 A inductive load @ 120 V (ac)
- 60 A make and 60 A break, inductive load @ 120 V (ac)

A resistance is an example of a *noninductive load.* This may be a resistance unit used as a heating element. An *inductive load* is basically a coil. This could be a solenoid, contactor coil, or motor starter coil.

The point to remember is that in determining the contact rating, it must be clear what rating is given.

Relay contacts are usually silver or a silver-cadmium alloy. This material is used because of the excellent conductivity of silver. Also, silver oxide, which forms on the contacts, is a good conductor. Adding a small amount of cadmium to the silver increases the interrupting capacity of the contacts. Some manufacturers now offer gold-plated contacts. This improves shelf life and provides improved contact reliability in low-energy, low-duty cycle applications. There are several points that the circuit designer should consider in the use of control relays:

1. Changing contacts normally open to normally closed, or vice versa. Most machine tool relays have some means of making this change, ranging from a simple flip-over contact to removing the contacts and relocating with spring changes.
2. Universal contacts. This refers to relays where the contacts may be NO or NC. Only one type may be selected in the use of each contact, but not both types.
3. Split, or bifurcated, contacts. As the name implies, the contact is divided into two parts. This provides for double the contact make points, improving reliability of contact and reducing contact bounce.
4. Contact bounce. All contacts bounce on closing. The problem is to reduce the bounce to a minimum. In rapid operating relays, contact bounce can be a source of trouble. For example, an interlock in a transfer circuit can be lost by a contact opening momentarily. The use of split contacts, overlapping contacts, other type of antibounce contacts, or a circuit change may help.
5. Overlap contacts. In this case, one contact can be arranged to operate at a different time relative to another contact on the same relay. For example, the NO contact can be arranged to close before the NC contact opens. This is called "make before break."
6. Contact wipe. This results from the relative motion of the two contact surfaces after they make contact. That is, one contact wipes against the other. One advantage to be gained from this action is that it tends to provide a self-cleaning action on the contacts.

Panel mounting space has been a problem for the user of relays. With increasing complexity of control, the number of relays used has grown. This requires more panel space. Several years ago, a line of 300-V relays was developed. The size was reduced from the old 600-V relay. This change, along with a modular form of

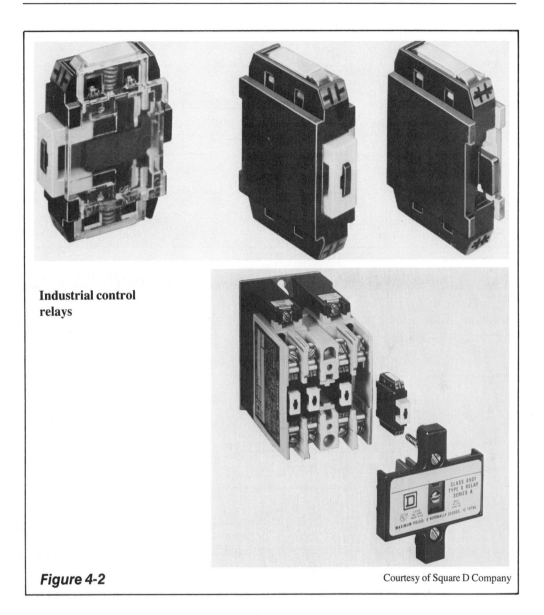

Industrial control relays

Figure 4-2   Courtesy of Square D Company

construction, was very helpful. More recently, a new line of 600-V relays were introduced. These relays are comparable with the 300-V line. Relays shown in this book are of the new 600-V design. Figure 4-2 shows a typical four-pole, 600-V control relay, with plug-in contact cartridges.

The reference to 300-V and 600-V relays applies to the voltage being carried by the contacts. The standard relay coil voltage in the machine control field is 120 V.

These relays are generally available with up to 12 contacts in combinations

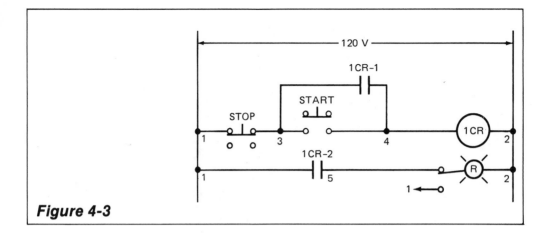

Figure 4-3

of NO and NC contacts. They may be either convertible or fixed. In most cases, the relays can be combined with a pneumatic timing head or mechanical latch. However, when adding the pneumatic timing head, the relay is restricted to four instantaneous contacts.

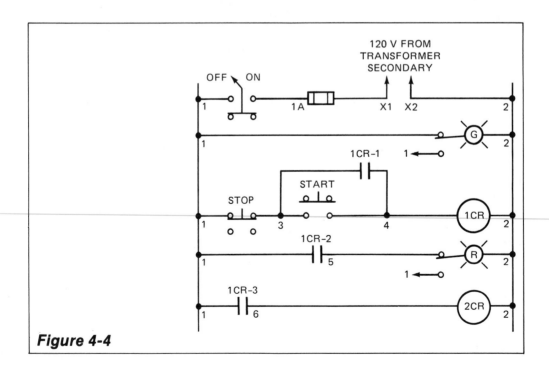

Figure 4-4

Figure 4-3 shows a circuit diagram using a relay with two NO contacts. One contact, 1CR-1, is used as an interlock around the START push button. Thus, an *interlock circuit* is a path provided for electrical energy to the load after the initial path has been opened. The second relay contact, 1CR-2, is used to energize a light. Remember that when a relay coil is energized, the NO contacts close. The circuit can be deenergized by operating the STOP push-button switch.

Figure 4-4 shows the addition of a selector switch, fuse, pilot light, and a second relay. When the selector switch is operated to the ON position, electrical energy is available at the two vertical sides of the circuit. The green light is energized, indicating that the operation has been completed. One additional relay contact is added in the circuit from relay 1CR. This contact closes when the relay 1CR is energized and it, in turn, energizes a second relay coil 2CR. The operating circuit can be deenergized by operating the STOP push-button switch.

Starting in this chapter it can be observed that a numbering system has been introduced in the circuits. This practice will continue in all the remaining circuits in the text.

The two vertical sides of the diagram, representing the source of electrical energy (control voltage), are numbered 1 and 2. While practice may vary with the manufacturer, this text will use 1 for the hot (protected) side and 2 for the common side (generally grounded). There also may be another variation of this system in industry. This is covered in Chapters 14 and 15.

Note that starting at one side or the other on each line of the circuit, the number changes each time you move through a contact or coil. Thus, each contact (push-button switch, selector switch, limit switch, pressure switch, temperature switch, and so on) will have two numbers, one for the incoming conductor and one for the outgoing conductor. This is also true for all coils (relays, timers, contactors, motor starters, and so on).

While this practice does not seem to have any importance in the operation of the circuit, the practice is invaluable when the components are wired on a machine in accordance with the specific circuit design.

This practice is not only a must for electromechanical control but also for solid-state and programmable controllers.

4.2 TIMING RELAY OR PNEUMATIC TIMER

The pneumatic timing head can be added to the control relay. The pneumatic timing relay is also available either as an add-on unit to the 300-V or 600-V relay line or as a separate unit.

In circuit work it is an advantage to have a timing contact as well as instan-

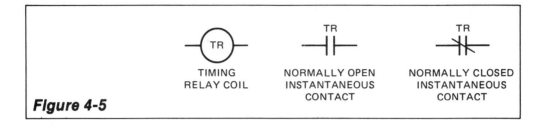

Figure 4-5

TIMING RELAY COIL   NORMALLY OPEN INSTANTANEOUS CONTACT   NORMALLY CLOSED INSTANTANEOUS CONTACT

taneous contacts from the same energized relay coil. The timing contact can be arranged to delay after energizing or deenergizing the coil.

The symbols for the timing relay coil and the instantaneous contact are shown in Figure 4-5. The final schematic description comes with the symbols for the timing contact. There are four separate symbols:

1. NO time delay after energizing
    NC time delay after energizing
    (These contacts are sometimes referred to as *"on" time-delay contacts.*)
2. NO time delay after deenergizing
    NC time delay after deenergizing
    (These contacts are sometimes referred to as *"off" time-delay contacts.*)

With the time-delay coil energized, the on time-delay contacts remain in their unoperated state through the preset time period. At the end of this period, the contacts change to their operated state. They remain in this state until the time-delay coil is deenergized.

Off time-delay contacts change to their operated state immediately when the time-delay relay coil is energized. They remain in the operated state during the time the time-delay coil is energized.

The contacts return to their unoperated condition after a preset time delay following the deenergizing of the time-delay coil. See Figure 4-6 for the symbols.

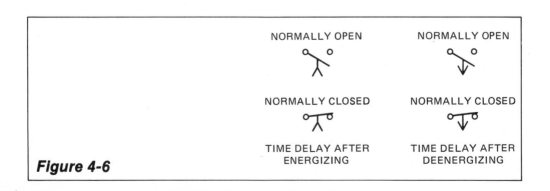

Figure 4-6

NORMALLY OPEN   NORMALLY OPEN

NORMALLY CLOSED   NORMALLY CLOSED

TIME DELAY AFTER ENERGIZING   TIME DELAY AFTER DEENERGIZING

A. Time-delay attachment for a pneumatic time-delay unit

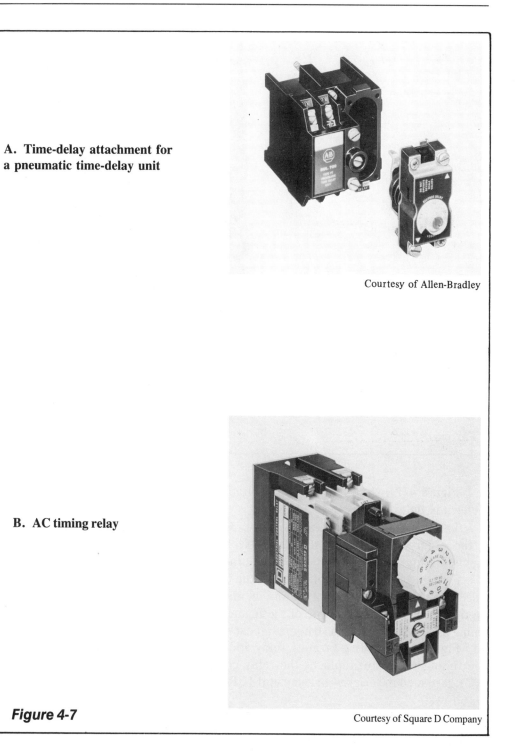

Courtesy of Allen-Bradley

B. AC timing relay

Figure 4-7

Courtesy of Square D Company

C. Solid-state timing relay

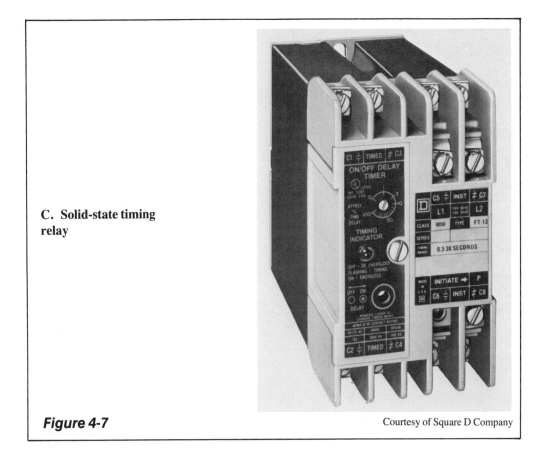

Figure 4-7

Courtesy of Square D Company

Figure 4-7A shows a time-delay attachment and Figure 4-7B shows a complete time-delay relay.

The adjustable timing range on the 300 volt or 600 volt relay line will be in the order of 0.1 seconds(s) to 60 seconds. Accuracy varies quite widely, depending upon the size, type and manufacturer. It is in the range of 10%.

Figures 4-7C and D display solid-state timing relays. Timing indication is provided by a LED that flashes during timing, glows steadily after timing and is off when the timer is deenergized. Some of the features include convertible ON-OFF delay, timing from 0.05 seconds to 10 hours, repeat cycle version, ± 1% repeat accuracy.

Figure 4-8A shows a timing relay added to the circuit shown in Figure 4-4. The timing relay coil replaces the relay coil 2CR. An ON TIME delay contact (NC), is placed in series with relay coil 1CR. The sequence of operations proceeds as follows:

1. Operate the START pushbutton switch.

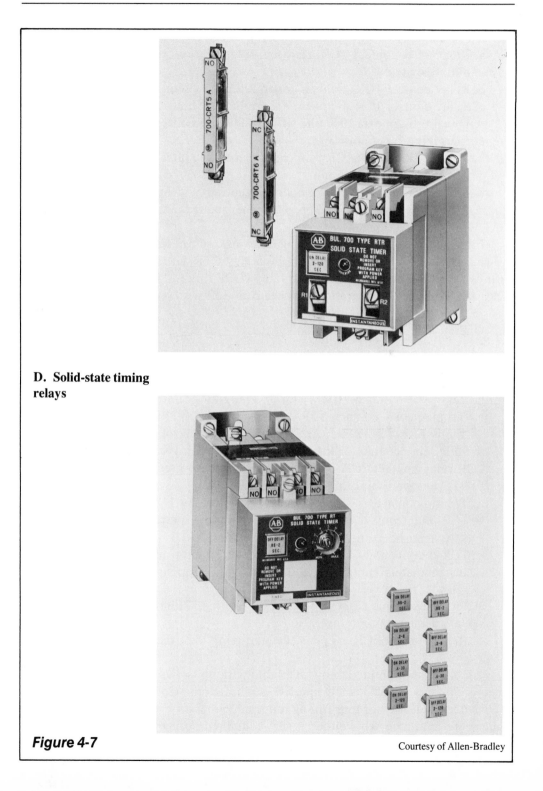

D. Solid-state timing relays

Figure 4-7

Courtesy of Allen-Bradley

2. Relay coil 1CR is energized.
    a. Relay 1CR contact 1CR-1 closes, interlocking around the START push-button switch.
    b. Relay contact 1CR-2 closes, energizing the green pilot light.
    c. Relay contact 1CR-3 closes, energizing the time-delay relay coil 1TR.
3. After a preset time on 1TR, the NC ON-TIME time-delay contact 1TR opens, deenergizing relay coil 1CR.
    a. Relay 1CR contact 1CR-1 opens, opening the interlock circuit around the START pushbutton switch.
    b. Relay 1CR contact 1CR-2 opens, deenergizing the green pilot light.
    c. Relay 1CR contact 1CR-3 opens, deenergizing the time-delay coil 1TR.

Figure 4-8B shows another timing relay and control relay added to the circuit shown in Figure 4-8A. The new timing relay 2TR will have one NO contact arranged for time delay after deenergizing (OFF time). The second relay coil is 2CR. The sequence of operations proceeds as follows:

1. Operate the START pushbutton switch.
2. Relay coil 1CR is energized.
    a. Relay 1CR contact 1CR-1 closes, interlocking around the START push-button switch.
    b. Relay 1CR contact 1CR-2 closes, energizing timing relay coil 1TR.
    c. Instantaneous contact 1TR-1 on timing relay 1TR closes, energizing timing relay coil 2TR.
    d. OFF delay contact 2TR closes, energizing relay coil 2CR.
3. 2CR relay contact 2CR-1 closes, energizing the green pilot light.
4. Preset time on timing relay 1TR expires.
    a. Timing contact 1TR opens, deenergizing relay coil 1CR.
    b. 1CR relay contact 1CR-1 opens, opening the interlock circuit around the START pushbutton switch.
    c. 1CR relay contact 1CR-2 opens, deenergizing timing relay 1TR.
    d. Instantaneous contact on timing relay 1TR opens, deenergizing timing relay 2TR.
5. After a preset time delay as set on 2TR expires, OFF time-delay contact 2TR opens, deenergizing relay 2CR.
    a. Relay contact 2CR-1 opens, deenergizing the green pilot light.

Another timing relay is now available with up to two timed and two instantaneous contacts in various NO and NC combinations. Convenient program keys (8) provide a selection of four timing ranges in either on-delay or off-delay mode. Timers are available with either a self-contained adjustment potentiometer or

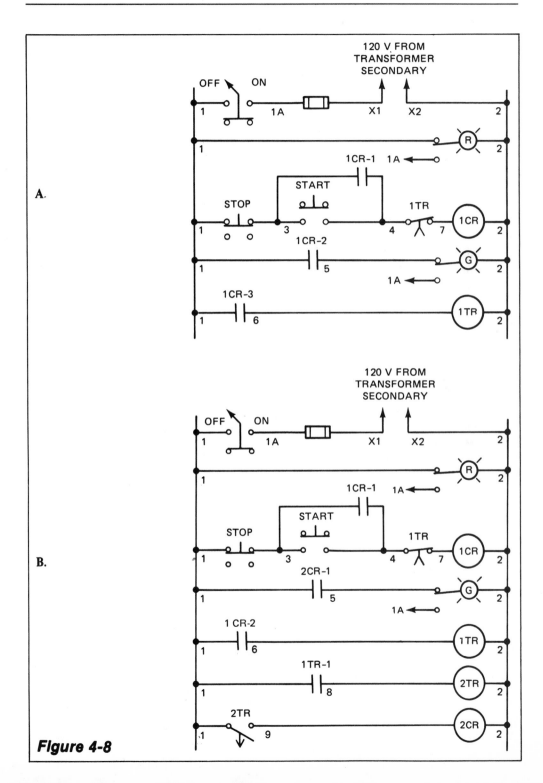

Figure 4-8

with provision for remote adjustment. Timers feature an LED indicator light which is off, flashing, or on to indicate timer operating status. Figure 4-7D illustrates this timing relay.

4.3 LATCHING RELAY

A mechanical latching attachment can be installed on the control relay. See Figure 4-9A. This obtains an interesting variation of the control capabilities of the relay. The latching relay is electromagnetically operated. It is held by means of a mechanical latch. By energizing a coil of the relay, called the *latch coil,* the relay operates. This results in the relay NO contacts closing and the NC contacts opening. The electrical energy can now be removed from the coil of the relay (deenergized), and the contacts remain in their operated condition. Now a second coil on the relay, the *unlatch coil,* must be energized in order to return the contacts to their unoperated condition. This arrangement is often referred to as a *memory relay.*

The latching relay has several advantages in electrical circuit design. For example, it may be necessary to open or close contacts early in a cycle. At the same time, it may also be desirable to deenergize the section of the circuit responsible for the initial energizing of the relay latch coil. Later in the cycle, the unlatch coil can be energized to return the contacts to their original or unoperated condition. The circuit is then set up for the next cycle. The symbol for the latching relay is shown in Figure 4-9B.

To further explain the symbols:

1. Contacts should be shown in the unlatched condition; that is, as if the unlatch coil were the last one energized.
2. The latch and unlatch coils are on the same relay and always have the same reference number.
3. The contacts are always associated with the latch coil as far as reference designations are concerned.

Another use for the latching-type relay involves power failure. Here it may be necessary that the contacts remain in their operated condition during the power-off period. Conditions in this case are the same after the power failure as they were before.

If quietness of operation is desired in a long cycle, this can be provided. The coil can be deenergized, thus eliminating the usual hum. Figure 4-9C illustrates a typical latching relay. The conventional two-coil circuit for the latching relay is shown in Figure 4-9D.

A. Latching attachment for a relay

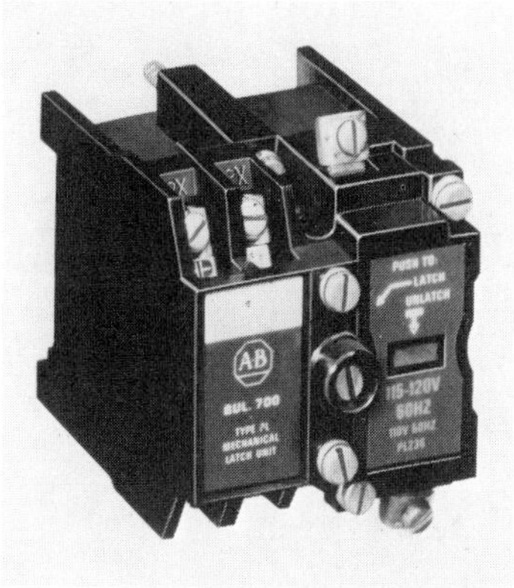

Courtesy of Allen-Bradley

B.

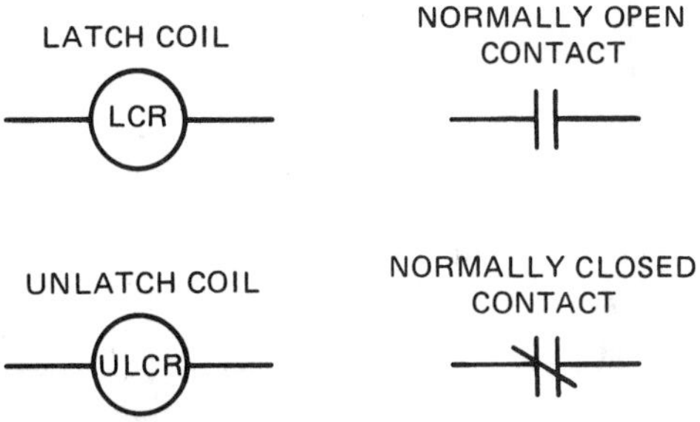

LATCH COIL

(LCR)

NORMALLY OPEN CONTACT

UNLATCH COIL

(ULCR)

NORMALLY CLOSED CONTACT

C. AC latching relay

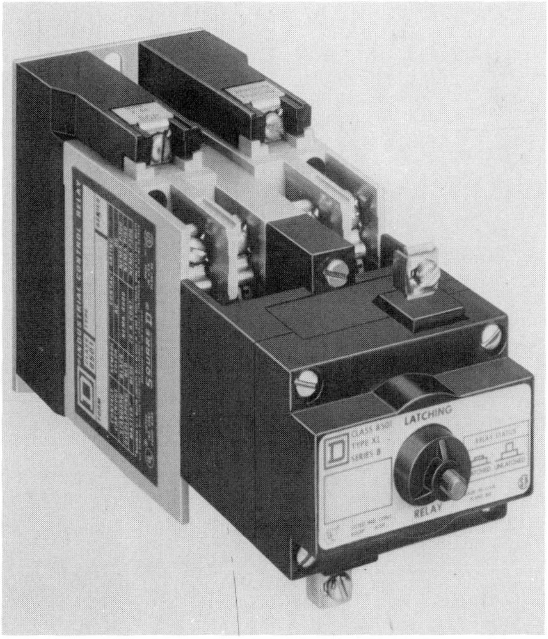

Figure 4-9

Courtesy of Square D Company

D.

E.

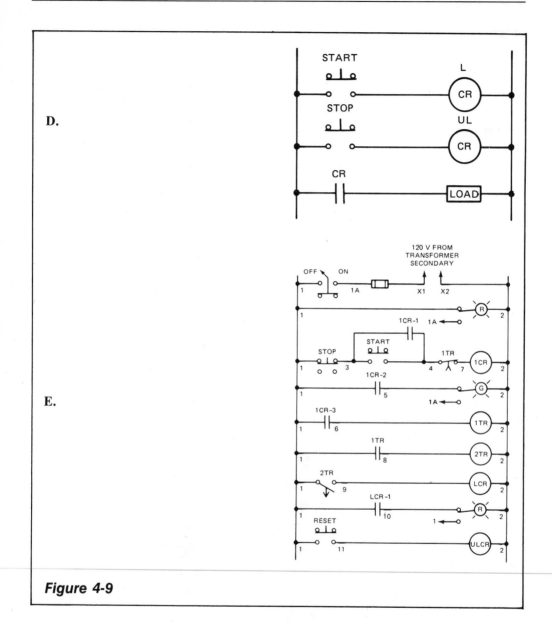

Figure 4-9

Refer again to the circuit shown in Figure 4-8B. The latch coil of a latching relay replaces the coil of control relay 2CR.

In Figure 4-9E, the closing of timing contact 2TR energizes the latch coil LCR. NO contact LCR-1 closes, energizing a red pilot light. When the timing relay 2TR times out and the timing contact opens, the latch coil is deenergized. However, the light remains energized until the RESET push-button switch is operated, energizing the unlatch coil ULCR.

4.4 PLUG-IN RELAYS

Some industrial machine relays are available in plug-in types. They are designed for multipole switching applications at or below 240 V. Coil voltages span standard levels from 6 V to 120 V. They are available for ac or dc. Mounting can be obtained with a tube-type socket, square-base socket mounting, or flange mounting using slip-on connectors.

A typical plug-in relay is shown in Figure 4-10. These relays are designed for multipole switching applications at or below 240 V.

The plug-in relay has a distinct advantage when changing relays without disturbing the circuit wiring. In critical operations, where the relay service is very hard

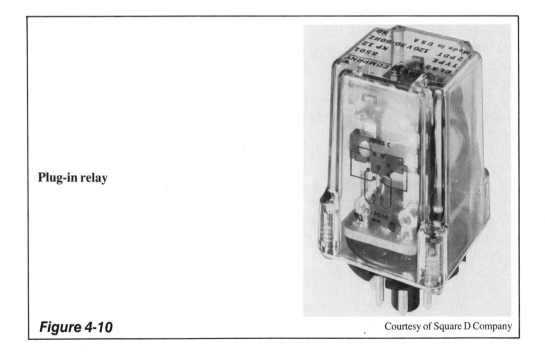

Plug-in relay

Figure 4-10   Courtesy of Square D Company

and downtime is a premium, the plug-in relay may have some advantages. Assessing the actual operating conditions in specific cases is the best way to determine their need.

4.5 CONTACTORS

The *contactor* is, in general, constructed in a similar fashion to the relay. Like the relay, it is an electromechanical device. The same coil conditions exist, in that a high inrush current is available when the coil of the contactor is energized. The current level drops to the holding or sealed level when the contacts are operated. Generally the contactor is supplied in two-, three-, or four-pole arrangements. The coil of the contactor is generally energized at 120 V. The major difference is in the size range available with contactors. Contactors capable of carrying current in the range of 9 A through approximately 2250 A are available. For example, a size 00 contactor is rated at 9 amperes (200-575 V); a size 9 contactor is rated at 2250 A (200-575 V).

One normally open auxiliary contact is generally supplied as standard on most contactors. This contact is used as a holding contact in the circuit. For example, around a normally open pushbutton switch. Additional normally open and normally closed auxiliary contacts can be obtained as an option. They can be supplied with the contactor from the manufacturer or ordered as a separate unit and mounted in the field.

Figure 4-11A shows two typical electromechanical contactors. The bottom one is a size 1 with a 27-ampere rating and the top one is a size 9 with a 2250-ampere rating.

The use of the auxiliary contact, which is generally rated at 10 amperes, is shown in Figure 4-11B. Here the contactor is added to the circuit shown in Chapter 3, Figure 3-9A.

It can be seen that the auxiliary contact is used like the relay interlock or holding circuit.

In the circuit in Figure 4-11B, the push-button switch used to energize the light in Figure 3-9A has been removed. The light is connected across the source of electrical energy (circuit lines 1-2). The light in the circuit in Figure 4-11B, therefore, is energized when the control ON-OFF selector switch is operated to the ON position.

The balance of the sequence is as follows:

1. Operate the HEAT ON push-button switch.
2. Coil of contactor 10CR is energized.

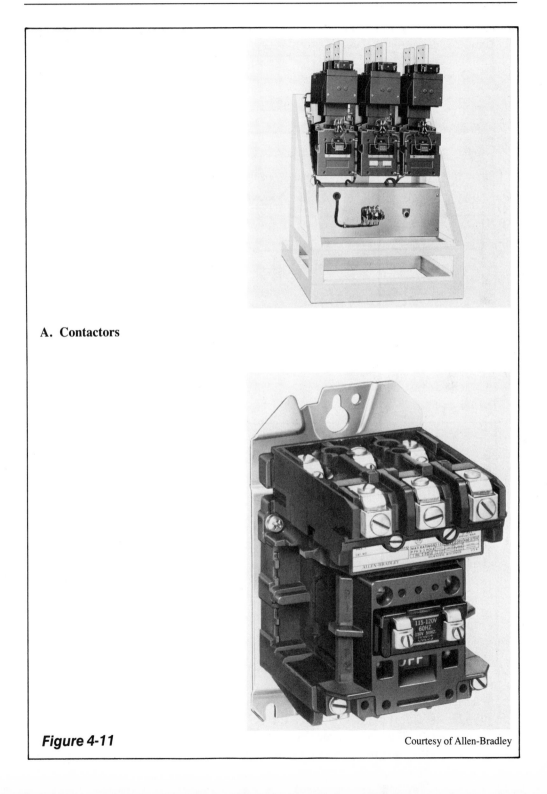

A. Contactors

Figure 4-11

Courtesy of Allen-Bradley

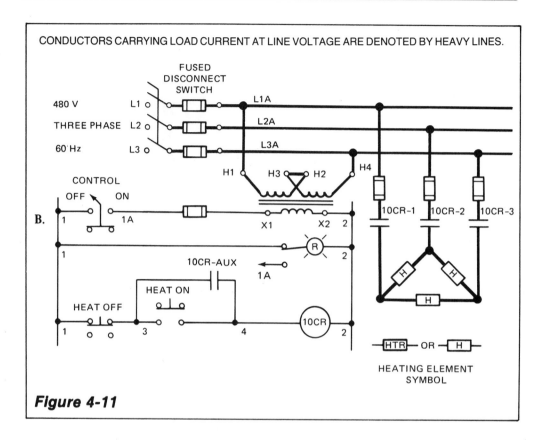

Figure 4-11

3. Power contacts 10CR-1, 10CR-2, 10CR-3 close, energizing the heating elements at line voltage.

4. Auxiliary contact 10CR aux. closes, interlocking around the HEAT ON pushbutton switch.

5. Operate the HEAT OFF push-button switch, deenergizing the coil of 10CR contactor.

    a. Contacts 10CR-1, 10CR-2, 10CR-3 open, deenergizing the heating elements.

    b. Auxiliary contact 10CR aux. opens, opening the interlock circuit around the HEAT ON push-button switch.

In addition to the conventional electro-mechanical contactor is the mercury-to-metal contactor. See Figure 4-12 showing a cross section through the contactor. The load terminals are isolated from each other by the glass in the hermetic seal. The plunger assembly which includes the ceramic insulator, the magnetic sleeves and related parts float on the mercury pool. When the coil is powered, causing a magnetic field, the plunger assembly is pulled down into the mercury pool,

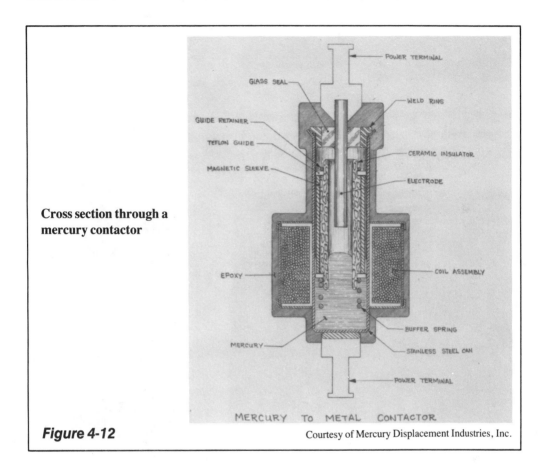

Cross section through a mercury contactor

Figure 4-12

Courtesy of Mercury Displacement Industries, Inc.

which, in turn, is displaced and moved up to make contact with the electrode. This closes the circuit between the top and bottom load terminal which is connected to the stainless steel can.

Figure 4-13 shows a group of one-, two-, and three-pole contactors. They are available up through 100 A capacity.

The basic use for the contactor is for switching of power in resistance heating elements, lighting, magnetic brakes or heavy industrial solenoids. Contactors can also be used to switch motors if separate overload protection is supplied.

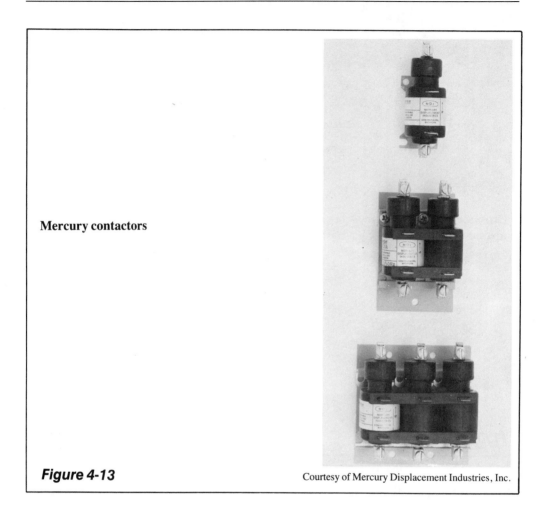

Mercury contactors

Figure 4-13

Courtesy of Mercury Displacement Industries, Inc.

ACHIEVEMENT REVIEW

1. What is the difference between inrush current and holding current in the coil of a control relay?

2. What are the three ratings generally published concerning control relay contacts?

3. Explain contact wipe, and explain how contact bounce can affect the operation of an electrical control circuit?

4. Draw the symbols for a control relay coil, normally open relay contact, and normally closed relay contact.

5. Show the use of a control relay contact in an interlock circuit. Use a control relay, two push-button switches, and a red push-to-test pilot light.

6. What advantages can be obtained from the use of a time-delay relay as compared to a standard control relay?

7. Identify the following time-delay relay contacts as to whether they operate before or after the relay coil is energized and whether they are normally open or normally closed.

A.

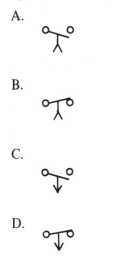

B.

C.

D.

8. Follow the instructions in question 5, but substitute a time-delay relay for the standard control relay. In this case, the red pilot light is to energize at some predetermined time after the relay coil is energized.

9. Describe how a latching relay operates.

10. How many poles are generally available on contactors? What is the range in current-carrying capacity?

11. What is one advantage in using the plug-in relay?
    a. No circuit numbers are required.
    b. The best application of the plug-in relay is in power circuits.
    c. Relays can be changed without disturbing the circuit wiring.

12. Coils on relays are designed not to drop out (deenergize) until the voltage drops to
    a. 60%.
    b. 85%.
    c. 100% of the rated voltage.

CHAPTER 5

SOLENOIDS

OBJECTIVES

After studying the chapter, the student will be able to:

- Explain why it is necessary for the plunger in a solenoid to complete its stroke.

- Discuss the two important problems to consider in the application of a solenoid.

- Describe the application of solenoids to operating valves.

- Draw the symbol for the solenoid.

- Draw a control circuit showing the energizing of a solenoid through the closing of a relay contact, using a control relay, two push-button switches, and a solenoid.

- Know the difference between sealed current and inrush current in a solenoid.

5.1 SOLENOID ACTION

The general principle of the solenoid action is very important in machine control. Like the relay and contactor, the *solenoid* is an electromechanical device. In this device, electrical energy is used to magnetically cause mechanical movement.

Review the explanation given in Chapter 4, Section 4.1 for the relationship between plunger travel and coil current. A similar condition exists with the solenoid. It is important that the plunger completes its stroke when the solenoid is energized. Otherwise, the current in the coil will be high, resulting in damage to the coil. Figure 5-1 shows a typical curve of the relationship of current and stroke. What

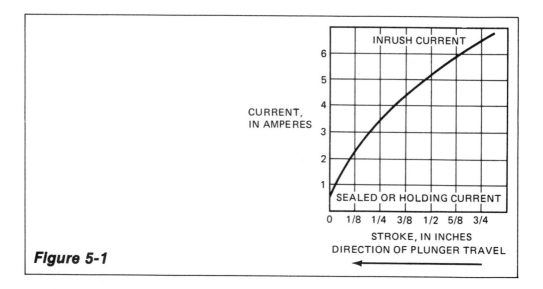

Figure 5-1

happens to the current when the stroke is not completed to zero can be seen.

The current in amperes at the open position is called *inrush current*. The current in amperes at the closed position is called *sealed*, or *holding*, *current*. The ratio of inrush current to sealed current generally varies from approximately 5:1 in small solenoids to as much as 15:1 in large solenoids.

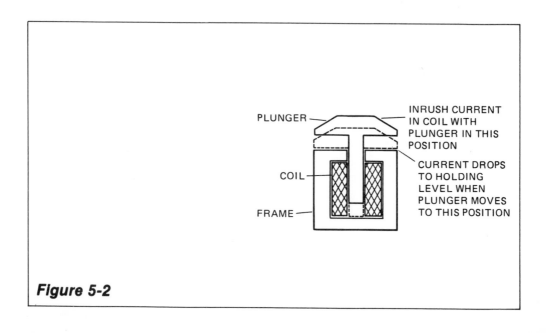

Figure 5-2

When the solenoid coil is energized, the plunger is in an out position. Due to the open gap in the magnetic path (circuit), the initial current in the coil is high. As the plunger moves into the coil, closing the gap, the current level drops to a lower value.

As shown in Figure 5-2, the solenoid is made up of three basic parts:

1. Frame   2. Plunger   3. Coil

**Cutaway sections of typical
solenoids**

Figure 5-3   Courtesy of Detroit Coil Company

The frame and plunger are made up of laminations of a high-grade silicon steel. The coil is wound of an insulated copper conductor.

Cutaway sections of typical solenoids are shown in Figure 5-3.

Solenoids for alternating-current use are now available as oil-immersed types. Heat dissipation and wear conditions are improved with this design. They are also available with a plug-in base. A typical example of this design is shown in Figure 5-4.

When the coil of a solenoid is energized, a magnetic field is produced about the coil. This magnetic field produces a force that acts on the solenoid plunger. Due to this force, the plunger moves into the coil. This force on the plunger is called *pull*.

The pull in solenoids varies widely. The pull may be as low as a fraction of an ounce (oz) or as high as nearly 100 pounds (lb).

Connections to the coil may be supplied in the following ways:

- Pigtail leads

- Terminals on the coil

- Terminal blocks

- Plug-in connections

There are two important problems to be considered in the application of a solenoid.

1. The pull of the solenoid must at all times exceed the load. If the pull is a little less, the solenoid action will be sluggish and may not complete the stroke. There are also conditions that the user may not always be able to control, such as low voltage or increased loading through friction or pressure. Therefore, it is generally advisable to overrate the solenoid by 20% to 25%. **Caution:** Too much pull will result in the plunger slamming, resulting in damage to the plunger and frame.
2. The duty cycle of the work load should be known. Some applications require a duty cycle with only an occasional operation. Other cases require up to several hundred operations per minute. **Caution:** The operation of the solenoid above its maximum cycling rate will result in excessive heating and mechanical damage.

The solenoid as applied to valves is used several times in this book. Therefore, a brief explanation of simple valve action is in order. The pressure referred to can be either hydraulic (fluid) or pneumatic (air). The pressure *ported* (directed) to a cylinder-piston assembly is used to explain the control of position and pressure changes on machines.

**Plug-in, oil-immersed
solenoids**

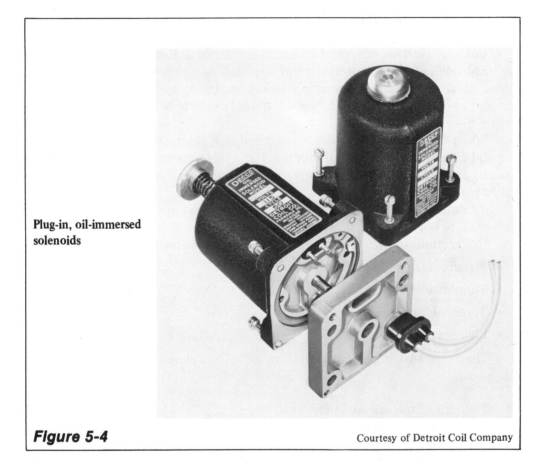

Figure 5-4   Courtesy of Detroit Coil Company

Figure 5-5A shows a single-solenoid, spring-return operating valve. It is shown in its normal deenergized condition. The symbols for the solenoid and valve are also shown.

In Figure 5-5B, the solenoid valve is shown in the energized condition. In the energized condition, note that the spring is loaded (depressed). When the solenoid is deenergized, the spring pressure returns the valve spool to its normal deenergized condition.

Note that when solenoid A is deenergized, pressure is free to flow to port A. Port B is open to tank. When solenoid A is energized, pressure is free to flow to port B, and port A is open to tank.

The double-solenoid valve is shown in Figure 5-5C. In the deenergized condition as shown, no pressure is available to either port A or port B, as it escapes around the land areas of the valve spool to the tank port. The centering springs keep the valve spool in this position, with both solenoids deenergized.

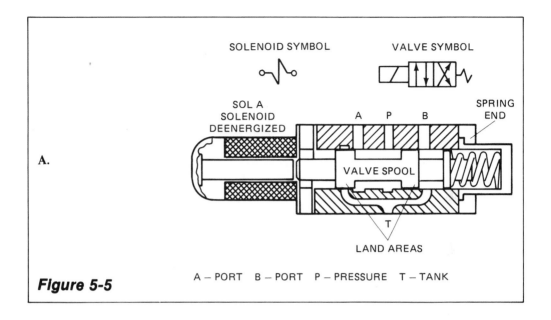

A.

Figure 5-5

SOLENOID SYMBOL

VALVE SYMBOL

SOL A
SOLENOID
DEENERGIZED

A P B

SPRING
END

VALVE SPOOL

T

LAND AREAS

A – PORT   B – PORT   P – PRESSURE   T – TANK

Figure 5-5D shows the double-solenoid operating valve with solenoid A energized. The force or pull exerted on the plunger in solenoid A moves the valve spool to the right. The spring is solenoid B is compressed (loaded). The land areas on the valve spool are now located so that pressure available at port P is free to flow into port A. Port B is open to the tank port T. As there is generally little or no pressure connected at the tank port, it follows that there will be little or no pressure at port B.

Figure 5-5E shows the double-solenoid operating valve with solenoid B energized. The force or pull exerted on the plunger in solenoid B moves the valve spool to the left. The spring in solenoid A is compressed (loaded). The land areas on the valve spool are now located so that pressure available at port P is free to flow into port B. Port A is open to the tank port T. As there is generally little or no pressure connected at the tank port, it follows that there will be little or no pressure at port A.

5.2 CIRCUIT APPLICATIONS

Refer again to the basic relay circuit in Chapter 4, in Figure 4-3. The same circuit is shown in Figure 5-6A, except that a solenoid is substituted for the red pilot light.

The sequence of operations proceeds as follows:

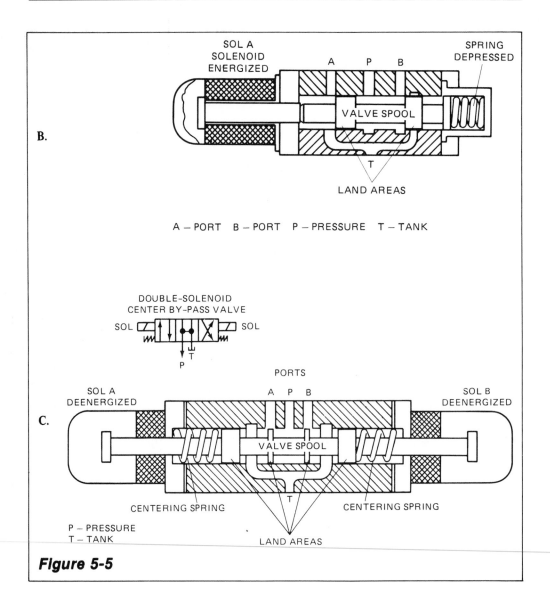

Figure 5-5

1. Operate the START push-button switch.
2. Coil of relay 1CR is energized.
3. Relay contact 1CR-1 closes, interlocking around the START push-button switch.
4. Relay contact 1CR-2 closes, energizing solenoid A.
5. Operate the REVERSE push-button switch.
6. Relay 1CR is deenergized.

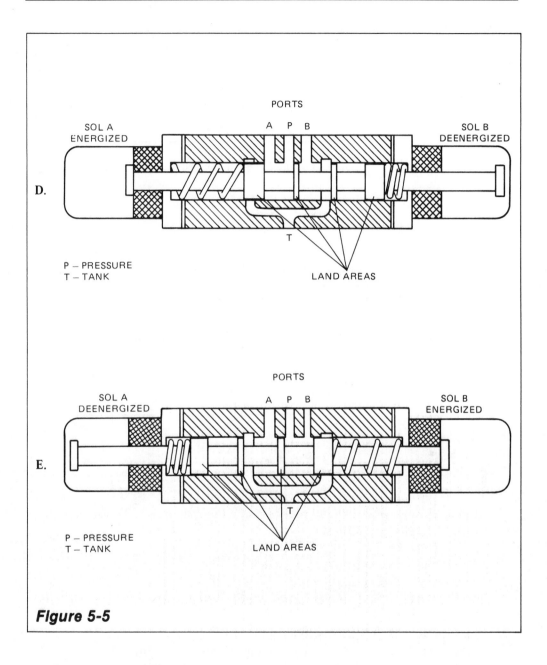

Figure 5-5

7. Relay contact 1CR-1 opens, opening the interlock circuit around the START push-button switch.
8. Relay contact 1CR-2 opens, deenergizing solenoid A.

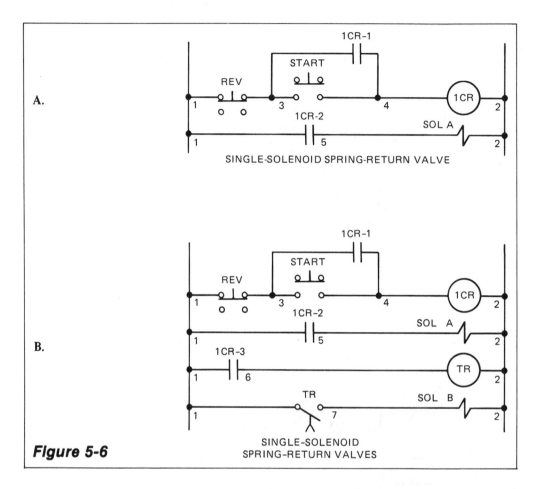

Figure 5-6

SINGLE-SOLENOID
SPRING-RETURN VALVES

In Figure 5-6B, a time-delay relay is added to the circuit. The sequence for this circuit proceeds as follows:

1. Operate the START push-button switch.
2. Coil of relay 1CR is energized.
3. Relay contact 1CR-1 closes, interlocking around the START push-button switch.
4. Relay contact 1CR-2 closes, energizing solenoid A.
5. Relay contact 1CR-3 closes, energizing timing-relay coil TR.
6. After a preset time on TR expires, the timing contact TR closes, energizing solenoid B.
7. Operate the REVERSE push-button switch, deenergizing relay coil 1CR.
8. Relay contact 1CR-1 opens, opening the interlock circuit around the START push-button switch.

9. Relay contact 1CR-2 opens, deenergizing solenoid A.
10. Relay contact 1CR-3 opens, deenergizing the coil of timing relay TR.
11. Timing relay contact TR opens, deenergizing solenoid B.

Note that if the REVERSE push-button switch is operated before the preset time expires on TR, solenoid B will not be energized.

ACHIEVEMENT REVIEW

1. Draw the symbol for the solenoid.
2. What happens in a solenoid if the plunger is prevented from completing its stroke?
3. At what position of the solenoid plunger does inrush current appear?
4. Why must the pull of a solenoid always exceed the load?
5. What will happen in a solenoid if the operation exceeds the solenoid design?
6. In the application of a solenoid to a spring-return operating valve, assume that the solenoid has been energized. What happens to the valve spool when the solenoid is deenergized?
7. Draw a control circuit using two push-button switches, a control relay, and a solenoid. The solenoid is to be energized when a normally open push-button switch is operated. It is to be deenergized when a normally closed push-button switch is operated.
8. The solenoid current in amperes with the plunger in the closed position is
   a. maximum current.
   b. no current will be indicated.
   c. sealed or holding current.
9. When a solenoid is energized in a single solenoid, spring-return valve, the spring
   a. remains in the same condition as it was before the solenoid was energized.
   b. is depressed (loaded).
   c. aids the solenoid in moving the valve piston.
10. The force acting on the plunger of a solenoid when the solenoid is energized is caused by
    a. hydraulic pressure.
    b. gravity.
    c. a magnetic field that is produced about the coil.

CHAPTER 6

LIMIT SWITCHES AND LINEAR POSITION-DISPLACEMENT TRANSDUCERS

OBJECTIVES

After studying this chapter, the student will be able to:

- List the three basic classes of limit switches.
- Explain where rotating-cam limit switches are used.
- Explain the following terms relative to limit switches:
    1. Operating force
    2. Release force
    3. Pretravel or trip travel
    4. Overtravel
    5. Differential travel
- Discuss the proximity limit switch and explain how it is used.
- Describe how the vane limit switch operates.
- Draw the limit switch symbols for four different conditions.
- Design an electrical operating circuit showing how the operation of a normally closed limit switch contact can be used to deenergize a solenoid, using two push-button switches, a normally closed limit switch contact, a relay, and a solenoid.
- List several methods for achieving proximity switching.
- List some uses for the linear transducer.

6.1 INDUSTRIAL USES

On the subject of electrical control of machines, position plays an important part. The problem is to accurately and reliably supply position information by providing an adequate electrical signal. The limit switch provides one of the best ways to accomplish this purpose.

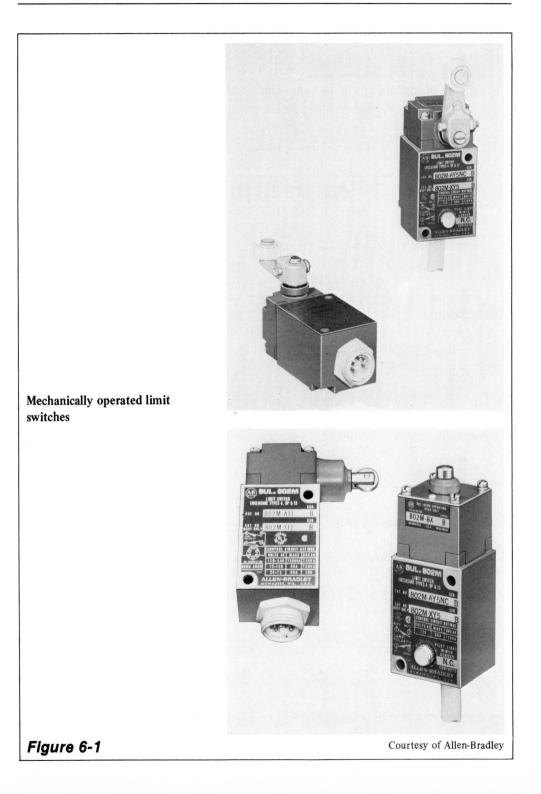

Mechanically operated limit switches

Figure 6-1

Courtesy of Allen-Bradley

To describe the various limit switches available on the market today and their uses would require an entire book. Therefore, this book covers only a few widely used units.

Industrial machine limit switches can be divided into three basic classes:

1. Mechanical   2. Proximity   3. Vane

6.2 MECHANICAL

Some typical mechanical limit switches are shown in Figure 6-1. Mechanical-type switches can be subdivided into those operated from linear motion, and those operated from rotary motion.

There are large types and small types of switches that operate from linear motion. Generally, the small class consists of the *precision limit switch.* This switch varies from the larger size mainly in a lower operating force and shorter stroke. The operating force may be as low as 1 pound. The stroke may be only a few thousandths of an inch.

Limit switches operated by rotary motion are generally called *rotating-cam limit switches.* These are control circuit devices used with machinery having a repetitive cycle of operation where motion can be correlated to shaft rotation. It is used to limit and control the movement of a rotating machine, and to initiate functions at various points in the repetitive cycle of the machine.

A rotating-cam limit switch arrangement is shown in Figure 6-2.

The switch assembly consists of one or more snap action switches. The cams are assembled on a shaft. The shaft in turn is either directly or through gearing driven by a rotary motion on the machine.

The cams are independently adjustable for operating at different locations within a complete 360-degree (360°) rotation. In some cases the number of total rotations available is limited. In other cases the rotation can continue and at speeds up to 600 revolutions per minute (r/min).

In selecting a limit switch, it is important to determine its application in the electrical circuit. The following factors must be considered:

- Contact arrangement
- Current rating of the contacts
- Slow or snap action
- Isolated or common connection

Rotating-cam limit switch

Figure 6-2

Courtesy of Allen-Bradley

- Spring return or maintained
- Number of NO and NC contacts required

In most cases the switch consists of double-break, snap-action, silver-tipped to solid silver contacts. The contact current rating will vary from 5 A to 10 A at 120 V ac continuous. The make contact rating will be much higher, and the break contact rating will be lower. Isolated NO and NC contacts are available. In some cases, multiple switches in the same enclosure operated by the same mechanical action are used.

A second important factor is the type of mechanical action available to operate the switch. Here, the *operator* is the major decision. Length of travel, speed, force available, accuracy and type of mounting possible are some of the considerations.

In discussing the action of limit switches, several terms are used. A knowledge of these terms is helpful.

1. Operating force — The amount of force applied to the switch to cause the "snap over" of the contacts
2. Release force — The amount of force still applied to the switch plunger at the instant of "snap back" of the contacts to the unoperated condition
3. Pretravel or trip travel — The distance traveled in moving the plunger from

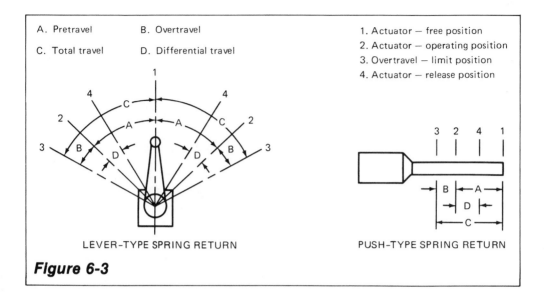

A. Pretravel   B. Overtravel

C. Total travel   D. Differential travel

1. Actuator — free position
2. Actuator — operating position
3. Overtravel — limit position
4. Actuator — release position

LEVER-TYPE SPRING RETURN   PUSH-TYPE SPRING RETURN

Figure 6-3

its free or unoperated position to the operated position

4. Overtravel – The distance beyond operating position to the safe limit of travel; usually expressed as a minimum value

5. Differential travel – The actuator travel from the point where the contacts snap over to the point where they snap back

6. Total travel – The sum of the trip travel and the overtravel.

Figure 6-3 shows the last four items in diagram form.

Most manufacturers of limit switches list this information for certain switches in their specifications. Accuracy of switch operators at the point of snap over varies with different types and manufacturers. In general, it is in the range of 0.001 inch (in) to 0.005 in.

The operator that has probably the greatest use is the roller lever. It is available in a variety of lever lengths and roller diameters.

The next most frequently used operator is the push rod. This operator can consist of only a rod or it can be supplied with a roller in the end.

In most cases, particularly with the oil-tight machine tool limit switch, the head carrying the operator can be rotated to four positions, 90° apart. It can also be either top or side mounted.

Two other operators used in machine control are the fork lever and the wobble stick.

Some of the various operators available are shown in Figure 6-4.

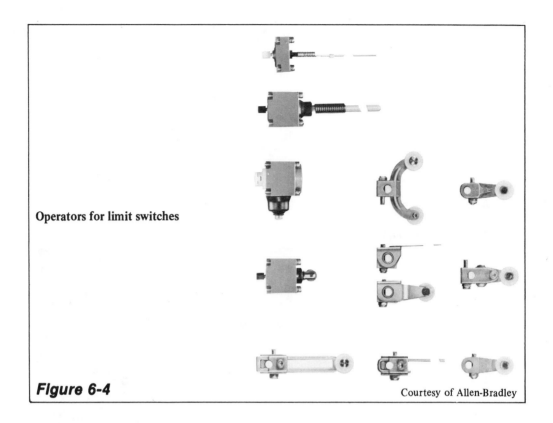

Operators for limit switches

Figure 6-4

Courtesy of Allen-Bradley

6.3 PROXIMITY SWITCHES

The proximity switch consists of a sensor used to detect the presence of an object without physical contact. This is the important difference when compared with the mechanical limit switch. See Figure 6-5A for typical proximity switches.

There are several methods used to achieve the operation of the proximity switch. For example, some methods have made use of the effect of objects on magnetic fields, RF (radio frequency) fields, capacitive fields, acoustic fields and light rays.

Magnetic fields have been used to close reed switches by bringing the magnet up close to the switch. Magnets have been used to alter electric fields in devices, making use of the Hall effect.

The Hall effect can be attributed to E. B. Hall who first noted the effect late in the last century. The effect can be stated that a magnetic field applied to a conductor carrying a current produces a voltage across the conductor. This voltage is known as the Hall voltage.

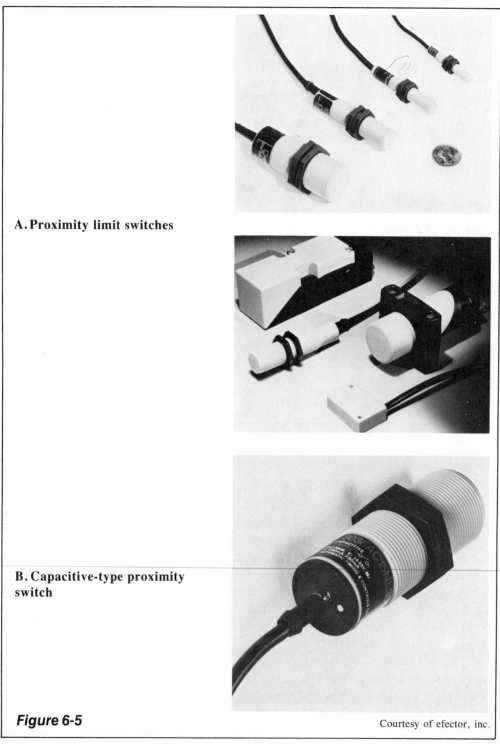

A. Proximity limit switches

B. Capacitive-type proximity switch

Figure 6-5

Courtesy of efector, inc.

The Hall voltage is proportional to the product of the current and the field. That is, a device that exhibits the Hall effect is a multiplier. With constant current, the Hall voltage will be proportional to the magnetic field. With the magnetic field constant, the Hall voltage will be proportional to the current flow.

More recently, the production of Hall effect integrated circuits (ICs) has eliminated the problems experienced with the use of discrete component circuit designs.

In addition to the use of Hall effect ICs in proximity switches, designers have obtained good results with the use of capacitive circuits, magnetic pick-up, and photoelectric switching.

A word of caution should be added about the use of Hall effect ICs. They are susceptible to mechanical damage. In hand soldering, a heat sink should be used on the leads.

RF fields are altered by the presence of ferrous materials. This material absorbs energy by any current produced in it by the field.

The capacitive switch makes use of a change in capacity that occurs when the object to be sensed acts as a plate of a capacitor. The sensor acts as the other plate.

Sonic devices utilize sound fields which are either interrupted by the object to be detected or detect the reflection of sound from objects.

Photoelectric devices operate in a similar manner. However, with these devices, light rays are detected rather than sound waves.

Most proximity switches are supplied in a rugged molded housing that protects the unit. The protection is against the dust, oil and dirt found in industrial installations. Metal housings are also available. For example, stainless steel is used in the food industry.

The detection range may vary from 0.1 mm to 50 mm with corresponding repeatability of 0.3 mm to 5 mm.

The inductive and capacitive type of proximity switches use oscillators in their output circuits. The speed of response for these switches is a function of the oscillator frequency. It follows then that the higher the frequency, the shorter the response time.

These devices have solid-state outputs. They may have NO or NC output. In some cases, this is programmable.

In many of the switches, an LED (light-emitting diode) is supplied. The light goes on when the solid-state output switch is closed. This feature aids the sensitivity adjustment procedure and provides a convenient check of the switch operational status.

Figure 6-5B shows a capacitive model of an electronic proximity switch. The magnet operated proximity limit switch is shown in Figure 6-6. This switch operates by passing an external magnet near the face of the sensing head. This

Magnet-operated limit switch

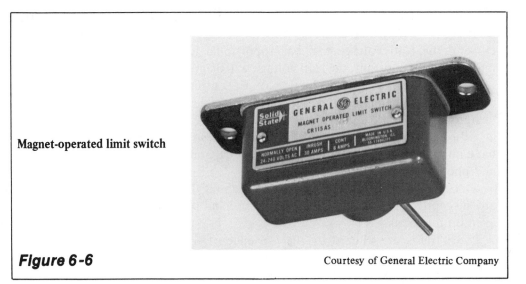

Figure 6-6

Courtesy of General Electric Company

ELECTRODE

Proximity switch

PLUNGER
FLOATING
ON MERCURY

PLUNGER
IS ATTRACTED
TO MAGNET

OPERATING
DISTANCE
1/8"

PERMANENT MAGNET
ABOUT 1,000 GAUSS

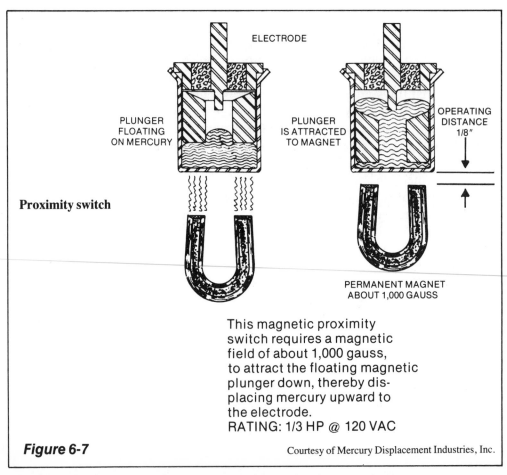

This magnetic proximity
switch requires a magnetic
field of about 1,000 gauss,
to attract the floating magnetic
plunger down, thereby dis-
placing mercury upward to
the electrode.
RATING: 1/3 HP @ 120 VAC

Figure 6-7

Courtesy of Mercury Displacement Industries, Inc.

actuates a small, hermetically sealed reed switch. The 120-volt ac pilot-duty model also includes an epoxy-encapsulated triac output.

One more proximity limit switch is the mercury switch. It operates by passing a permanent magnet of sufficient strength past the switch. The sketch shown in Figure 6-7 displays the internal working parts and the relationship between the strength of the magnet and the operating distance.

6.4 VANE SWITCHES

The *vane-operated limit switch* is actuated by passage of a separate steel vane through a recessed slot in the switch, Figure 6-8.

Either the vane or switch can be attached to the moving part of the machine. As the vane passes through the slot, it changes the balance of the magnetic field, causing the contacts to operate. The switch is available with either a normally open or normally closed contact.

The switch can detect very high speeds of vane travel without detrimental effects such as arm or mechanism wear or breakage. There is no physical contact between vane and switch. Therefore, the upper limit on vane speed is governed by factors other than the switch.

The vane switch offers excellent accuracy and response time. Providing the path of the vane through the slot is constant, repeatability is constant within ±0.0025 inch or less. The time required for the switch to operate after the vane has reached the operating point (response time) is less than a millisecond.

Figure 6-9 shows the oil-tight and dust-tight vane-operated limit switch. It provides high reliability and long life, with an electrical rating capable of handling high inductive loads.

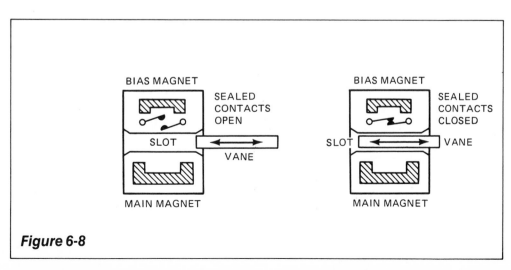

Figure 6-8

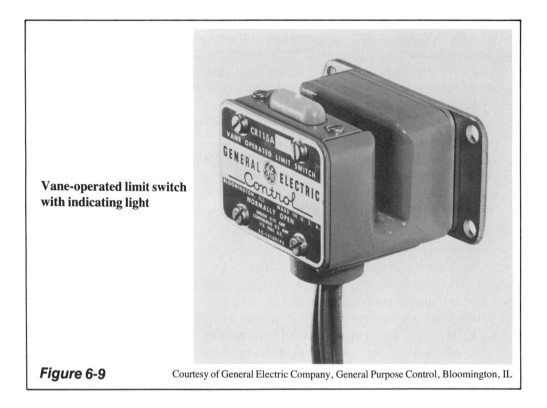

Vane-operated limit switch with indicating light

Figure 6-9   Courtesy of General Electric Company, General Purpose Control, Bloomington, IL

6.5 LINEAR POSITION-DISPLACEMENT TRANSDUCERS

By precisely sensing the position of an external magnet, the solid-state transducer (as shown in Figure 6-10), is able to measure linear displacements with infinite resolution. Since there is no contact between the magnet and the sensor rod, there is no wear, friction, or degradation of accuracy. Their outputs represent an absolute position, rather than an incremental indication of position change. There is an option of either a digital or analog output.

The measuring principle by which this transducer operates can be explained as follows: when a current pulse is sent through a wire (which has been threaded through a tube and returned outside), the resulting magnetic field is concentrated in the tube, which acts as a waveguide. If this tube is then passed through a doughnut-shaped magnet, the two magnetic fields interact. A tube made of magnetostrictive material will experience a local rotary strain where these fields interact. This strain will continue for the duration of the electrical pulse. The rotary strain pulse travels along the wave guide element at ultrasonic speed and can be detected at the end of the tube.

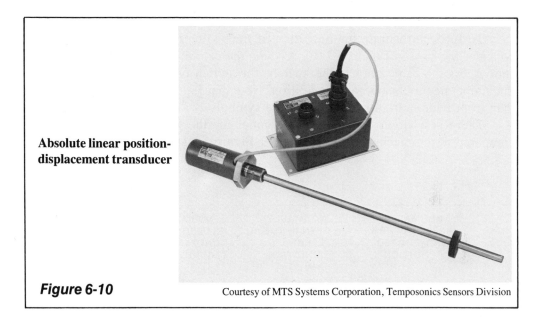

Absolute linear position-displacement transducer

Figure 6-10

Courtesy of MTS Systems Corporation, Temposonics Sensors Division

By measuring the time from the generation of the initial electrical pulse until the ultrasonic pulse is detected, one can determine the distance of the external magnet from a reference point.

This transducer is used for either position readout or closed loop control. It may be mounted inside hydraulic cylinders or externally on machines.

The applications are very broad, covering small winding machines, machine tools, plastic forming machines, etc.

6.6 LIMIT SWITCH SYMBOLS

The mechanical limit switch symbol is used in applying the limit switch to the schematic or elementary circuit diagram. There are four separate symbols, Figure 6-11A. These represent the switch in four different conditions. Basically, the switch has a normally open contact and/or a normally closed contact. In some switches now available, there are two NO and two NC contacts.

The switch symbol should be shown as it is in the unoperated condition of the machine. This may result in the switch being shown in either the operated or unoperated condition.

There are times when both the NO and NC contacts on a given limit switch may be used in a circuit. Under this condition, it helps to join the two contact symbols with a broken line. This indicates that they are contacts on the same limit switch, Figure 6-11B.

To further illustrate the operation of the switch in relation to the symbol, consider the case in Figure 6-11C. Assume the cam to be moving to the right as shown. As the cam contacts the switch, the switch operates and changes from the unoperated to the operated condition. Note that Figure 6-11C has no use other than to provide help in remembering the symbols.

There is a slight change in the symbol for the proximity limit switch as compared to the mechanical limit switch, Figure 6-11D.

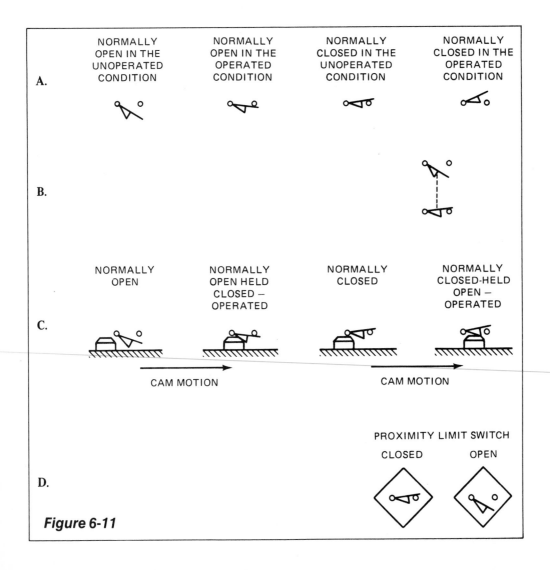

Figure 6-11

6.7 CIRCUIT APPLICATIONS

Limit switches in machine tool electrical control are used to gather information relative to the position of a machine part. To illustrate this, it is helpful to show a means of providing motion. In Chapter 5, Solenoids, the operating valve is introduced. The valve, along with a cylinder-piston assembly, is used in the limit switch application circuits.

At this point an explanation is given to help understand the solenoid operating valve symbol as it is used and shown in circuit problems.

Figure 6-12A shows the symbol for a single-solenoid, spring-return operating valve in the deenergized condition. Figure 6-12B shows the symbol for the same valve with solenoid A energized. Note the change in the flow of pressure through the valve to the piston-cylinder assembly.

Figure 6-12C shows the symbol for the double-solenoid operating valve with both solenoids deenergized. Figure 6-12D shows the symbol with solenoid A energized. Figure 6-12E shows the symbol with solenoid B energized. Here again, as with the single-solenoid operating valve, note the change that takes place when either solenoid is energized. Remember that when either solenoid is deenergized, centering springs in the valve return the piston to the center position.

The first circuit is built on the basic solenoid circuit shown in Figure 5-6A. The limit switch circuit is shown in Figure 6-13. A cylinder-piston assembly is shown as a means of moving a cam on the piston from position A to position B.

In Figure 6-13A, the piston is in position A. Solenoid A is deenergized. In Figure 6-13B, the piston is in position B. Solenoid A is energized. Limit switch 1LS contact shown in the circuit of Figure 6-13, is normally closed.

The sequence of operations proceeds as follows:

1. Operate the START pushbutton switch.
2. Relay coil 1CR is energized.
   a. Relay contact 1CR-1 closes, interlocking around the START pushbutton switch.
   b. Relay contact 1CR-2 closes, energizing solenoid A.

The valve spool shifts, permitting pressure to enter the main cylinder area (left-hand end). The pressure medium (water, oil, or air) in the right-hand end of the cylinder (rod end), is free to return to the tank. The piston moves from its start position at A to a new position at B.

3. At position B, limit switch 1LS is operated, opening its normally closed contact.

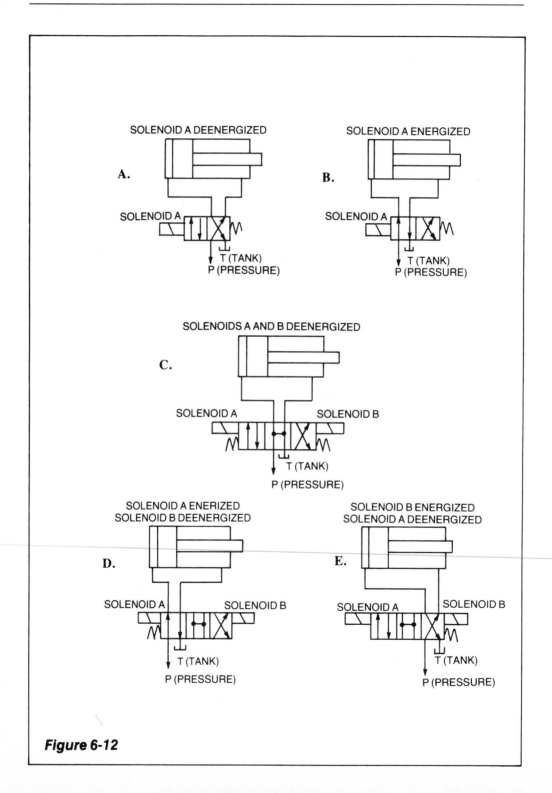

Figure 6-12

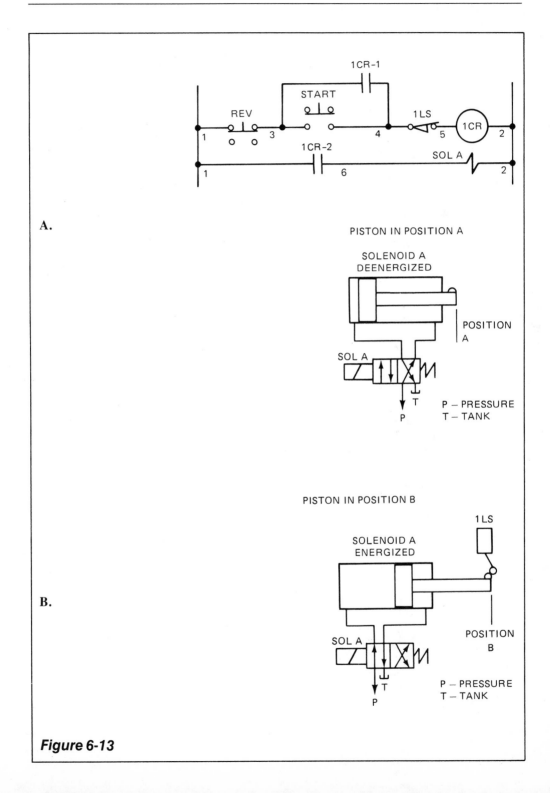

A.

B.

Figure 6-13

4. Relay coil 1CR is deenergized.
   a. Relay contact 1CR-1 opens, opening the interlock circuit around the START pushbutton switch.
   b. Relay contact 1CR-2 opens, deenergizing solenoid A.

   With solenoid A deenergized, the valve spring returns the valve spool to its initial position. This permits pressure to enter the rod end (right hand) returning the piston to position A.

   In the circuit shown in Figure 6-13, it is assumed that the piston was at position A to start. To ensure that the piston is at position A to start a cycle, a second limit switch, 2LS, is placed at position A. See Figure 6-14. A normally open limit switch contact is used. It is held operated (contact closed) by a cam on the piston. Note that 2LS limit switch contact is placed inside the START pushbutton interlock circuit formed by relay contact 1CR-1. Otherwise, as soon as the cam moves off limit switch 2LS, the limit switch contact opens, deenergizing the circuit.

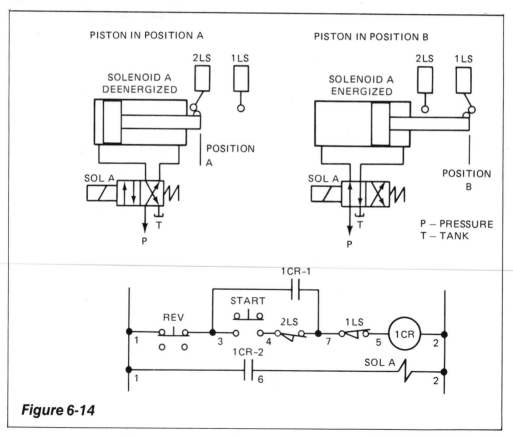

Figure 6-14

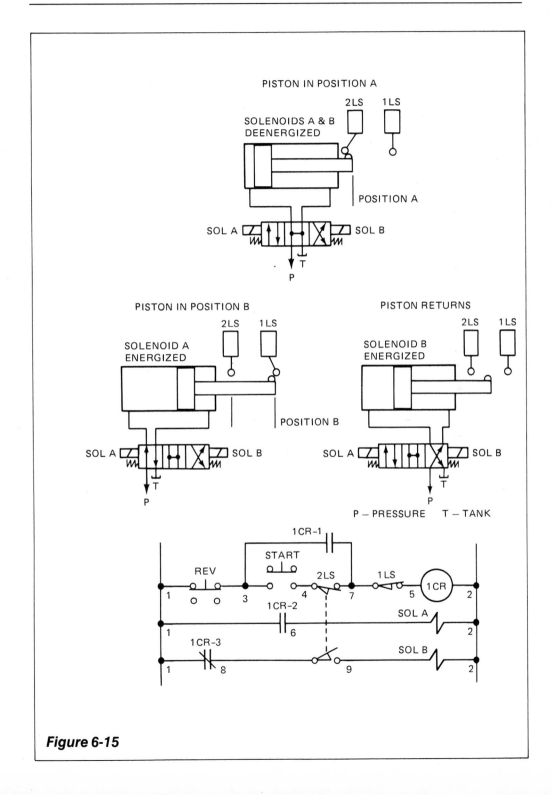

Figure 6-15

The operation of this circuit is identical to that shown in Figure 6-13, except the piston must be in position A, operating limit switch 2LS for start conditions.

A third circuit is shown in Figure 6-15. This circuit uses a double-solenoid, spring-return-to-center valve. In this circuit, a limit switch performs a double duty. The NO contact on limit switch 2LS, held closed in position A (start position), ensures that the piston is at position A for start conditions. The NC contact on limit switch 2LS is held open at the start position. This opens the circuit to solenoid B.

The sequence of operations for the circuit shown in Figure 6-15 is as follows:

1. Operate the START pushbutton switch.
2. With limit switch 2LS normally open contact held closed, relay 1CR is energized.
   a. Relay contact 1CR-1 closes, interlocking around the START pushbutton switch.
   b. Relay contact 1CR-2 closes, energizing solenoid A.
   c. Relay contact 1CR-3 opens.

As the piston moves off limit switch 2LS, the normally open contacts open and the normally closed contacts close. However, since relay contact 1CR-3 is now open, solenoid B remains deenergized.

3. Piston moves to position B, operating limit switch 1LS, opening its normally closed contact.
4. Relay 1CR is deenergized.
   a. Relay contact 1CR-1 opens, opening the interlock circuit around the START pushbutton switch.
   b. Relay contact 1CR-2 opens, deenergizing solenoid A.
   c. Relay contact 1CR-3 closes, energizing solenoid B.

The piston now moves back to position A, operating limit switch 2LS.

5. The normally closed limit switch contact 2LS opens, deenergizing solenoid B.

ACHIEVEMENT REVIEW

1. What are the three basic classes of industrial limit switches?
2. What is one use for the rotating-cam limit switch?

3. List some of the considerations when selecting a limit switch for operation in an electrical control circuit.

4. What is the difference between operating force and release force in a mechanical limit switch?

5. Draw a sketch to show the difference between pretravel and overtravel.

6. Discuss the accuracy and response time in a vane-type switch.

7. Draw the mechanical limit switch symbols for four different conditions.

8. Some of the methods used to achieve the operation of proximity limit switches are
   a. capacitive fields.
   b. light rays.
   c. low current input.

9. The speed of response for the proximity limit switch using an oscillator in the output circuit is
   a. the location of the switch.
   b. the speed of the object being sensed.
   c. dependent upon the oscillator frequency.

10. Draw an electrical control circuit, using two push-button switches and a limit switch, relay and solenoid operating valve. The solenoid operating valve is a single-solenoid, spring-return valve. The relay and solenoid are to be energized when the normally open push-button switch is operated. The relay and solenoid are to be deenergized when the limit switch is operated. The relay and solenoid can also be deenergized at any time during a cycle, before the limit switch is operated, by operating a second push-button switch.

CHAPTER 7

PRESSURE SWITCHES AND TRANSDUCERS

OBJECTIVES

After studying this chapter, the student will be able to:

- Define terms used with pressure switches.
- Discuss tolerance in a pressure switch.
- List four types of pressure switches.
- Explain how the piston-type pressure switch operates.
- Discuss why the rated operating pressure of a Bourdon tube pressure switch should not be exceeded during use
- Describe the operation of the diaphragm pressure switch.
- Explain how the pressure transducer operates.
- Draw the normally open and normally closed pressure switch symbols.
- Draw an electrical control circuit showing how a normally closed pressure switch contact can be used to deenergize the circuit.

7.1 IMPORTANCE OF PRESSURE INDICATION AND CONTROL

Pressure indication and control is important in many electrical control systems where air, gas, or a liquid is involved. The control takes several forms. It may be used to start or stop a machine on either rising or falling pressure. It may be necessary to know that pressure is being maintained.

The *pressure switch* is used to transfer information concerning pressure to an electrical circuit. The electrical switch unit can be a single normally open or normally closed switch. It can be both, with a common terminal or two independent circuits. Usually it is easier to use a switch with two independent circuits in the design of a control circuit.

Some terms associated with pressure switches are best explained in chart form. Figure 7-1 shows some of the terms we are concerned with and some of the most important points.

Tolerance in most switches is about 2%. Accuracy is about 1%. These are expressed in percent of the working range. In some cases it is necessary to know and use a differential pressure. As used here, the *differential* is the range between the actuation point and the reactuation point. For example, the electrical contacts operate at a preset rising pressure and hold operated until the pressure drops to a

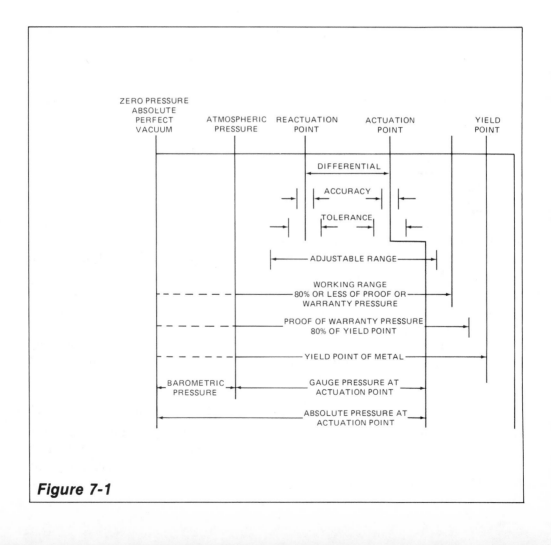

Figure 7-1

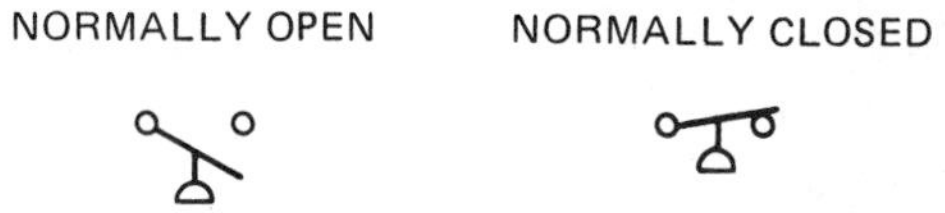

Figure 7-2

NORMALLY OPEN   NORMALLY CLOSED

lower level. This differential may be a fixed amount, generally proportional to the operating range, or it can be adjustable.

The pressure switch symbol is shown in Figure 7-2. The NC contact opens upon reaching a preset pressure. The NO contact closes upon reaching a preset pressure.

7.2 TYPES OF PRESSURE SWITCHES

There are several types of pressure switches available. Some of the most widely used types are the following:

- Sealed piston
- Bourdon tube
- Diaphragm
- Solid state

7.2.1 Sealed Piston Type   This type of pressure switch is actuated by means of a piston assembly which is in direct contact with the hydraulic fluid. Referring to Figure 7-3, the assembly consists of a piston sealed with an O-ring, direct acting on a snap-action switch. An extremely long life can be expected from this switch where mechanical pulsations exceed 25 per minute, and where high overpressures occur. The sealed piston saves the cost of installing return lines.

The piston-type pressure switch can withstand large pressure changes and high overpressures. These units are suitable for pressures in the range of 15 pounds per square inch (psi) to 12,000 psi.

7.2.2 Bourdon-tube Pressure Switch   The pressure operating element in this unit is made up of a tube formed in the shape of an arc. Application of pressure to the tube will cause it to straighten out. Care should be taken that pressures beyond the rating of the unit are not applied. Excessive pressures tend to bend the tube beyond its ability to return to its original shape.

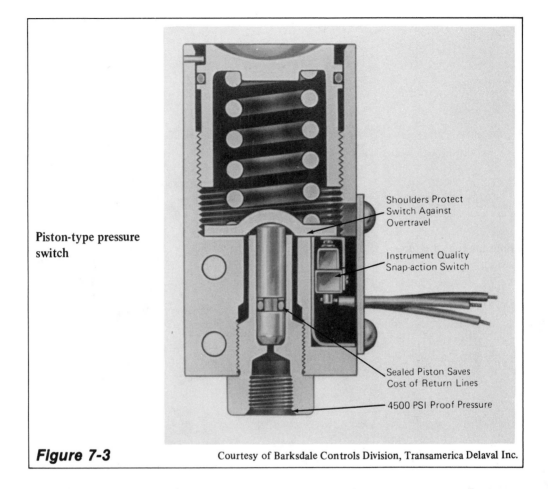

Piston-type pressure switch

Shoulders Protect Switch Against Overtravel

Instrument Quality Snap-action Switch

Sealed Piston Saves Cost of Return Lines

4500 PSI Proof Pressure

Figure 7-3   Courtesy of Barksdale Controls Division, Transamerica Delaval Inc.

The Bourdon-tube pressure switch unit is extremely sensitive. It senses peak pressures. Due to this condition, it may be necessary to use some form of dampening or snubbing of the pressure entering the tube.

This pressure switch is available for applications in a range from 50 psi to 18,000 psi. It is generally used more in indicating service than in switching. This is due to the inherent low work output of the unit.

Figure 7-4 shows a typical Bourdon-tube pressure switch.

7.2.3 Diaphragm Pressure Switch   This unit consists of a disk made with convolutions around the edge. The edge of the disk is fixed within the case of the switch. Pressure is applied against the full area of the diaphragm. The center of the diaphragm opposite the pressure side is free to move and operate the snap-action electrical contacts. Units are available in the range of vacuum to 150 psi.

Figure 7-5 shows a diaphragm pressure switch.

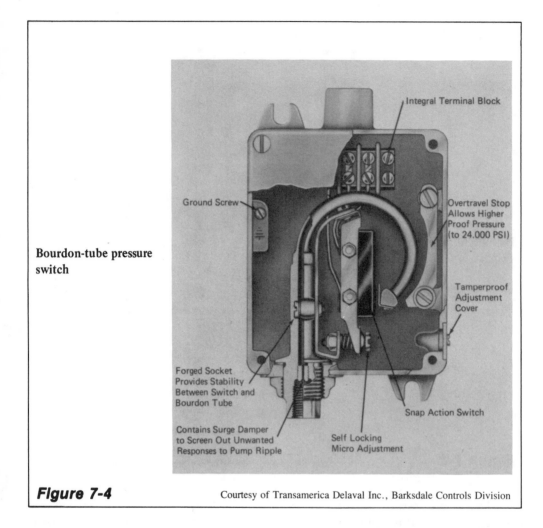

Bourdon-tube pressure switch

Integral Terminal Block

Ground Screw

Overtravel Stop Allows Higher Proof Pressure (to 24.000 PSI)

Tamperproof Adjustment Cover

Forged Socket Provides Stability Between Switch and Bourdon Tube

Contains Surge Damper to Screen Out Unwanted Responses to Pump Ripple

Self Locking Micro Adjustment

Snap Action Switch

Figure 7-4   Courtesy of Transamerica Delaval Inc., Barksdale Controls Division

7.2.4 Solid-state Pressure Switch   This switch is interchangeable with existing electromechanical pressure switches. Pressure sensing is performed by semiconductor strain gauges with proof pressures up to 15,000 psi acceptable. Switching is accomplished with solid-state triacs. An enclosed terminal block allows four different switching configurations:

1. Single-pole, single-throw (NC)
2. Single-pole, single-throw (NO)
3. Single-pole, double-throw
4. Double-make, double-break

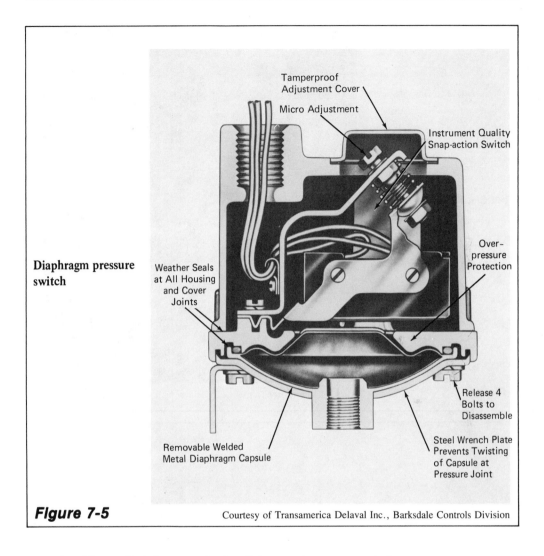

Diaphragm pressure switch

Figure 7-5   Courtesy of Transamerica Delaval Inc., Barksdale Controls Division

See Figure 7-6 for a typical solid-state pressure switch.

7.3.1 Pressure Transducers   With the present industrial use of programmable controllers (covered later in Chapter 19), the use of pressure transducers has become important.

A pressure transducer, such as shown in Figure 7-7, utilizes semiconductor strain gauges which are epoxy-bonded to a metal diaphragm. Pressure applied to the diaphragm through the pressure port produces a minute deflection which introduces strain to the gauges. The strain produces an electrical resistance change proportional to the pressure. Four gauges (or two gauges with fixed resistors) form a Wheatstone bridge, (Figure 7-8).

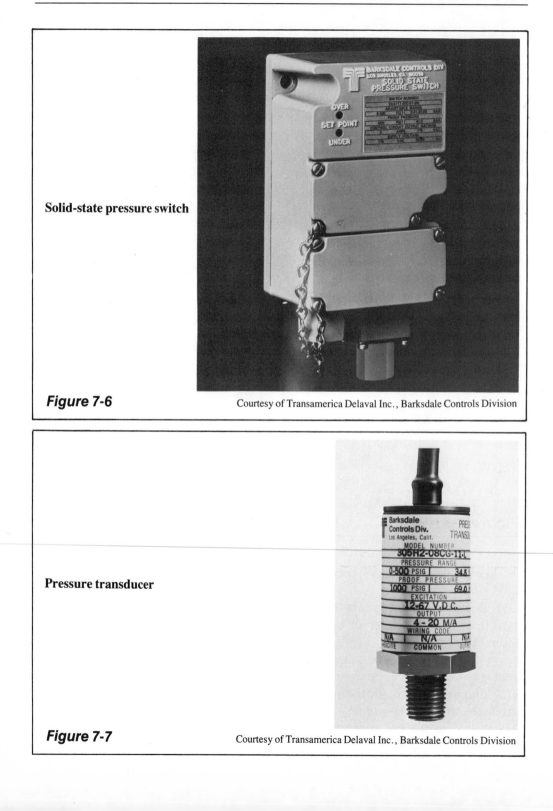

Solid-state pressure switch

Figure 7-6   Courtesy of Transamerica Delaval Inc., Barksdale Controls Division

Pressure transducer

Figure 7-7   Courtesy of Transamerica Delaval Inc., Barksdale Controls Division

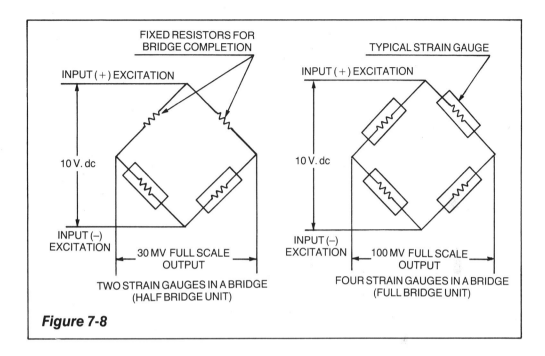

Figure 7-8

The differential resistance is measured by applying a constant voltage to the bridge. Diaphragm reflection results in an analog (millivolt) output which is proportional to the pressure.

The rigid sensing diaphragm, with bonded semiconductor strain gauges and low mass, provide a no-moving-parts transducer with excellent resistance to shock and vibration.

7.3 CIRCUIT APPLICATIONS

The pressure switch contact is now substituted for the limit switch contact. The basic circuit showing the limit switch as a means of reversing the piston is shown in Figure 6-11A. In Figure 7-9A and 7-9B, the pressure switch is connected into the fluid power pressure circuit. A normally closed contact is connected into the electrical circuit, 7-9C.

The function of this circuit is the same as that shown in Figure 6-11A, except now the piston reverses when a preset pressure is reached. Remember that a normally closed pressure switch contact opens on rising pressure.

In the circuit shown in 7-10A, 7-10B, 7-10C and 7-10D, the double-solenoid valve is used. Remember that in the deenergized condition of both solenoids,

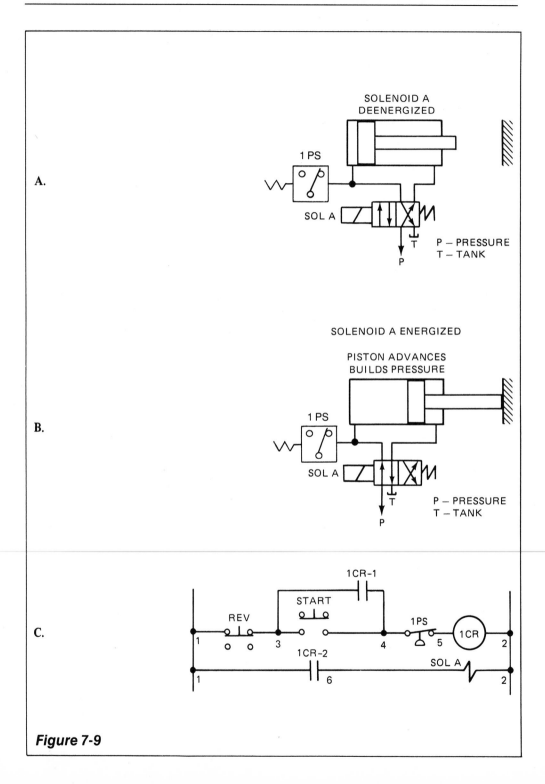

Figure 7-9

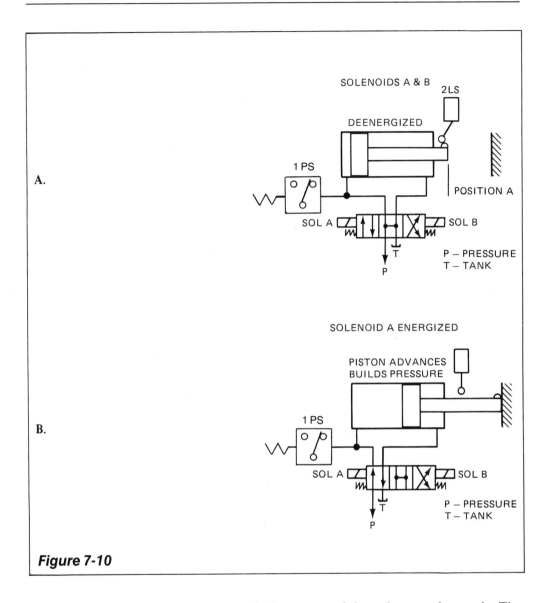

Figure 7-10

pressure is allowed to bypass through the center of the valve spool to tank. The piston must be in position A for start conditions.

The sequence of operations proceeds as follows:

1. Operate the START pushbutton switch.
2. Relay coil 1CR is energized.
   a. Relay contact 1CR-1 closes, interlocking around the START pushbutton switch.
   b. Relay contact 1CR-2 closes, energizing solenoid A.

C.

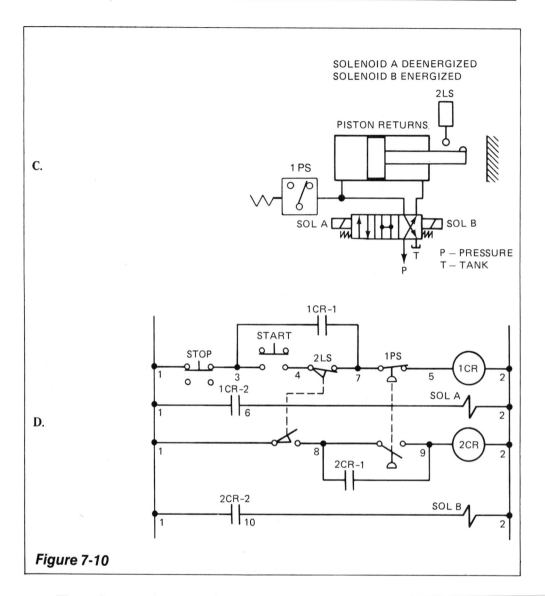

D.

Figure 7-10

The valve spool now shifts, allowing pressure to enter the main cylinder area (left-hand end). The piston moves forward. On reaching the work piece, pressure builds against the work piece. When a preset pressure is reached on pressure switch 1PS, its NC contact opens.

   3. Relay coil 1CR is deenergized.
      a. Relay coil 1CR-1 opens, opening the interlock circuit around the START pushbutton switch.
      b. Relay contact 1CR-2 opens, deenergizing solenoid A.

When the NC pressure switch contact opened, the NO pressure switch contact closes. Since the cam on the piston is away from position A, the NC contact on limit switch 2LS is allowed to close. This completes the circuit to relay 2CR.

4. Relay coil 2CR is energized.
   a. Relay contact 2CR-1 closes, interlocking around the NO pressure switch contact.
   (Note: This is necessary because the pressure drops when solenoid A is deenergized. The NO pressure switch contact opens.)
   b. Relay contact 2CR-2 closes, energizing solenoid B.

The valve spool now moves back, allowing pressure to enter the pull back area (right-hand end). The piston now returns to its initial position (A). On reaching this point, limit switch 2LS is operated, opening its NC contact.

5. Relay coil 2CR is deenergized.
   a. Relay contact 2CR-1 opens, opening the interlock circuit around the pressure switch 1PS contact.
   b. Relay contact 2CR-2 opens, deenergizing solenoid B.

Note the broken line connecting the two limit switch contacts. This indicates that they are two contacts in the same limit switch and operate at the same time. A similar explanation can be made for the broken line connecting the two contacts on 1PS. Both contacts are on the same pressure switch and operate at the same time.

The circuits in Figures 7-11A, 7-11B, 7-11C, and 7-11D use two separate cylinder-piston assemblies. Each assembly is powered through a single-solenoid, spring-return operating valve.

The sequence of operations for these circuits is as follows:

1. Operate the START push-button switch.
2. Relay coil 1CR is energized.
   a. Relay contact 1CR-1 closes, interlocking around the START push-button switch.
   b. Relay contact 1CR-2 closes, energizing solenoid A.
3. The piston moves forward.

On the forward stroke of piston A, limit switch 3LS, NO contact closes cam is no longer operating 2LS). This causes no circuit action as NC relay contact 1CR-3 is open.

4. The piston reaches the workpiece, builds pressure to a preset amount on 1PS, operating the pressure switch contacts.

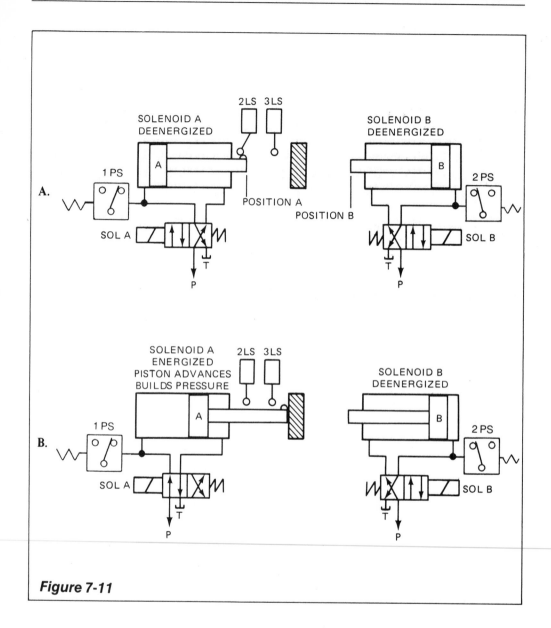

Figure 7-11

5. The normally closed contact on 1PS opens, deenergizing relay coil 1CR.
   a. Relay contact 1CR-1 opens, deenergizing relay coil 1CR.
   b. Relay contact 1CR-2 opens, deenergizing solenoid A.
   c. Relay contact 1CR-3 closes.

Piston A now returns to position A. On the return travel, the cam on the

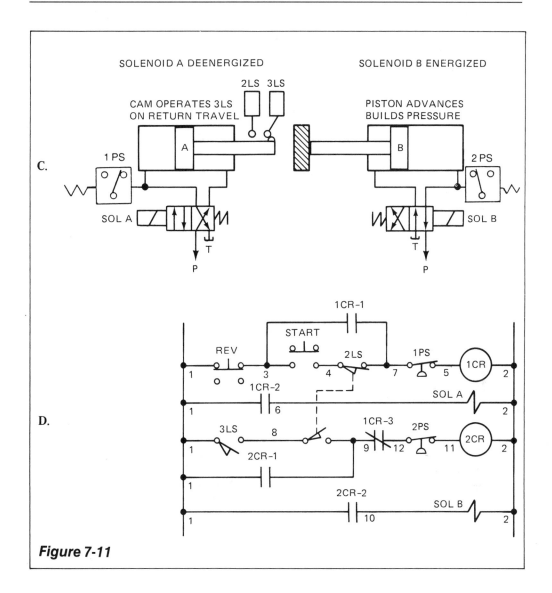

Figure 7-11

piston operates limit switch 3LS. With relay contact 1CR-3, limit switch 2LS NO contact, limit switch 3LS and pressure switch 2PS closed,

6. Relay coil 2CR is energized.
   a. Relay contact 2CR-1 closes, interlocking around 2LS NC contact and 3LS NO contact.
   b. Relay contact 2CR-2 closes, energizing solenoid B.

Piston B now moves forward, meeting the work and building pressure to a preset amount set on 2PS.

7. Normally closed 2PS contact opens, deenergizing relay coil 2CR.
   a. Relay contact 2CR-1 opens, opening the interlock circuit around 3LS and 2LS.
   b. Relay contact 2CR-2 opens, deenergizing solenoid B.

Piston B now returns to its initial position at B.

ACHIEVEMENT REVIEW

---

1. Draw the normally open and normally closed pressure switch symbols.

2. Explain differential in pressure switch operation.

3. List four major types of pressure switches.

4. In a piston-type pressure switch, what is used to prevent oil leakage past the piston?

5. Why is a form of dampening sometimes used with the Bourdon-tube type pressure switch?

6. What is the usual pressure operating range for the diaphragm-type pressure switch?

7. Design and draw an electrical control circuit, showing how the operation of a normally closed pressure switch contact can be used to de-energize an output solenoid. Use two pushbutton switches, a relay, pressure switch contact and a solenoid.

8. To transfer pressure information, a pressure transducer can use
   a. a low resistance coil.
   b. a specially designed plug-in relay.
   c. semiconductor strain gauges.

9. How is switching accomplished in the solid-state pressure switch? List four different switching configurations available in the solid-state pressure switch.

10. The differential resistance in a pressure transducer is measured by
    a. applying a constant voltage to the bridge.
    b. calculating using data provided by the manufacturer.
    c. estimating the change on rising pressure.

CHAPTER 8

TEMPERATURE CONTROLLERS AND SWITCHES

OBJECTIVES

After studying this chapter, the student will be able to:

- Identify five important factors to consider when selecting a temperature controller.
- Explain the difference between controller sensitivity and controller operating differential.
- Describe how the thermocouple, thermister, and resistance unit obtain temperature information that can be used in a control circuit.
- Explain how time proportioning is used in a temperature controller.
- List the advantages and disadvantages of the potentiometric-type controller.
- Discuss band width, automatic reset, and rate control in pyrometers.
- Name two types of temperature switches or thermostats and explain how each operates.
- Show how a temperature switch contact can be used in an electrical control circuit to prevent the operation of the circuit unless the temperature of the part being sensed is at or above a preset temperature.

8.1 TEMPERATURE CONTROLLERS

Temperature controllers basically consist of two parts: a sensing element that responds to temperature, and a switch consisting of normally open and/or normally closed contacts. The contacts operate from the temperature sensing element. The operating point is preset. The switch operation is usually accomplished when the temperature at the sensing point changes from any given level to the preset point. The main difference among temperature switches is the means by which the temperature information is transferred from the sensing element to the switching element.

There are several factors to consider when selecting a temperature controller.

1. Temperature range available. Temperature range is the overall operating range of the controller. Not all controllers cover the entire temperature range used in industrial control. The controller is generally selected after the range is determined. The sensing element often becomes an important factor.
2. Type of sensing element. There are three basic types: 1) electric, 2) differential expansion of metal, and 3) expansion of fluid, gas, or vapor.
3. Response time. The response to a temperature change may vary between one type of sensing element and another. Here, the application may be the deciding factor. For example, if rapid response is not required, the slower-response and generally less expensive types may be used. Speed of response is a measure of the elapsed time from the instant the temperature change occurs at the sensing element until it is converted into controller action.
4. Sensitivity. Units vary with the amount of temperature change required to operate the switch. This amount is fixed in some switches. In others it is adjustable. It is usually desirable to have a controller that has a relatively high sensitivity. Resolution sensitivity is the amount of temperature change necessary to actuate the controller.
5. Operating differential. This is the difference in temperature at the sensing element between make and break of the controller's contacts when the controller is cycled.

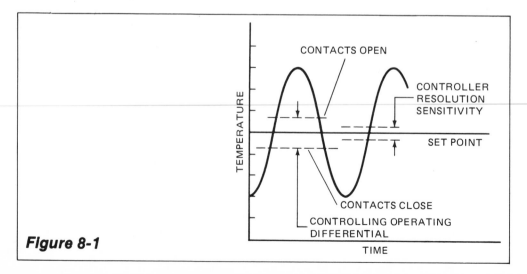

Figure 8-1

Figure 8-1 is a sketch illustrating some of these operating characteristics in temperature controllers.

8.1.1 Electronic Temperature Controller (Pyrometer) The electronic controller, or pyrometer, may have one of three different temperature sensing elements:

- Thermocouple (approximate temperature range 200°F–5000°F)

- Thermister (approximate temperature range 100°F–600°F)

- Resistance temperature detector (RTD) (approximate temperature range 300°F–1200°F)

The *thermocouple* operates on the principle of joining two dissimilar metals A and B at their extremities, Figure 8-2A. When a temperature difference exists between the two extremity points, a potential is generated. This is in proportion to the temperature difference. Such combinations as copper-constantan, iron-constantan, and chromel-alumel are used in thermocouple construction.

When the temperature at T1 is different than at T2, electromotive force (emf) exists. By the use of an indicating device that indicates emf or the flow of current, the temperature difference can be determined.

The combination of iron and constantan form the type J thermocouple. The relationship between temperature in degrees Fahrenheit and millivolts output is shown in Figure 8-2B.

The *thermistor* is a semiconductor whose resistance decreases with increasing temperature. This element is connected into a null reading type ac bridge circuit and the output fed to an amplifier. The output of the amplifier operates a relay that is used in the control circuit.

The *RTD unit* consists of a tube made of stainless steel or brass. It contains an element made in the form of a coil. The coil is generally wound of a fine nickel, platinum, or pure copper wire. As the temperature changes, the resistance of the coil changes proportionally. This change is converted to a voltage by its application in the electrical control circuit. Therefore, as the

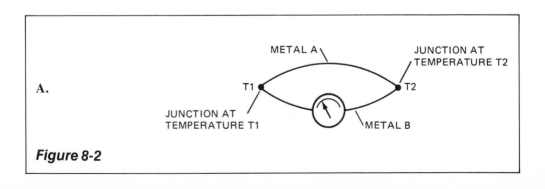

Figure 8-2

B.

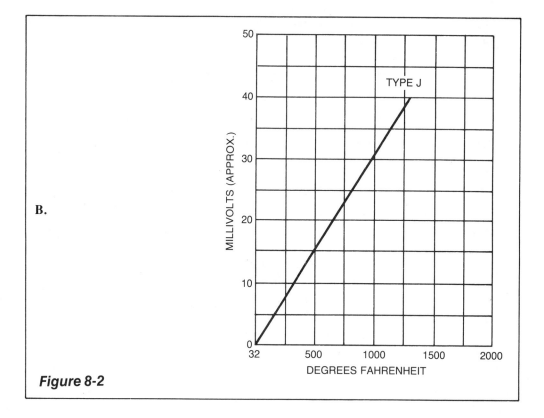

Figure 8-2

voltage changes with a temperature change, the voltage level is compared to a reference. Thus, an output proportional to the temperature change is obtained.

Three different physical arrangements of thermocouples are shown in Figures 8-3A, 8-3B, and 8-3C. Figure 8-3A shows the dual-element thermocouple. It can be used to sense temperature at two different physical levels. Figure 8-3B shows the single thermocouple. Both thermocouples are spring loaded. The single thermocouple uses an adaptor and small steel ferrule on a flexible armored cable to obtain the spring loading. The dual-element thermocouple uses a small steel plate that is secured to the deep couple with an adaptor and locking nut. The springs on both thermocouples are loaded against the steel plate. Figure 8-3C shows an adjustable-depth plastic melt type with a plug attachment.

Figure 8-4A shows a resistance temperature detector (RTD unit).

Figure 8-4B shows a line drawing of a thermistor.

The electronic temperature controller is usually the most expensive type. It is more sensitive and has a faster response. Generally, electronic controllers have a smaller sensing device. The sensing device can be remotely located with greater ease. Normal control accuracies of one-fourth of 1% to 5% of full-scale reading can be expected.

A. Dual-element thermocouple

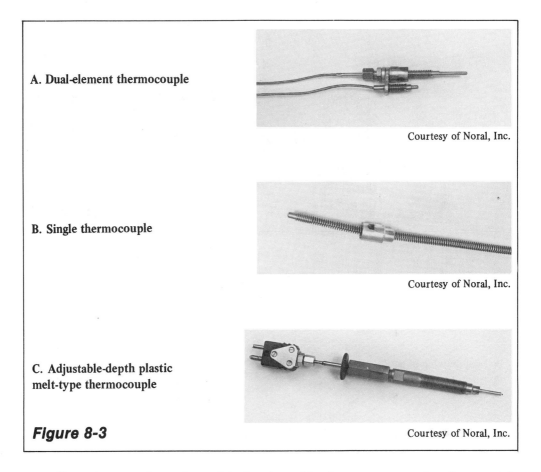

Courtesy of Noral, Inc.

B. Single thermocouple

Courtesy of Noral, Inc.

C. Adjustable-depth plastic melt-type thermocouple

Figure 8-3

Courtesy of Noral, Inc.

Recent years have brought about rapid advancement and refinement in electronic temperature controllers. Thus, some terms and expressions that may not be too familiar have been developed. Explanations of these terms and expressions are included in the following sections.

Electronic controllers can be divided into two groups as to their means of obtaining control.

1. The first group includes on-off and proportioning.
   A. *On-off* is the most simple and oldest method of temperature control. Assuming no time thermal lags in the object being heated or cooled, the on-off control produces a fluctuating temperature. The limits are within the heating or cooling band. Depending on the time lag, the amplitude of the temperature swing may exceed good operating control.
   B. *Proportioning.* The time proportioning was the next major development in controllers. This method turns the heating elements on and off in accordance with the demands of the temperature set point. There is a point

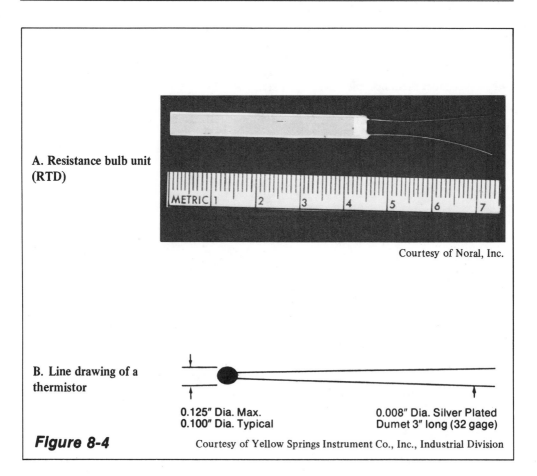

A. Resistance bulb unit (RTD)

Courtesy of Noral, Inc.

B. Line drawing of a thermistor

0.125″ Dia. Max.
0.100″ Dia. Typical

0.008″ Dia. Silver Plated
Dumet 3″ long (32 gage)

Figure 8-4   Courtesy of Yellow Springs Instrument Co., Inc., Industrial Division

below the set point where the power to the heating elements is on continuously. Likewise, there is a point above the set point where the power is off. Between these two points the power is on at a proportional rate. For example, at a temperature 20% through the proportional band, the power may be 80% on; 50% of the way, it will be 50% on. This method differs from on-off control in that it does not require an actual temperature change to cycle.

To accomplish on-off and proportioning, pyrometers can be divided into two types. These are the millivoltmeter controller and the potentiometric controller.

The *millivoltmeter controller* is the oldest type on the market. It uses a millivoltmeter actuated directly from a temperature sensor such as a thermocouple. Setting the desired temperature, positions either a photo cell and light source or a set of oscillator coils. The indicating pointer carries a small flag. As the pointer approaches the set point, the flag moves between the photo cell and light source or between the oscillator coils. Control action now starts, cutting back on heat.

One of the major sources of trouble with this type of controller is the adverse effect of shock and vibration. Another problem involves the sensing device. For good indication accuracy, the external resistance of the thermocouple circuit should match that of the instrument. Some manufacturers offer controllers with adjustable external resistance circuits. Such instruments can be calibrated after they are installed.

Potentiometric controller differs from the millivoltmeter type in that the signal from the temperature sensor is electronically controlled and compared to the set point temperature. This instrument can be supplied with or without indication, because with this control, the indicating method is completely separate.

The potentiometric controller has the following advantages:

- No moving parts

- Does not have to be calibrated for external resistance

- Not affected by shock and vibration

    The potentiometric controller has the following disadvantages:

- More electronic circuitry that is not easily serviced; it is generally advisable to replace the entire plug-in controller

Figure 8-5 shows two potentiometric controllers.

2. The millivoltmeter and potentiometric controllers both generally use a contactor to energize the heating element load. The contactor is energized and deenergized in response to a demand for heat. For example, when heat is required as sensed by a thermocouple, the contactor is energized by the controller output. Electrical power is then connected to the heating element load. When there are no heat requirements, the heating elements are completely isolated from the power source by the opening of the contacts on the contactor.

A more recently developed controller does not use a contactor to energize and deenergize the heating element load. With this latest type of controller, the power remains connected to the load at all times. Just enough power is supplied to the heating elements so that the temperature requirements are satisfied. This type of controller is called a *stepless* or *proportional controller.* The stepless or proportional controller has a current output. It is generally used with silicon-controlled rectifiers (SCR) or Triac solid-state power packs.

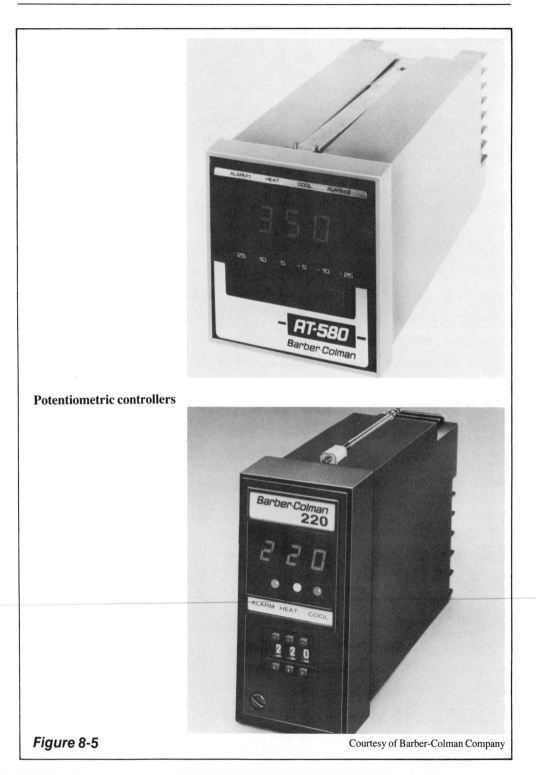

Potentiometric controllers

Figure 8-5

Courtesy of Barber-Colman Company

Two different outputs are available with this type of controller. Power is applied to the heating element load through:

- Phase control (phase angle fired)
- Zero crossing (zero angle fired)

With phase control, the output is continuous at the amount required for consistent temperature. This is accomplished by "firing" the SCR at some point through the wave, depending on the amount of power required.

One problem with phase fired control is the radio frequency interference (RFI) generated. This is caused by the SCR switching at any place in the wave at an extremely fast rate. This is on the order of one-half microsecond (μ s). Another problem is the resulting distorted wave shape. This creates a problem for the power company.

These problems can be avoided by the use of zero crossing SCR circuitry. Its characteristics are similar to the time proportioning results. However, the power is always switched off and on when the power voltage is zero.

8.2 CONTROLLER OUTPUTS

The relay in pyrometer outputs generally is rated at 5 A to 10 A. This means it can handle a small heating element load of approximately 500 watts (W) to 1000 W at 120 V. For larger loads, the relay is used to energize a contactor.

Controllers are now available with a solid-state output capacity capable of 20 A inrush and 1 A continuous. This control output can be used with load contactors up to sizes 2 or 3.

8.3 ADDITIONAL TERMS

In discussing the use of pyrometers in the control of temperature, there are several terms that need explanation.

A. **Band Width.** This is sometimes referred to as the *proportioning band.* In most controllers this band is adjustable, Figure 8-6. If the band width is

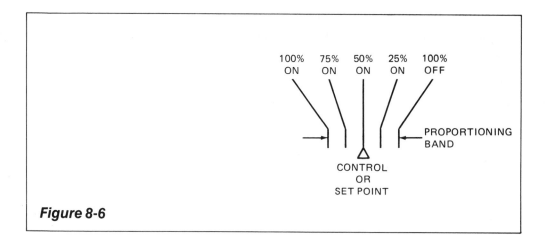

Figure 8-6

set for 3%, this is 24°F on an 800°F instrument. If the set point is at 450°F, the band extends from 438°F to 462°F. In controlling it is quite possible for the temperature to settle out at some temperature other than the set point. For example, assume the temperature settles out at 454°F. This represents an offset of +4°. This can be corrected by adjusting the set point at 446°F. Resetting of the control point in this way is called *manual reset.*

B. **Automatic Reset.** The addition of the auto reset function in a controller automatically eliminates the offset. Circuitry in the controller recognizes the offset. It will electronically shift the band up or down scale as required to remove an error. The auto reset feature is sometimes referred to as the integral function.

C. **Rate.** The application of rate control is valuable in applications having rapid changes in temperature caused by external heating and cooling. Rate control works in the opposite direction of reset, and at a faster speed. For example, a sudden cooling causes the controller to turn the heaters full on. This is done by an upward shift of the band. Rate is sometimes referred to as the *derivative* function.

D. **Mode.** A *two-mode controller* is one having proportioning and auto reset or proportional and auto reset. A *three-mode controller* is one having proportioning, auto reset, and rate or proportional, auto reset, and rate.

E. **Analog and Digital Set Point.** With an analog set point, the temperature is set on a scale. With the digital set point controller, the temperature is displayed in numbers, Figure 8-7.

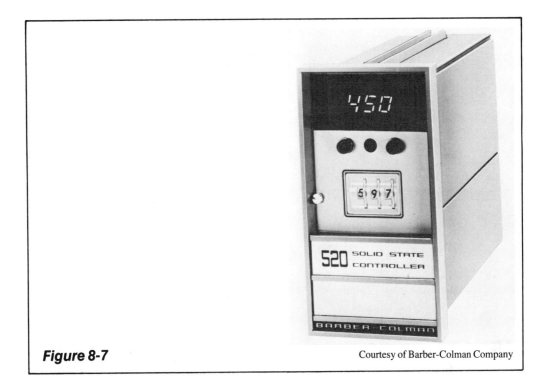

Figure 8-7

Courtesy of Barber-Colman Company

8.4 TEMPERATURE SWITCHES (THERMOSTATS)

Temperature switches using differential expansion of metals may be of two different types. One uses a mechanical link. The other uses a fused bimetal.

The mechanical link has a temperature range of 100°F–1500°F. The bimetal type has a temperature range of 40°F–800°F.

The mechanical-link type has one metal piece directly connected or subjected to the part where the temperature is to be detected. The metal expands or contracts due to temperature change. This produces a mechanical action that operates a switch.

The bimetal type operates on the principle of uneven expansion of two different metals when heated. With two metals bonded together, heat will deform one metal more than the other. The mechanical action resulting from a temperature change is used to operate a switch. In bimetal units the sensing element and switch are generally enclosed in the same container or are located adjacent.

The bimetal types is usually the smallest and least expensive type of controller. It is not suited for high precision work. However, its compact, rugged construction

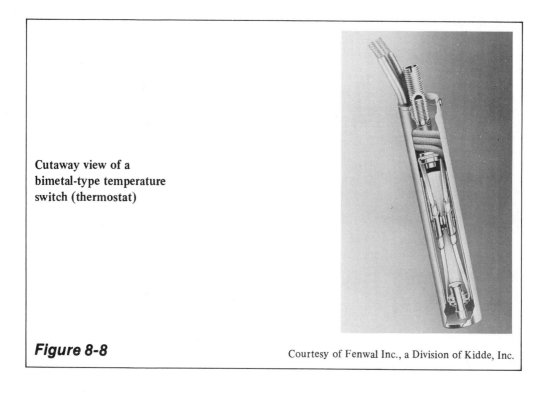

**Cutaway view of a
bimetal-type temperature
switch (thermostat)**

Figure 8-8   Courtesy of Fenwal Inc., a Division of Kidde, Inc.

lends it to many uses. Accuracies of 1% to 5% of full scale can be expected.

Figure 8-8 is a cutaway view of a bimetal unit.

With temperature switches using liquid, gas, or vapor, the sensing elements are generally located remote from the switch. The temperature ranges for these units are as follows:

> Liquid filled: 150°F to 2200°F (these may be self contained)
> Gas filled: 100°F to 1000°F
> Vapor filled: 50°F to 700°F

These units operate on the principle that when the temperature increases, expansion of the medium (fluid, gas, or vapor) takes place. Thus, a force is exerted on a device which in turn operates a switch.

The liquid-filled type that is self-contained (switch and sensing element together) has a relatively fast response time.

A factor to consider in using units with a remote sensing head is the problem with the tube connecting the sensing bulb with the switching mechanism. The tube is easily damaged by mechanical abuse. Also, the response is slower with long lengths of connecting tube.

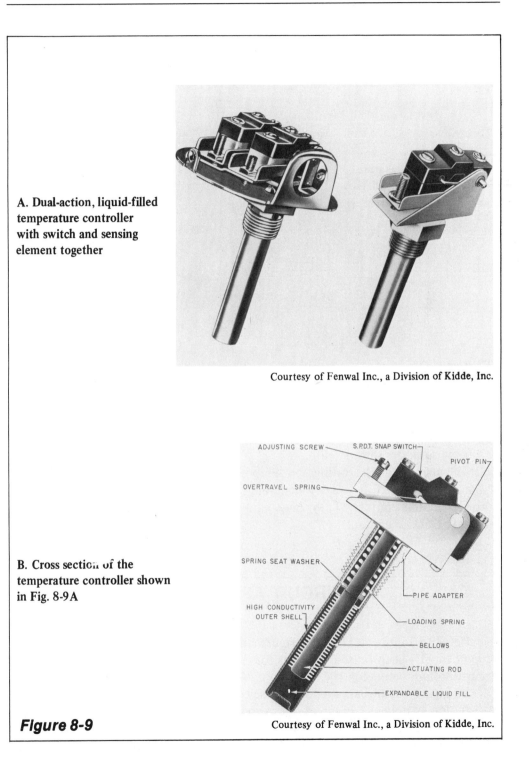

A. Dual-action, liquid-filled temperature controller with switch and sensing element together

Courtesy of Fenwal Inc., a Division of Kidde, Inc.

B. Cross section of the temperature controller shown in Fig. 8-9A

Figure 8-9

Courtesy of Fenwal Inc., a Division of Kidde, Inc.

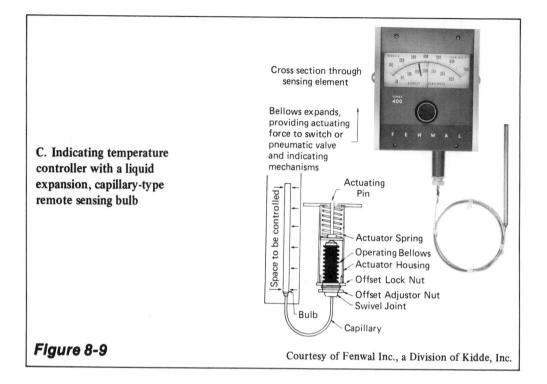

C. Indicating temperature controller with a liquid expansion, capillary-type remote sensing bulb

Cross section through sensing element

Bellows expands, providing actuating force to switch or pneumatic valve and indicating mechanisms

Actuating Pin

Actuator Spring
Operating Bellows
Actuator Housing
Offset Lock Nut
Offset Adjustor Nut
Swivel Joint

Space to be controlled

Bulb

Capillary

Figure 8-9

Courtesy of Fenwal Inc., a Division of Kidde, Inc.

An example of a self-contained unit is shown in Figure 8-9A. A cross section through the unit is shown in Figure 8-9B. Figure 8-9C is an example of the unit using a remote sensing bulb.

The switch contact ratings on temperature switches (thermostats) vary from 5 A to 25 A at 120 V. They generally consist of one normally open and one normally closed contact. They may have a common connection point, or they may have isolated contacts.

8.5 CIRCUIT APPLICATIONS

Refer to Figure 4-11. This figure shows a heating circuit with only heat OFF – heat ON push-button switches for control. These switches control the heat circuit through contactor 10CR

In Figure 8-10A, a two-position selector switch is substituted for the two push-button switches in Figure 4-11. A pyrometer is inserted in the circuit to the contactor coil 10CR. Note that the symbol for the pyrometer shows only the relay contact in the pyrometer and the thermocouple connections. More details on internal circuit construction can be obtained from the manufacturer of the pyrometer.

When the heat OFF-ON selector switch is operated to the ON position, the circuit is complete to the NO relay contact. If the temperature of the part being sensed is below the set point on the pyrometer, the relay contact will be closed. This completes the circuit to the coil of contactor 10CR. The coil of contactor 10CR is energized. Contactor contacts 10CR-1, 10CR-2, and 10CR-3 close, energizing the heating elements at power line voltage.

When the temperature of the part being sensed by the thermocouple reaches the set point, the pyrometer relay contact opens. This deenergizes the coil of

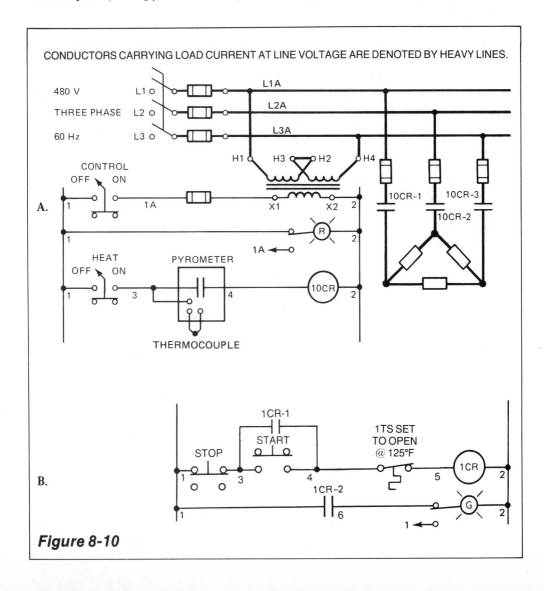

Figure 8-10

contactor 10CR. Contactor contacts 10CR-1, 10CR-2, and 10CR-3 open, deenergizing the heating elements. The pyrometer relay contact continues to close and open, depending on the thermocouple input. The rate of opening and closing depends on the type of pyrometer.

The temperature switch (thermostat) is generally used at lower temperatures than the pyrometer. It is also used where accuracy and sensitivity are not important factors in temperature control. For example, a tank containing a fluid of some type (oil, asphalt, or brine) is to be heated to a given temperature. Depending on the size of the tank, it may take considerable time to heat it safely to a given temperature. If the preset temperature is 200°F, it generally is not important whether the temperature is actually 204°F or 196°F.

Another use of the thermostat is to indicate and/or control the temperature of fluid used in a machine operation. For example, it may be required that the fluid in a hydraulic system be at a minimum temperature of 70°F for the machine to operate, and the temperature should not exceed 120°F.

The circuits that follow illustrate temperature control of a fluid that is used in a machine for machine operation.

In Figure 8-10B, a temperature switch is added to the circuit shown in Figure 4-3. The sequence of operations for the circuit in Figure 8-10B are as follows:

1. Temperature below set point of 125°F, circuit ready to operate.
2. Operate the START pushbutton switch.
3. Energize relay coil 1CR.
   a. Relay contact 1CR-1 closes, interlocking around the START pushbutton switch.
   b. Relay contact 1CR-2 closes, energizing the green pilot light.
4. Temperature exceeds 125°F, NC temperature switch contact opens.
5. Relay coil 1CR deenergized.
   a. Relay contact 1CR-1 opens, opening the interlock circuit around the START pushbutton switch.
   b. Relay contact 1CR-2 opens, deenergizing the green pilot light.

The temperature must drop below 125°F, allowing the NC 1TS contact to close before relay coil 1CR can again be energized.

Figure 8-10C shows the control for two heating elements. With the temperature of the medium being heated (oil, asphalt, brine), below 80°F:

1. Contactor coil 10CR is energized.
2. Contactor contacts 10CR-1 and 10CR-2 close, energizing the two heating elements.
3. Contactor auxiliary contact #2 opens, preventing the energizing of relay coil 1CR through the operation of the START pushbutton switch.

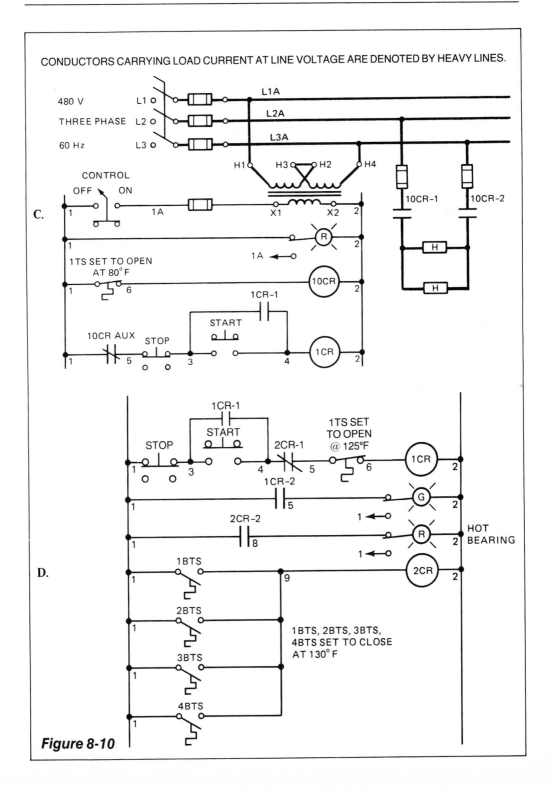

Figure 8-10

4. The heated medium now reaches 80°F, deenergizing 10CR contactor coil by the opening of 1TS.
5. Contactor contacts 10CR-1 and 10CR-2 open, deenergizing the two heating elements.
6. Contactor auxiliary contact closes.
7. The START pushbutton switch can now be operated, energizing relay coil 1CR.
    a. Relay contact 1CR-1 closes, interlocking around the START pushbutton switch.

Relay coil 1CR now remains energized until such time as the temperature of the heated medium drops below 80°F. At that time contactor coil 10CR will again energize, energizing the heating elements and opening the 10CR auxiliary contact. Relay coil 1CR will be deenergized.

In the circuit shown in Figure 8-10D, hot bearing control is added to the circuit as shown in Figure 8-10B.

Assume that a machine has four bearings. The machine should not operate with any of the bearings at a temperature in excess of 130°F. A thermostat can be placed on each of the four bearings. The thermostat contacts will be normally open, set to close (operate) at 130°F.

1. With bearing temperature below 130°F, relay coil 2CR not energized.
2. When the temperature of one or more of the bearings exceeds 130°F, relay coil 2CR energized.
    a. Normally closed relay contact 2CR-1 opens.
       a-1. Relay coil 1CR deenergized.
       a-2. Relay contact 1CR-1 opens, opening the interlock circuit around the START pushbutton switch.
       a-3. Relay contact 1CR-2 opens, deenergizing the green pilot light.
    b. Relay contact 2CR-2 closes, energizing the hot bearing light (red).

When the bearing problem has been corrected and all of the normally open thermostat contacts are open, the 1CR relay circuit can be energized through the START pushbutton switch.

ACHIEVEMENT REVIEW

1. What are the three basic types of temperature sensing devices?
2. What is speed of response in a temperature controller?

3. Explain the difference between controller resolution sensitivity and controller operating differential.

4. Explain the thermocouple operating principle.

5. Explain the function of time proportioning control.

6. What size of loads can be handled with a relay output controller? Can this load capacity be increased? If so, how?

7. If the band width in a 800°F temperature controller is set for 2%, what is the band width in degrees?

8. Explain how the mechanical-link temperature switch operates.

9. What are the disadvantages or problems found in the use of a liquid-filled sensing element as used with a temperature switch?

10. Draw the symbols for a normally open and normally closed temperature switch contact.

11. What is the approximate millivolt output of a type J thermocouple at 500°F?
    a. 10
    b. 15
    c. 20

12. The bimetal type of temperature switch operates on the principle of
    a. chemical reaction of the two metals.
    b. uneven expansion of two different metals when heated.
    c. expansion or air, gas or vapor.

CHAPTER 9

TIMERS

After studying this chapter, the student will be able to:

- Describe the differences between a timer and a time-delay relay.
- Explain where the timer has a definite use.
- List three major types of timers discussed in this chapter.
- Explain the arrangement of the contacts in a reset-type timer under the conditions of reset, timing and timed out.
- Draw simple control circuits showing the:
  1. Clutch and motor energized at the same time
  2. Clutch energized before the motor
  3. Motor energized before the clutch
- Explain how the multiple-interval timer operates.
- Discuss the operation of the repeat-cycle timer.
- Describe some of the design features of the solid-state timers.
- Explain how the dashpot-type pneumatic timer operates.

9.1 SELECTED OPERATIONS

Many types of timers are available for use on industrial machines. Their major function is to place information about elapsed time into an electrical control circuit. The method of accomplishing this varies with the type of timer used.

Time range, accuracy, and contact arrangement vary among types of timers. A few suggestions are made here for the proper use of timers.

It is true that time elapses during nearly every action taking place on a machine. This does not necessarily mean that a timer is always the best means of control.

For example, a machine part is to move from point X to point Y. This is a definite distance. It is observed that this motion requires approximately 5 seconds (s) to complete. If the position of the machine part is important, a position control device (such as a limit switch) should be used to indicate that the part did arrive at point Y, not dependent on elapsed time. Due to variables in the machine, one cycle could require 5.0 seconds, the next 5.1 seconds, and a third cycle 4.9 seconds. Therefore, if a timer is used the machine part will not always stop at point Y.

In Figure 9-1, the timer is set for 5 seconds. In case A, time runs out before the part reaches point Y. In case B, time runs out after the part has passed point Y.

A similar situation may exist when a timer is substituted for a pressure indication. In one case the pressure may build to a preset value before a preset time has elapsed. Also, time may elapse before the preset pressure is reached. A like example is when a timer is substituted for a temperature controller, to obtain temperature information.

The important point is to determine the critical condition to be met. Is it time, position, pressure, or temperature? Timers are very useful tools, but they must be applied properly.

Another point to clarify is reference to timers and time-delay relays. Generally, when reference is made to *time-delay relays*, this indicates devices having a timing function after the timer coil has been energized or deenergized. Time-delay relays have a normally open and a normally closed timing contact. In some cases the contacts are isolated. In other cases they have a common terminal. Sometimes time-delay relays have several NO and/or NC instantaneous contacts which operate immediately when the time-delay coil is energized.

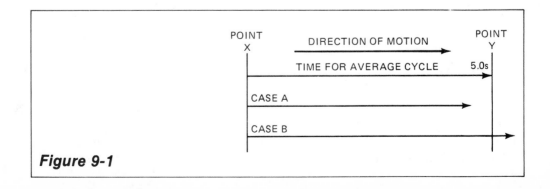

Figure 9-1

When reference is made to *timers,* the time function may start on one or more of the contacts upon energization, or at any time after energization during a preset time cycle. Likewise, the timing function may stop on one contact or more during the cycle after timing has been started on the particular contact(s).

In general, a timer opens or closes electrical circuits to selected operations according to a timed program. Instantaneous contacts are also available on some timers.

9.2 TYPES OF TIMERS

In grouping timers according to their method of timing, there are three types that are applied most frequently on industrial machines:

1. Synchronous motor driven
2. Solid state
3. Dashpot

Three other types of timers that have industrial applications are the mechanical, electrochemical, and thermal types. Since these have only limited use on industrial machines, they are not covered in this text.

9.3 SYNCHRONOUS MOTOR-DRIVEN TIMERS

The synchronous motor-driven timer can well be termed the "workhorse" of the timer field. Its many applications make it a very useful tool.

Under the general heading of synchronous motor-driven timers, the following subdivisions can be made:

• Reset timers

• Repeat-cycle timers

• Manual-set timers

9.3.1 Reset-type Timers The *reset timer* depends on the clutch and the synchronous motor for its operation. The symbols for the clutch and synchronous motor are shown in Figure 9-2.

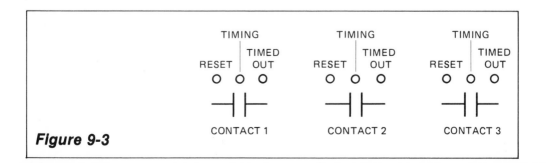

TIMER CLUTCH   TIMER MOTOR

Figure 9-2

To provide an output for the timer to control electrical circuits, most reset-type timers have a set of three contacts. Each contact can be adjusted to perform in the same way or in a different way, as required. Figure 9-3 shows that there are three conditions for each contact during a complete cycle:

1. Reset or deenergized — clutch and motor deenergized
2. Timing — clutch energized; motor may or may not be energized, depending on its circuit application
3. Timed out — motor deenergized; clutch may or may not be deenergized, depending on its circuit application

When the contact is open, the ○ symbol above the contact is shown open, (○). When the contact is closed, the symbol is shown solid (●).

For example, in a given timer contact arrangement, assume contact 1 is open in the reset condition, closed in the timing condition, and open in the timed-out condition. The complete symbol appearing above that contact would be ○ ● ○.

Remember that the contact changes from its reset condition to the timing condition when the clutch is energized. The contact changes from the timing condition to the timed-out condition when the motor has run for the preset amount of time. The contact then changes back to the reset condition when the clutch is deenergized.

There are several arrangements available for each of the three contacts. A few of the arrangements are shown in Figure 9-4.

In some reset-type timers, the motor connections are not brought out for external connection. In this case they are generally wired so that the motor is

TIMING   TIMING   TIMING
                    |TIMED |TIMED |TIMED
RESET|OUT   RESET|OUT   RESET|OUT
○   ○   ○   ○   ○   ○   ○   ○   ○

—| |—   —| |—   —| |—

CONTACT 1   CONTACT 2   CONTACT 3

Figure 9-3

SEQUENCE NUMBER	RESET Clutch deenergized; motor deenergized	TIMING Clutch energized; motor energized	TIMED OUT Clutch engaged until solenoid is deenergized; motor deenergized
1	○	○	●
2	○	●	○
3	●	●	○
4	●	○	●
5	○	●	●
6	●	○	○

Figure 9-4   ○ CONTACT OPEN   ● CONTACT CLOSED

energized from a normally open circuit contact which closes when the clutch is energized. On other timers, all connections are brought out. In this way, the motor as well as the clutch can be energized from an external source at any time during a cycle.

The circuit in Figure 9-5A illustrates conditions where the motor is energized directly from the clutch being energized. This is the normal usage.

Figure 9-5B illustrates the motor energized at a later time, after the clutch is energized. Increased timing accuracy is possible with this arrangement, where the timing period is relatively short in an overall long cycle.

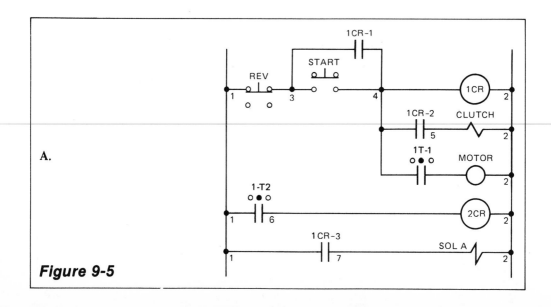

A.

Figure 9-5

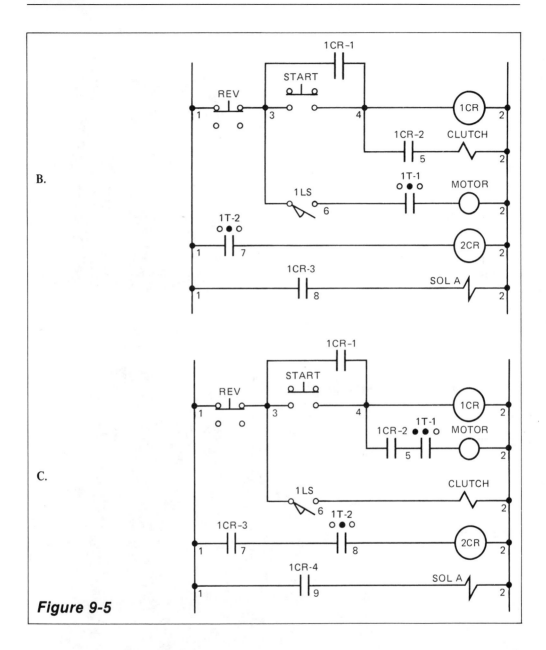

B.

C.

Figure 9-5

A third condition is shown in Figure 9-5C. Here the motor is energized before the clutch is energized. Increased accuracy can be obtained if the timing period is nearly that of the overall cycle.

Figure 9-6 shows an example of the reset-type timer.

Some of the more simple reset timers have only a snap-action switch that operates at the end of a preset time period. When the clutch is deenergized, the timer resets and the switch is released.

Reset-type timer

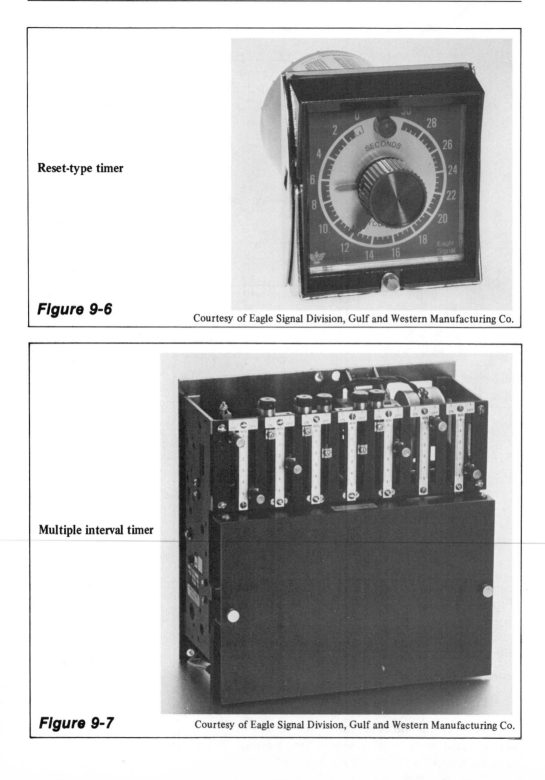

Figure 9-6

Courtesy of Eagle Signal Division, Gulf and Western Manufacturing Co.

Multiple interval timer

Figure 9-7

Courtesy of Eagle Signal Division, Gulf and Western Manufacturing Co.

Another form of the reset-type timer is often called a *multiple-interval timer.* It is used extensively in programming-type control. Multiple time periods can be controlled by the adjustment of each individual ON and OFF setting.

The timer displayed in Figure 9-7 consists of a series of contacts. Each contact has two fingers which ride on a cam plate. During the timing interval, a synchronous motor drives the plate downward through a solenoid-operated clutch. The contacts are closed and opened in the sequence set for the fingers to drop off the edge of the cam plate. When the clutch solenoid is deenergized to disengage the clutch, the cam plate is reset by a spring. Simultaneously, the contact fingers are lifted to their normally open or normally closed RESET position.

9.3.2 Repeat-cycle Timers The *repeat-cycle timer* is used to control a number of electrical circuits in a predetermined sequence.

The timer shown in Figure 9-8 consists of a synchronous motor driving a cam shaft. The cam shaft rotates continuously as long as the motor is energized. Adjustable cams determine the point of closing and opening a switch during each cam-shaft revolution.

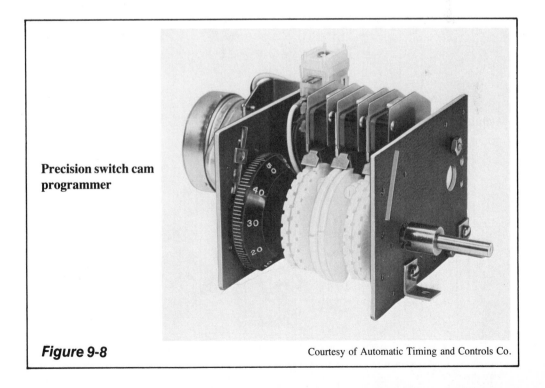

Precision switch cam programmer

Figure 9-8

Courtesy of Automatic Timing and Controls Co.

In the unique split-cam design, each side of the cam is separately screwdriver-adjustable in either direction. Either side determines the precise instant during the cycle when the switch will actuate; the other side determines how long the switch remains actuated. Adjustments are easy and precise. One-quarter turn of the adjusting screw equals .5% of cycle time. A setting disc, calibrated in 1% increments, facilitates program set-up and indicates cycle progress.

9.3.3 Manual-set Timers   The *manual-set timer* requires manual operation of the timer to start operation. The timer then runs a selected time and stops automatically.

In the timers illustrated in Figure 9-9, timing begins when the START switch is closed. At the same time, the timing light-emitting diode (LED) goes on. A

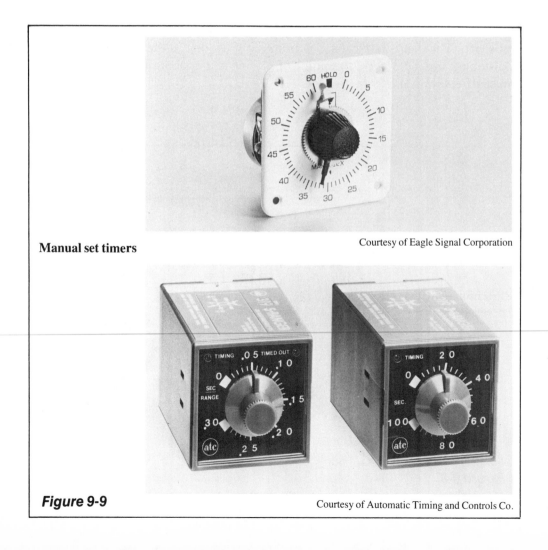

Manual set timers   Courtesy of Eagle Signal Corporation

Figure 9-9   Courtesy of Automatic Timing and Controls Co.

relaxation oscillator starts to run at a rate determined by the set point. The timer times out and the timing LED turns off when the oscillator count is to the level set by the range switch. At time out, the load relay is energized, transferring its contacts, and the timing circuit is automatically deenergized. Reset occurs when the START switch is opened or when power is interrupted.

9.4 SOLID-STATE TIMERS

Great advancement has been made in timer design in the last ten years. A thorough review of the many types and variations of solid-state timers would more than fill this entire book. For our purposes, then, we shall discuss the following important design features:

- Plug-in modules
- Digital set and optional digital readout
- Provision for external set

Some recent developments in the solid-state timer field are pictured in Figures 9-10A, 9-10B, and 9-10C.

The ATC Series 365 Long-Ranger computing timer is shown in Figure 9-10A. This compact and versatile timer is controlled by a built-in microcomputer. It is totally adjustable between 0.01 second (s) and 999 hours (h). The timer has a repeat accuracy of ±10 milliseconds (ms) at all settings. It is easily programmed to time up or down from the set point. This timer can also be programmed to stop or go (for example, continue displaying time) after time out. A built-in self-test program verifies proper operations without test instruments.

The Eagle 531 series shown in Figure 9-10B has a digital readout using light-emitting diodes. The CT530 series has all the design and capabilities provided in the CT531 except the digital readout.

The CT530 series consists of two basic parts: the solid-state (MOS/LS1) timing circuit, and two plug-in output relays rated at 10 A. The one internal relay provides instantaneous contact action as soon as the unit is energized. The second relay operates after a set delay. Digital timing is obtained through the use of an integrated chip. The chip times by counting line frequency through one thousand transistors.

Repeat accuracy is 8.33 ms maximum on all time ranges. Reset time is 20 ms maximum. The temperature range is 32°F to 158°F.

The solid-state timer shown in Figure 9-10C is the SCI model TWT. Its function is to provide eight externally adjusted time delays. The delay time of each section is determined by the resistance connected to it. Resistance

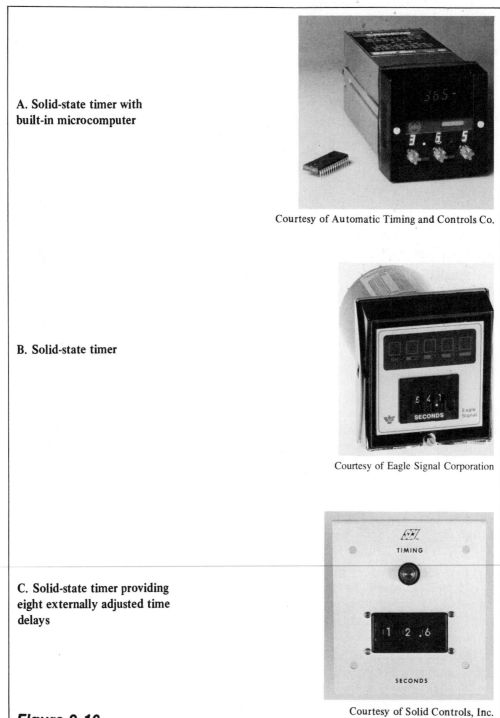

A. Solid-state timer with built-in microcomputer

Courtesy of Automatic Timing and Controls Co.

B. Solid-state timer

Courtesy of Eagle Signal Corporation

C. Solid-state timer providing eight externally adjusted time delays

Courtesy of Solid Controls, Inc.

Figure 9-10

comes from external thumbwheel heads connected to an 18-pole terminal block located on the top edge of a TWT card. The 18-pole terminal block on the TWT card has one pole for each timing resistance connection, one pole for each ○ ● ○ output, and one pole each for the +12Vd.c. and D.C. common. These connections are used in the operation of the thumbwheel timer heads. The ○ ● ○ outputs are also accessible by cable connection. Minimum time is 10 ms.

9.5 DASHPOT TIMERS

One type of dashpot timer is the *pneumatic timer.* This timer works on the principle of transferring air from one chamber to another through an adjustable orifice. This orifice controls the rate of air flow, in turn controlling the movement of a diaphragm and contact assembly. A cross section of a pneumatic timing head is shown in Figure 9-11. The timing head is shown in the timed-out position. It must be reset before any timing can occur.

To reset the timing head, the plunger (A) is rapidly pushed upward, generally by the motion of the armature of the device. An operating lever attached to the

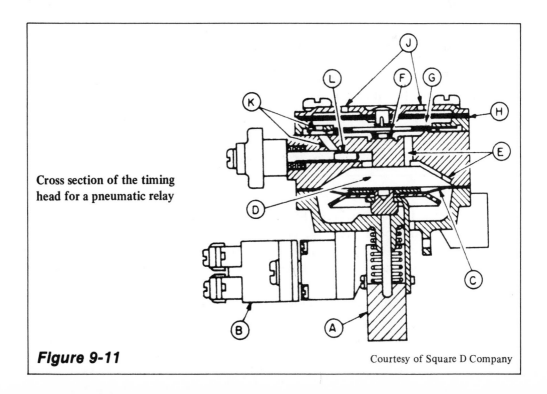

Cross section of the timing
head for a pneumatic relay

Figure 9-11   Courtesy of Square D Company

plunger (A) causes the snap switch (B) to reset. (The contacts transfer from an open to a closed position or from a closed to an open position.) The diaphragm (C) is attached to the plunger and is also moved upward, compressing the air in the chamber (D). The compressed air in the chamber (D) is then forced through the outlet (E) and against the valve seat (F). The pressure of the air on the reset stroke is great enough to force the valve seat (F) upward. This allows the air to escape into chamber (G) through filter (H) and out into the surrounding atmosphere through opening (J). Valve seat (F) then returns to its normal position. All of the above happens in a few milliseconds. The timing head is now in the RESET position.

To begin the timing period, plunger (A) is when the device armature moves downward and away from it. Diaphragm (C) tries to move downward under spring pressure, but a vacuum is created in chamber (D), holding the diaphragm (C) and plunger (A) in position. Air is then drawn in from the surrounding atmosphere through opening (J) and filter (H), through the inlet hole (K), past a needle valve (L), and into the air chamber (D). As air enters the chamber (D), the diaphragm and plunger move downward. The timing period ends when the plunger (A) and the operating lever cause the snap switch (B) to operate. The above reset and timing cycles can then be repeated. The length of the timing period can be adjusted by the needle valve (L). Adjusting the needle valve in restricts the air flow and lengthens the timing period. Adjusting it out allows more air to flow past at a faster rate and shortens the timing period.

The diaphragm (C) and valve seat (F) are generally made of silicon rubber. The timing head shown in Figure 9-11 is an open-air system. External air is drawn in and expelled through a filter. Closed-air systems recirculate the air internally and do not use external air. (Closed-air systems, however, depend on a seal to keep out external air and dust.)

9.6 CIRCUITS

Refer again to Chapter 7. In Figure 7-6C, the ram advanced until it reached the work piece and built pressure. On reaching a preset pressure, the normally closed pressure switch contact opened, and the ram reversed.

In Figure 9-12, a reset-type timer is added. The pressure switch contact is changed from normally closed to normally open.

The sequence of operations for the circuit shown in Figure 9-12 is as follows:

1. Operate the START pushbutton switch.
2. Relay coil 1CR is energized.

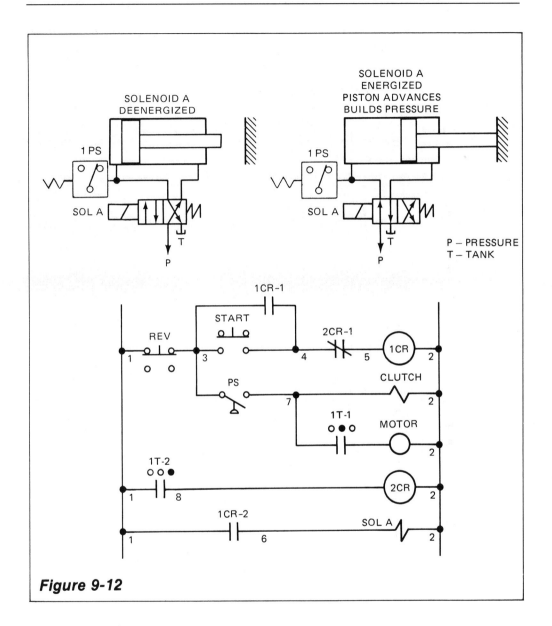

Figure 9-12

    a. Relay contact 1CR-1 closes, interlocking around the START pushbutton switch.

    b. Relay contact 1CR-2 closes, energizing solenoid A. The piston advances to the work and builds pressure.

3. The normally open pressure switch contact closes, energizing the timer clutch.

a. Timer contact 1T-1 closes, energizing the timer motor.
4. When a preset time elapses, timer contact 1T-1 opens, deenergizing the timer motor.
5. Timer contact 1T-2 closes, energizing relay 2CR.
    a. Relay contact 2CR-1 opens, deenergizing relay coil 1CR.
    b. Relay contact 1CR-1 opens, opening the interlock circuit around the START pushbutton circuit.
    c. Relay contact 1CR-2 opens, deenergizing solenoid A.
6. Pressure drops, opening pressure switch contact PS.
7. Timer clutch deenergized.
8. Timer contact 1T-2 opens, deenergizing relay 2CR.
9. Normally closed relay contact 2CR-1 closes, setting up the circuit for the next cycle.

In the circuit shown in Figure 9-13, an additional cylinder-piston assembly is added with an additional pressure switch contact 2PS.

The function of this circuit is to start piston #2 with a time delay after piston #1 has built pressure to a preset level on the work piece. The sequence of operations proceeds as follows:

1. Operate the START pushbutton switch.
2. Relay coil 1CR is energized.
    a. Relay contact 1CR-1 closes, interlocking around the START pushbutton switch.
    b. Relay contact 1CR-2 closes, energizing solenoid A.

Piston #1 advances and builds pressure on the work piece. When a preset pressure is reached, pressure switch contact 1PS operates.

3. Normally closed pressure switch contact 1PS opens, deenergizing relay coil 1CR.
    a. Relay contact 1CR-1 opens, opening the interlock around the START pushbutton switch.
    b. Relay contact 1CR-2 opens, deenergizing solenoid A.

Piston #1 now returns to its initial start position.

4. Normally open pressure switch contact 1PS closes energizing relay coil 2CR.
    a. Relay contact 2CR-1 closes, interlocking around 1PS pressure switch contact.
5. Timer clutch energized.
    a. Timer contact 1T-1 closes, energizing the timer motor.

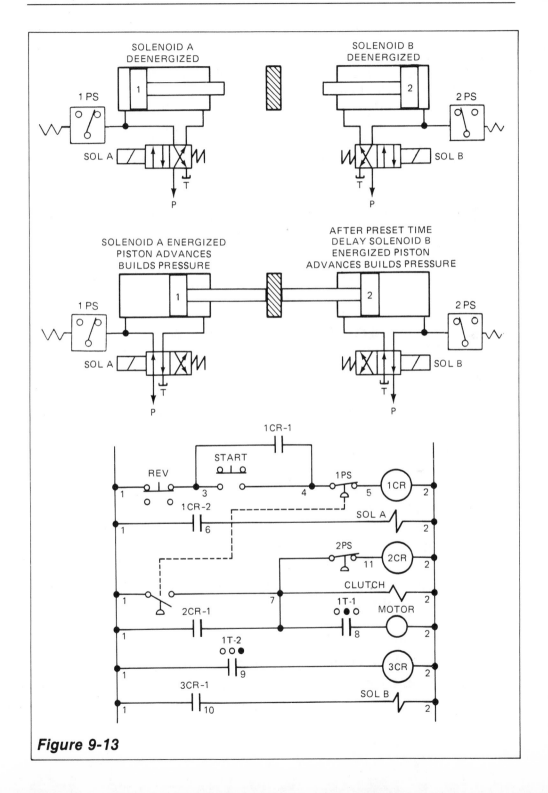

Figure 9-13

6. Timer times out.
   a. Timer contact 1T-1 opens, deenergizing the timer motor.
   b. Timer contact 1T-2 closes, energizing relay coil 3CR.
7. Relay contact 3CR-1 closes, energizing solenoid B.

Piston #2 advances and builds pressure on the work piece.

8. Normally closed pressure switch contact 2PS opens, deenergizing relay coil 2CR.
   a. Relay contact 2CR-1 opens, deenergizing timer clutch.
9. Timer contact 1T-2 opens, deenergizing relay coil 3CR.
   a. Relay contact 3CR-1 opens, deenergizing solenoid B.

Piston #2 now returns to its initial start position.

ACHIEVEMENT REVIEW

1. Describe the differences between a time-delay relay and a timer.

2. List the three types of timers that are applied most frequently in the industrial field.

3. What are the two major components responsible for the operation of the reset-type time?

4. List the conditions of the two major components in a reset-type timer under the following conditions:
   A. Reset or deenergized
   B. Timing
   C. Timed out

5. In a control circuit using two push-button switches, two relays, a reset-type timer, and a solenoid, show how the clutch and motor of the reset-type timer can be energized at the same time. The cycle starts with the operation of a push-button switch.

6. What is the major use of the multiple-interval timer?

7. What controls the opening and closing of a switch in the repeat-cycle timer?

8. What controls the movement of the diaphragm in a pneumatic timer?

9. In the control circuit shown in Figure 9-12, the timer clutch is energized when
   a. relay 1CR is energized.
   b. relay 2CR is deenergized.
   c. pressure builds, operating pressure switch 1PS.

10. In the control circuit shown in Figure 9-13, solenoid B is deenergized when
    a. relay 1CR is deenergized.
    b. the timer times out.
    c. pressure builds, operating pressure switch 2PS.

CHAPTER 10

COUNTERS

OBJECTIVES

After studying this chapter, the student will be able to:

- Explain the basic difference between an electromechanical reset-type timer and an electromechanical reset-type counter.
- Discuss how the clutch and count motor operate through one complete cycle.
- Use the counter contact symbol in each of three conditions: deenergized, counting, and counted out.
- Follow through a typical control circuit using an electromechanical reset-type counter.

10.1 PRESET ELECTRICAL IMPULSES

The electromechanical control counter is similar to the electromechanical reset-type timer, except that the synchronous motor of the timer is replaced by a stepping motor. The stepping motor advances one step each time it is deenergized. After a preset number of electrical impulses, a contact is opened or closed.

The electromechanical counter requires a minimum of approximately 0.5 seconds off time between input impulses to reset.

As in the electromechanical timer, there are three output contacts. Each contact can be arranged for a sequence. The conditions for each contact can change as shown in Figure 10-1.

With the clutch deenergized, the contacts are in the reset condition. When the clutch is energized, the contacts go to the counting condition. When the

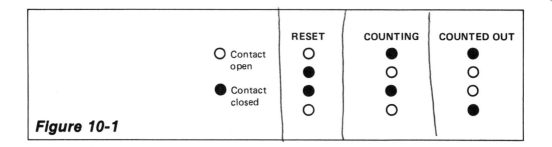

Figure 10-1

count motor has received the same number of count impulses as set on the dial, the contacts go to the counted-out condition. When the clutch is deenergized, the contacts return to the reset condition.

Figure 10-2A shows a typical electromechanical counter. These devices are generally available with analog set and readout dials.

Considerable advancement has been made in solid-state counter design. Solid-state counters are available with digital set and digital readout. High-speed pulse operation with 100% accuracy is available. Figure 10-2B shows a typical solid-state counter.

A recent development in the solid-state timer/counter field is the programmable timer/counter. This unit has the following features:

- One to six circuits
- Single or dual outputs per circuit
- One cycle
- Automatic repeat cycle
- Visual readout
- Internal or external cycle time base
- 1200 counts per minute
- Frequency counts up to 30 kilohertz (kHz)

The programmable timer/counter is shown in Figure 10-2C.

10.2 CIRCUIT APPLICATIONS

A typical control circuit using a reset-type counter is shown in Figure 10-3. In this example a single piston-cylinder assembly is used. The piston is to travel to the right as shown until it engages and operates limit switch 1LS at position P1. The pis-

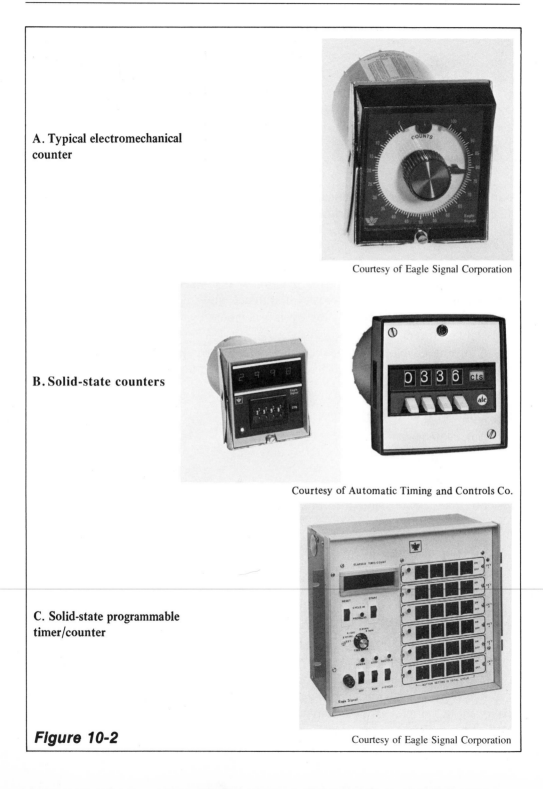

A. Typical electromechanical counter

Courtesy of Eagle Signal Corporation

B. Solid-state counters

Courtesy of Automatic Timing and Controls Co.

C. Solid-state programmable timer/counter

Figure 10-2

Courtesy of Eagle Signal Corporation

ton returns to position P2, operating limit switch 2LS. The piston now travels to the right. This reciprocating motion continues until the preset number of counts has been reached. The counter is now in the counted-out condition. The piston continues to the right until a work piece is engaged and preset pressure builds to the setting of pressure switch 1PS. The piston now returns to the start position.

A single-solenoid, spring-return operating valve is used to supply fluid power to the cylinder. The circuit is arranged to return the piston to the start position at any time by operating the REVERSE push-button switch.

The sequence of operation proceeds as follows:

1. Operate the START push-button switch.
2. Relay coil 1CR is energized.
   a. Relay contact 1CR-1 closes, interlocking around the START push-button switch and energizing the counter clutch.
   b. Relay contact 1CR-2 closes, energizing solenoid A.

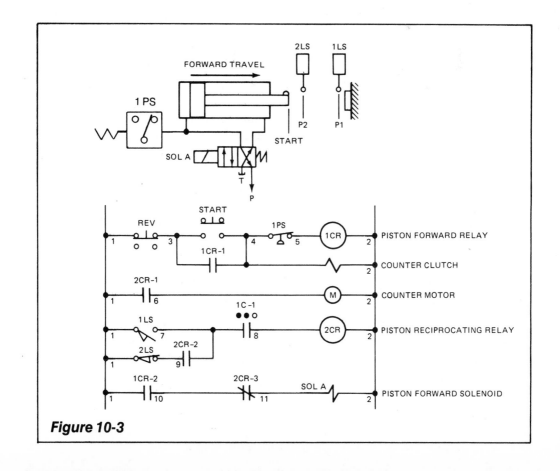

Figure 10-3

The piston travels to the right. At position P2, limit switch 2LS is operated. No action results as relay contact 2CR-2 is open. The piston continues to travel to position P1, operating limit switch 1LS.

3. Relay coil 2CR is energized.
   a. Relay contact 2CR-1 closes, energizing the counter motor.
   b. Relay contact 2CR-2 closes, forming an interlock circuit with normally closed limit switch contact 2LS around normally open limit switch contact 1LS.
   c. Relay contact 2CR-3 opens, deenergizing solenoid A.

The piston travels back (to the left) until it operates limit switch 2LS.

4. Relay coil 2CR is de-energized.
   a. Relay contact 2CR-1 opens, deenergizing the counter motor. The counter has now completed one count.
   b. Relay contact 2CR-2 opens.
   c. Relay contact 2CR-3 closes, energizing solenoid A.

The piston now travels forward (to the right) until it operates limit switch 1LS.

5. Relay coil 2CR energized.
   a. Relay contact 2CR-1 closes, energizing the counter motor.
   b. Relay contact 2CR-2 closes, interlocking limit switch 1LS.
   c. Relay contact 2CR-3 opens, deenergizing solenoid A.

The piston travels back (to the left) until it operates limit switch 2LS.

6. Relay coil 2CR is deenergized.
   a. Relay contact 2CR-1 opens, deenergizing the counter motor. The counter has now completed two counts.
   b. Relay contact 2CR-2 opens.
   c. Relay contact 2CR-3 closes, energizing solenoid A.

The piston continues to shuttle between limit switches 1LS and 2LS until the counter has counted the number of preset counts. When the counter has counted out, counter contact 1C-1 opens. The next time the piston advances and operates limit switch 1LS, relay coil 2CR is not energized.

7. Solenoid A remains energized.
8. Piston continues past limit switch 1LS, meeting the work piece and builds pressure to a preset amount on pressure switch 1PS.

9. The normally closed pressure switch contact 1PS opens.
10. Relay coil 1CR is deenergized.
   a. Relay contact 1CR-1 opens, opening the interlock circuit around the START push-button switch and deenergizes the counter clutch.
   b. Contact 1CR-2 opens, deenergizing solenoid A.

The piston now returns to its initial start position.

ACHIEVEMENT REVIEW

1. How does the electromechanical reset-type counter differ from the electromechanical reset-type timer?

2. A given counter contact has been arranged to be open in the deenergized condition of the counter, closed in the counting condition, and closed in the counted-out condition. Show the contact and symbol for these conditions.

3. What accuracy can you expect to receive when using a solid-state counter?

4. Given the following counter contact symbol, explain the condition of the contact in the deenergized condition of the counter, the counting condition, and the counted-out condition.

5. In the electromechanical counter, what is the condition of the contacts with the clutch deenergized?

6. In modern solid-state counters, the following accuracy can be expected.
   a. 75%
   b. 90%
   c. 100%

7. What are some of the features available in the solid-state timer-counters?

8. In the circuit shown in Figure 10-3, what operational change would occur if the counter contact 1C-1 had a sequence of ● ○ ● (closed-open-closed)?

9. In the same circuit (Figure 10-3), what is the function of the NC relay contact 2CR-3?

10. The stepping motor advances one step each time the motor is
    a. energized.
    b. deenergized.

CHAPTER

MOTOR STARTERS

11

OBJECTIVES

After studying this chapter, the student will be able to:

- Discuss the difference between a contactor and a motor starter.
- Explain why overload relays are used on motor starters.
- Describe how the normally closed contact in the overload relay is connected into the motor starter control circuit.
- Explain how the ambient compensated overload relay operates.
- Draw the basic power and control circuit for a magnetic full-voltage motor starter.
- Explain why both mechanical and electrical interlocks are used on the reversing motor starter.
- Describe the relationship between applied voltage on a motor and the resulting torque.
- Draw the basic power and control circuit for a resistor-type, reduced-voltage motor starter and explain how it operates.
- Explain why a reduced-voltage motor starter should be used in some cases rather than a full-voltage motor starter.

11.1 OVERLOAD RELAYS

The motor starter is similar to the contactor in design and operation. Both have one important feature in common: contacts operate when the coil is energized. The important difference is the use of overload relays on the motor starter.

Manual motor starters are available. In these units the contacts are closed by manual operation. The end results obtained are energizing and protecting a motor. However, most machines today require additional control, safety, and convenience of remote control on one motor or multiple motors. Thus, very limited use is made of the manual motor starter on industrial machines.

The *magnetic motor starter* has three main contacts in the form used most frequently. These three contacts are normally open. This arrangement is used

NEMA SIZE	MOTOR STARTERS HORSEPOWER		VOLTS
	SINGLE PHASE	TWO AND THREE PHASE	
00	1	1 1/2 2 2	200–230 460 575
0	2	3 5 5	200–230 460 575
1	3	7 1/2 10 10	200–230 460 575
2	7 1/2	15 25 25	200–230 460 575
3	15	30 50 50	200–230 460 575
4		50 100 100	200–230 460 575
5		100 200 200	200–230 460 575
6		200 400 400	200–230 460 575
7		300 600 600	200–230 460 575
8		450 900 900	200–230 460 575

Figure 11-1

in the starting of three-phase motors. Most industrial plants in this country have three-phase power available as standard. Sometimes two-phase, four-wire power is used. In such cases, four normally open contacts are supplied on the starter.

Like the power contactor, magnetic motor starters are available in many sizes. They start at approximately the relay size and extend up in capacity. Figure 11-1 lists motor starters, showing size, rated horsepower, and rated voltage.

The important concern with motor starters comes with the closing of contacts. These connect the motor to the source of electrical power. Unfortunately, it is not always possible to control the amount of work applied to the motor. Therefore, the motor may be overloaded, resulting in serious damage. For this reason, overload relays are added to the motor starter.

The goal is to protect the motor from overheating. The current drawn by the motor is a reasonably accurate measure of the load on the motor and thus a measure of its heating.

Most overloads today use a thermally responsive element. That is, the same current that goes to the motor coils (causing the motor to heat) also passes through the thermal elements of the overload relays.

The thermal element is connected mechanically to a normally closed contact. When an excessive current flows through the thermal element for a long enough time period, the contact is tripped open. This contact is connected in series with the control coil of the starter. When the contact opens, the starter coil is deenergized. In turn, the starter power contacts disconnect the motor from the line.

Figure 11-2 shows a typical thermal overload relay.

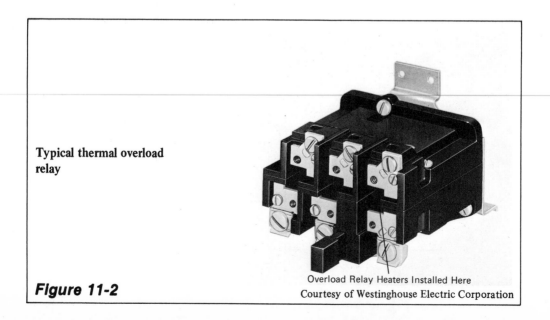

Typical thermal overload relay

Overload Relay Heaters Installed Here
Courtesy of Westinghouse Electric Corporation

Figure 11-2

A motor can operate on a slight overload for a long period of time or at a higher overload for a short period of time. Overheating of the motor will not result in either case. Therefore, the overload heater element should be designed to have heat-storage characteristics similar to those of the motor. However, they should be just enough faster so that the relay will trip the normally closed relay contact before excessive heating occurs in the motor.

The *ambient* (surrounding air) temperature in the location of the motor and starter also has some effect. It is necessary to specify the rating of a given temperature base plus the allowable temperature rise due to the load current. For example, an open motor rating is generally based on 40°C (104°F). The motor nameplate will specify the allowable temperature rise from this base.

The motor and starter are usually located in the same general ambient temperature. Thus, the overload heater elements are affected by the same temperature conditions. The overloads will open the motor starter control circuit either through excessive motor current, a high ambient temperature, or a combination of both.

Ambient compensated overload relays are now available. These are used when the control is located in a varying ambient temperature and the motor which it protects is in a constant ambient temperature.

The ambient compensated overload relay operates through a compensating bimetal relay. The relay maintains a constant travel-to-trip distance, independent of ambient conditions. Operation of this bimetal relay is responsive only to heat generated by the motor overcurrent passing through the heater element.

All starter manufacturers list the size of overload heaters for a given starter application. The lists show the range of motor currents with which they should be used. These may be in increments of from 3% to 15% of full load current. The smaller the increment, the closer can be the selection to match the motor to its actual work. Since the heater varies with the enclosure, this information is also included.

All overload relays should be *trip free*. This means it should be impossible for anyone to hold or block a RESET button down, resulting in damage to the motor. After a relay has tripped, the cause of the overload should be investigated. The problem should be solved before the RESET button is depressed to put the starter back into operation.

There is one point about overload relays that is often misunderstood. Overload relays are not intended to protect against short-circuit currents. Short-circuit protection is the function of fuses and circuit breakers.

Another feature of the motor starter is the auxiliary contact normally used in the starter control circuit. In some small starters, an additional load contact can be used. This is not usually the case in large starters.

Auxiliary contacts may be normally open or normally closed and have a 10-ampere rating. In most cases, one NO interlock contact is supplied as standard. Additional contacts (up to four) can be obtained. These can be purchased with the original starter or can be added later. A kit is available from the manufacturer containing all hardware necessary to make the installation.

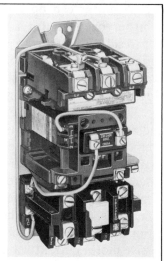

Courtesy of Allen-Bradley

A. Full voltage motor starter

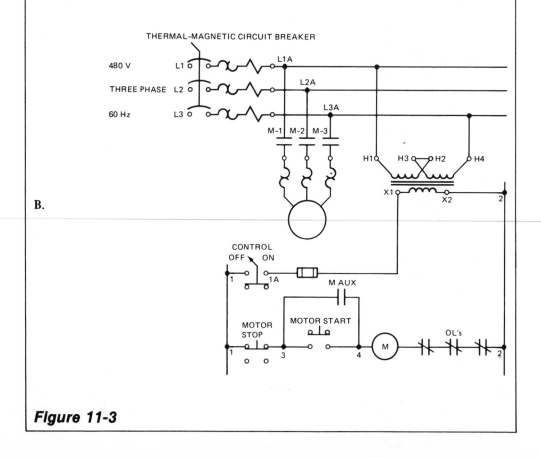

B.

Figure 11-3

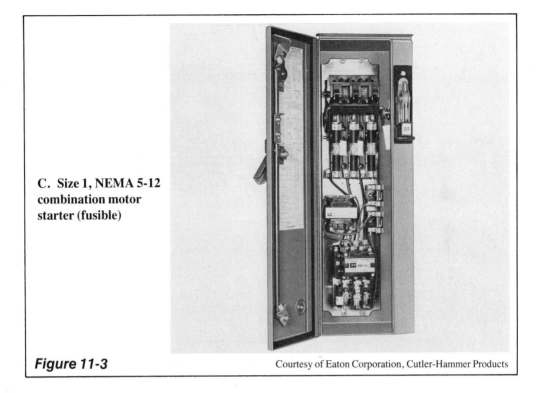

C. Size 1, NEMA 5-12 combination motor starter (fusible)

Figure 11-3   Courtesy of Eaton Corporation, Cutler-Hammer Products

11.2 ACROSS-THE-LINE (FULL-VOLTAGE) STARTERS

The *across-the-line (full-voltage) starter* is often referred to as the "simple" type. This starter has one set of contacts that close when the starter coil is energized. When the contacts close, the motor is connected directly to the line voltage. This type of motor starter has control that is relatively simple, inexpensive, and easy to maintain.

Figure 11-3A shows a typical across-the-line motor starter. Such items as the line terminals, load terminals, motor starter coil, overload relays, and auxiliary contact can be seen.

Figure 11-3B is a complete circuit diagram. Here, as in most machine control, the voltage for control is reduced to 120 V.

Figure 11-3C shows an across-the-line motor starter assembled in a NEMA 12 enclosure with a control transformer and disconnecting and protective devices. The NEMA 12 enclosure is designed to be oil-tight and dust tight. Enclosures are discussed in more detail in Chapter 14, Use of Electrical Codes and Standards. Note that the disconnecting device is mounted in the dead front of the enclosure. This allows the disconnecting device to be locked in the open position with the door open or closed. This arrangement is referred to as a *combination starter*.

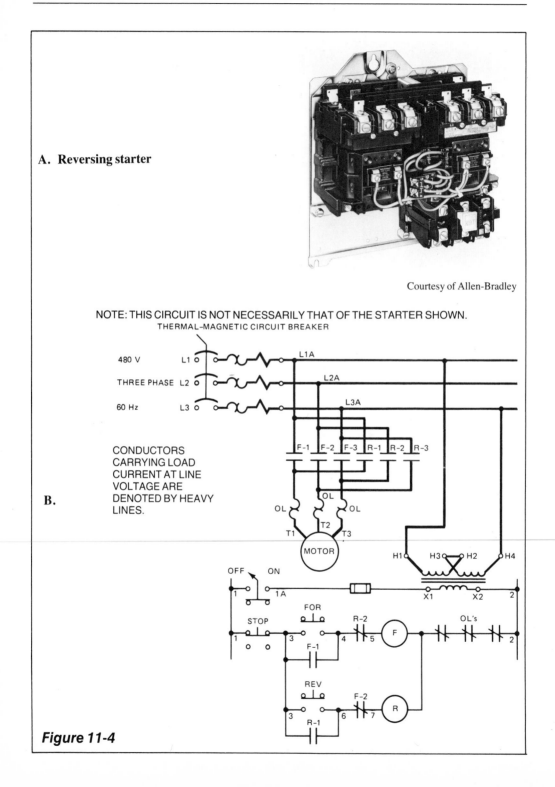

A. Reversing starter

Courtesy of Allen-Bradley

NOTE: THIS CIRCUIT IS NOT NECESSARILY THAT OF THE STARTER SHOWN.

THERMAL–MAGNETIC CIRCUIT BREAKER

480 V — L1 — L1A

THREE PHASE — L2 — L2A

60 Hz — L3 — L3A

B.

CONDUCTORS CARRYING LOAD CURRENT AT LINE VOLTAGE ARE DENOTED BY HEAVY LINES.

F-1   F-2   F-3   R-1   R-2   R-3

OL   OL   OL

T1   T2   T3

MOTOR

H1   H3   H2   H4

OFF   ON
1   1A   X1   X2   2

STOP

FOR   R-2   F   OL's
1   3   4   5   2

F-1

REV   F-2   R
3   6   7

R-1

Figure 11-4

11.3 REVERSING MOTOR STARTERS

Reversing and multispeed motor starters may be considered as special applications of across-the-line starters.

In the *reversing starter,* there are two starters of equal size for a given horsepower motor application. The reversing of a three-phase, squirrel-cage induction motor is accomplished by interchanging any two line connections to the motor. The concern is to properly connect the two starters to the motor so that the line feed from one starter is different from the other. Both mechanical and electrical interlocks are used to prevent both starters from closing their line contacts at the same time. Only one set of overloads is required as the same load current is available for both directions of rotation.

A reversing starter is shown in Figure 11-4A. Figure 11-4B is the circuit diagram. Note the change of connections from the line to the contacts from one unit to the other (lines L2 and L3).

11.4 MULTISPEED MOTOR STARTERS

Many industrial applications require the use of more than one speed in their normal operation. While many new methods for adjustment of speed are available, the *pole changing* method is still used. For example, in a squirrel-cage induction motor, the speed is dependent upon the number of poles. This is obtained by the design of the stator winding. At 60 Hz, which is standard in the United States, a two-pole motor operates at approximately 3600 revolutions per minute (r/min). A four pole operates at 1800 r/min, a six pole at 1200 r/min, an eight pole at 900 r/min, and a ten pole at 720 r/min. By having two or more sets of winding leads brought out into the terminal connection box, the number of effective poles can be changed. Also, one winding can be connected in more than one way.

With a change in speed, the horsepower will also change. For example, in a two-speed motor where the slower speed is one-half the higher speed, the horsepower will also be one-half the horsepower of the higher speed. This means that two sets of overload relays must be used to provide adequate protection.

Figure 11-5A shows a two-speed reconnectable multispeed starter. The circuit of a two-speed motor and starter is shown in Figure 11-5B.

Additional across-the-line starter circuits are shown in Figures 11-5C through 11-5N. To start this group of circuits, it is important to become acquainted with two more expressions in the electrical field. These are *no-voltage or low-voltage release* and *no-voltage or low-voltage protection.*

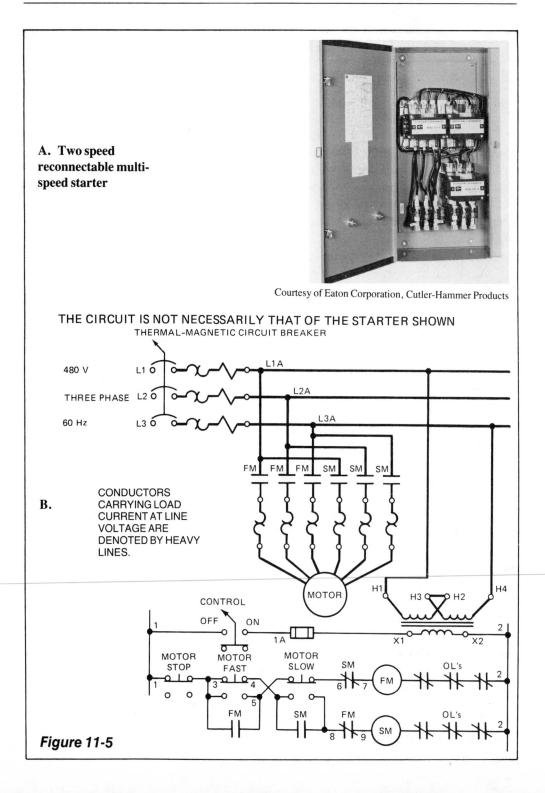

A. Two speed reconnectable multi-speed starter

Courtesy of Eaton Corporation, Cutler-Hammer Products

THE CIRCUIT IS NOT NECESSARILY THAT OF THE STARTER SHOWN

THERMAL–MAGNETIC CIRCUIT BREAKER

480 V

THREE PHASE

60 Hz

B.

CONDUCTORS
CARRYING LOAD
CURRENT AT LINE
VOLTAGE ARE
DENOTED BY HEAVY
LINES.

Figure 11-5

No-voltage or low-voltage release means that when there is a voltage failure, the starter will open its contacts, or "drop out." However, the contacts will close again, or "pick up," as soon as the voltage returns. A typical wiring diagram showing no-voltage or low-voltage release is shown in Figure 11-5C. Note that this is a wiring diagram, as compared to a schematic or elementary diagram. The elementary diagram is shown in Figure 11-5D.

A *wiring diagram* shows the actual wiring on a group of components in a control center or system. In this case, it is an across-the-line motor starter. This approach in explaining the circuit is helpful when physically checking the wires and their connections. Most manufacturers of motor starters include a wiring diagram in the starter enclosure.

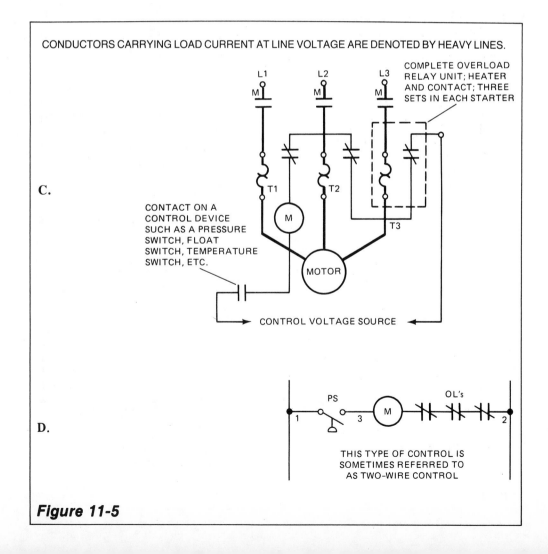

CONDUCTORS CARRYING LOAD CURRENT AT LINE VOLTAGE ARE DENOTED BY HEAVY LINES.

COMPLETE OVERLOAD RELAY UNIT; HEATER AND CONTACT; THREE SETS IN EACH STARTER

C.

CONTACT ON A CONTROL DEVICE SUCH AS A PRESSURE SWITCH, FLOAT SWITCH, TEMPERATURE SWITCH, ETC.

CONTROL VOLTAGE SOURCE

D.

THIS TYPE OF CONTROL IS SOMETIMES REFERRED TO AS TWO-WIRE CONTROL

Figure 11-5

No-voltage or low-voltage protection means that when there is a voltage failure, the starter contacts will open or drop out. However, they will not reclose, or pick up, automatically when the voltage returns. Figure 11-5E is a typical wiring diagram showing this arrangement. The elementary diagram is shown in Figure 11-5F.

There are times when a motor starter must be energized and deenergized from several different locations. Figure 11-5G shows such an arrangement, using four motor STOP-START push-button units. Note that each set of STOP-

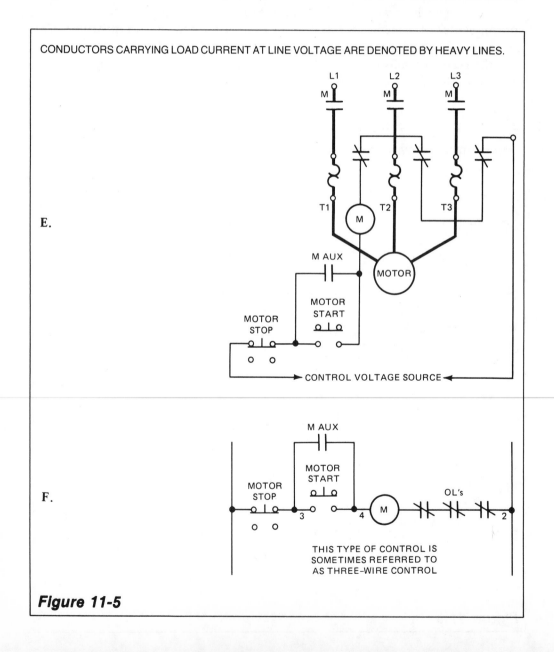

Figure 11-5

START push-buttons may be located remote from any of the other sets. Each has complete control of the starter coil. For safety, the START push-buttons may have individual locks to prevent unauthorized operation of any units.

The use of a pilot light to indicate that a motor is energized is a convenience. Many times it is also a safety factor. For example, a motor may be located remote from its START-STOP push-button and therefore cannot be seen. Also, sometimes the shop noise level is so high that it is hard to know if a motor is operating. Figure

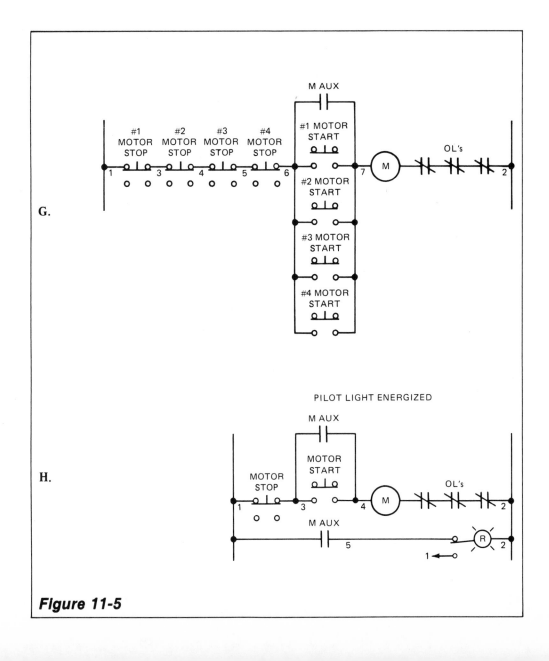

Figure 11-5

11-5H shows a motor starter control. The pilot light indicates its energized condition. The motor is energized in Figure 11-5H. The motor is deenergized in Figure 11-5I.

In addition to the start-stop function of control for the motor starter, there is also *jogging.* The starter is energized only as long as the JOG push button is operated. The use of the relay in this circuit provides a safety factor to prevent the starter from "locking in" during a jogging operation. The circuit for the jogging operation is shown in Figure 11-5J.

The sequence of operation for the jogging circuit is as follows:

For normal motor operation:
1. Operate the motor START push-button switch.

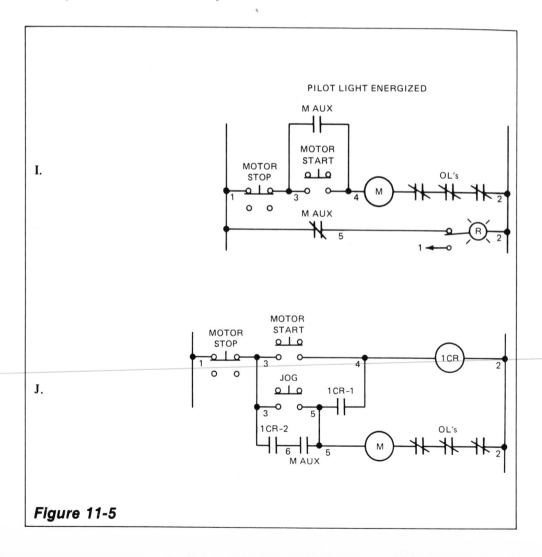

Figure 11-5

2. Relay coil 1CR is energized.
   a. Relay contact 1CR-1 closes, energizing the motor starter coil.
   b. Relay contact 1CR-2 and motor auxiliary contact M Aux. both close. Together with the closed 1CR-1 contact, they form an interlock circuit around the motor START push-button switch.
3. Operating the motor STOP push-button switch deenergizes the motor starter coil and relay coil 1CR.

For jogging operation:
1. Operate the JOG push-button switch.
2. The motor starter coil M will remain energized only as long as the JOG push-button switch is held operated. Note that relay coil 1CR is not energized in this operation.

Some machines have multiple motors. Due to power line conditions, it may not be advisable to start all the motors at the same time.

An arrangement of three motors starting from a single START push button but energizing at different times is shown in Figure 11-5K. Two time-delay relays are used. They can be set to start #2 and #3 motors at predetermined time intervals after #1 motor is energized. The STOP push-button switch will deenergize all three motors. Additional motors can be started from this arrangement by adding time-delay relays for each motor starter. They should be added in the same way that #1 and #2 time-delay relays are used.

The sequence of operations for the circuit shown in Figure 11-5K is as follows:

1. Operate the motor START push-button switch.
2. Motor starter coil 1M is energized.
   a. Time delay relay coils 1TR and 2TR are energized.
3. Motor starter auxiliary normally open contact 1M Aux. closes, interlocking around the motor START push-button switch.
4. After a preset time delay as set on timing relay 1TR, timing relay contact 1TR closes.
5. Motor starter coil 2M is energized.
   a. Motor starter auxiliary normally open contact 2M Aux. closes, interlocking around the 1TR timing relay contact.
6. After a longer preset time as set on timing relay 2TR, timing relay contact 2TR closes.
7. Motor starter coil 3M is energized.
   a. Motor starter auxiliary normally open contact 3M closes, interlocking around the 2TR timing relay contact.
8. All motors are now energized.

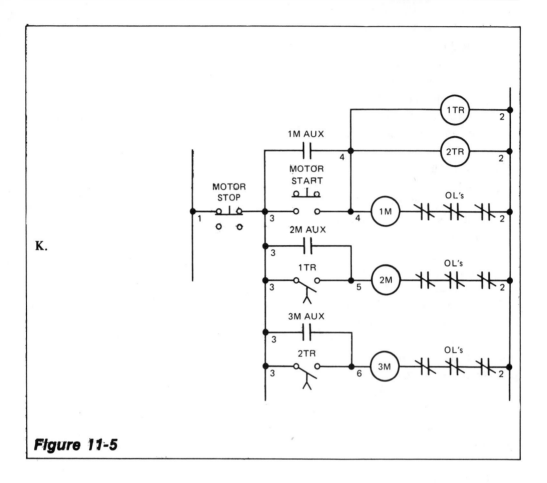

K.

Figure 11-5

9. Operation of the motor STOP push-button switch deenergizes all motor starter coils and the timing relay coils.

A multiple number of motors can be sequenced into operation using the auxiliary contact of the preceding motor starter. This gives a short fixed time delay of only the starter closing time. The opening of any of the motor starters through overload (opening overload contacts) will deenergize all the motors in sequence. All the motors will be deenergized through operation of the one common STOP push-button switch. This circuit is shown in Figure 11-5L. Note that the START push-button switch must be held operated until #3 motor starter coil is energized and the 3M auxiliary contact closes.

The sequence of operations for the circuit shown in Figure 11-5L is as follows:

1. Operate the motor START push-button switch.

2. Motor starter coil 1M is energized.
   a. 1M motor starter auxiliary contact 1M Aux. closes.
3. Motor starter coil 2M is energized.
   a. 2M motor starter auxiliary contact 2M Aux. closes.
4. Motor starter coil 3M is energized.
   a. 3M motor starter auxiliary contact 3M Aux. closes, interlocking around the motor START push-button switch.
5. Operating the motor STOP push-button switch at any time deenergizes all motor starter coils.

Additional motors can be added to this sequence circuit. The auxiliary contact on the last motor starter is generally used to interlock the START push button.

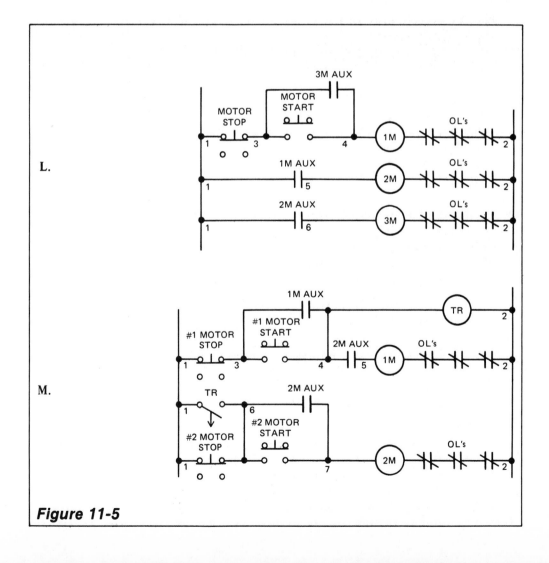

Figure 11-5

Occasionally it is necessary to prevent one motor from stopping until some time has elapsed after the stopping of another motor. This arrangement is used in fluid power applications where a large motor is driving a main pressure pump. If the motor driving the main pressure pump and a small motor driving a pilot pressure pump were deenergized at the same time, the small motor would come to rest before the main pressure pump motor. This would cause control pressure on the large main pressure pump to be lost during the time the large motor was coming to rest.

Figure 11-5M shows a circuit using two motor starters with a time delay. This prevents #2 motor starter from being deenergized until after a preset time has elapsed from the time #1 motor starter is deenergized. The sequence of operations for the circuit shown in Figure 11-5M uses two motor starters with a time delay. Motor starter coil 2M must be energized first.

1. Operate 2M motor START push-button switch.
2. 2M motor starter coil is energized.
    a. 2M Aux. #1 contact closes, interlocking around the #2 motor START push-button switch.
    b. 2M Aux. #2 contact closes, setting up the start circuit for the #1 motor.
3. Operate the #1 motor START push-button switch.
4. Motor starter coil #1M is energized.
5. Time delay relay coil TR is energized.
6. OFF delay-time delay contact TR closes, interlocking around the #2M motor STOP push-button switch.

Motor starter coil #1M must be deenergized before #2M.

7. Operate the #1M motor STOP push-button switch.
    a. Motor starter coil #1M and time delay relay coil TR are deenergized.
8. After a preset time delay as set on TR, the OFF time delay contact opens.
9. Operate #2M motor STOP push-button switch.
10. Motor starter coil 2M is deenergized.

The circuit shown in Figure 11-5N is similar to that shown in Figure 11-5K. The main difference is in the operation of the motor START push-button switch to initially energize the #2 motor starter coil 2M. In this circuit, the 2M motor starter coil can not be energized until the preset time on timing relay has expired. The sequence proceeds as follows:

1. Operate the #1 motor START push-button switch.
2. Motor starter coil 1M and timing relay coil TR are energized.
    a. 1M motor starter auxiliary contact 1M Aux. closes, interlocking around the motor START push-button switch.

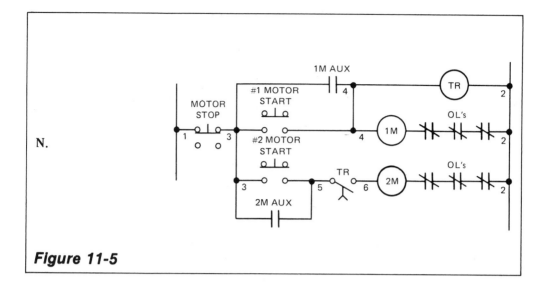

N.

Figure 11-5

3. After a time delay as set on timing relay TR has expired, timing contact TR closes.
4. Operate the #2 motor START push-button switch.
5. Motor starter coil # 2M is energized.
   a. 2M motor starter auxiliary contact 2M Aux. closes, interlocking around the #2 motor START push-button switch.

Note that in this circuit, 2M motor starter coil can be deenergized through the opening of an overload relay contact without deenergizing 1M motor starter coil. However, if 1M motor starter coil is deenergized through the opening of an overload contact, timing relay coil TR is deenergized. This opens the circuit to 2M motor starter coil, deenergizing the motor starter coil 2M.

The sequencing of several motors is shown in Figure 11-5O. The sequence of operations for the circuit shown in Figure 11-5O is as follows:

1. Operate the motor START push-button switch.
2. Motor starter coil 1M is energized.
   a. Green pilot light is energized.
   b. 1M motor starter auxiliary contact 1M Aux. closes.
3. Motor starter coil 2M is energized.
   a. 2M motor starter auxiliary contact 2M Aux. closes.
4. Motor starter coil 3M is energized.
   a. 3M motor starter auxiliary contact 3M Aux. closes.
5. Motor starter coil 4M is energized.
   a. Amber pilot light is energized.

b. 4M motor starter auxiliary contact 4M Aux. closes, interlocking around the motor START push-button switch.

6. Operating the motor STOP push-button switch will deenergize all motor starter coils and pilot lights.

Note that the motor START switch must be maintained operated until 4M Aux. closes.

Note that this circuit is similar to that shown in Figure 11-5L. The advantage of the arrangement in Figure 11-5O is that the STOP push-button switch has

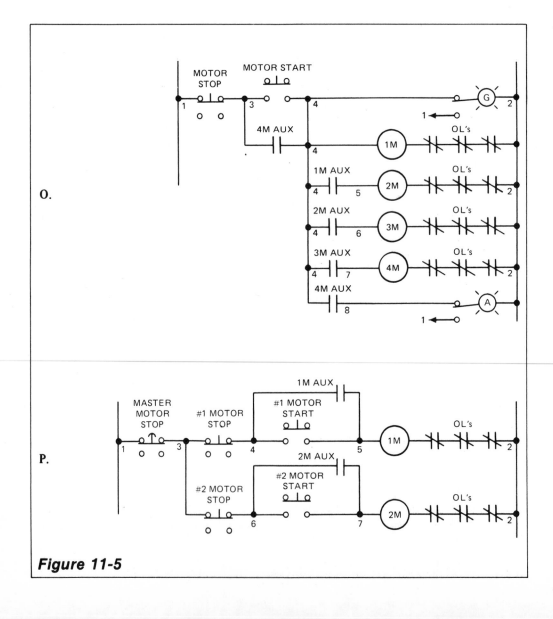

Figure 11-5

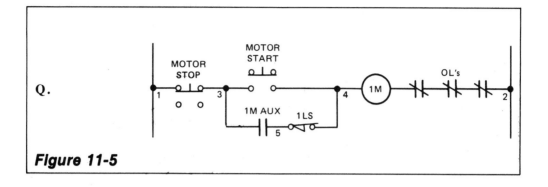

Figure 11-5

complete control in opening the circuit (deenergizing) to all the starter coils. In the case of the circuit in Figure 11-5L, the starters drop out (deenergized) in sequence from the preceding starter. The disadvantage of the circuit in Figure 11-5O is that it may limit the number and/or size of starters used. The current used to energize the starter coils comes through the push-button switches. Therefore, the start circuit is controlled to a degree by the rating of the switches.

The use of a master motor STOP push-button switch can be arranged with each motor-starter circuit having an individual STOP push-button switch. The master motor STOP push-button switch can be used as a convenience or as emergency safety. This switch is generally supplied with a mushroom head operator. The circuit is shown in Figure 11-5P.

As a safety precaution, it is often necessary to provide an overstroke limit switch to deenergize a motor starter coil. For example, a machine part such as a moving table may travel a given distance under safe operating conditions. However, if this distance is exceeded, damage to equipment or injury to personnel may result. The limit switch is located to operate if the proper travel distance is exceeded. When the limit switch is operated, deenergizing the motor starter coil, the motor must then be energized on an emergency basis to back the operating cam off the limit switch. This is accomplished by switching the operating circuit to reverse, and backing the cam off the limit switch by inching the motor START push-button switch. This circuit is shown in Figure 11-5Q.

11.5 REDUCED-VOLTAGE STARTERS

Reduced voltage-type motor starters can be divided into several different designs:

- Autotransformer (or Compensator)
- Primary Resistor
- Wye (or Star) Delta
- Part Winding

In connecting a motor directly across the line, the resulting current at start condition may be in the order of 4 to 10 times the full load rating of the motor. This current is sometimes referred to as *locked rotor current*. The high current often causes line disturbances if the power supply is inadequate or the motor is large.

The basic principle of the reduced-voltage starter is to apply a percentage of the total voltage to start. After the motor starts to rotate, switching is provided to apply full line voltage. For example, in the autotransformer type, starting steps may be 50%, 65%, or 80% of full voltage.

A timer is provided in the starter control circuit so that the time between starting and running may be adjusted to suit actual operating conditions.

It should be pointed out that at reduced voltage, the torque available from the motor will be reduced. This may cause concern where the motor is starting under load. Torque varies as the square of the impressed voltage.

For example, starting on the

- 50% voltage tap — torque will be 25% of normal.
- 65% voltage tap — torque will be 42% of normal.
- 80% voltage tap — torque will be 64% of normal.

Each type of reduced-voltage motor starter is best explained by referring to its circuit diagram. Therefore, a circuit diagram and photo accompany the description of each type of reduced-voltage starter.

Detail control may vary between manufacturers and with the use of the starter. Therefore, the circuit is not necessarily that of the photo shown. The manufacturer lists the maximum horsepower for a range of 5 through 1000 for voltages of 200, 230, 380, 460 and 575. The NEMA size is given for each rating.

11.5.1 Autotransformers (or Compensator). See Figures 11-6A and 11-6B

Two contactors are used in the control. One is a five pole. It is called the *start contactor* (SC). The second contactor is a three-pole contactor, the *run contactor* (RC). A typical starter is shown in Figure 11-6A.

An autotransformer of sufficient size to handle the horsepower rating of the motor is shown connected to the line leads and motor through the open contacts. Taps are available for a starting voltage of 80%, 65%, or 50% of full line voltage.

On operating the motor START push-button switch, the start contactor is energized and the SC contacts close. This connects the motor to the line through

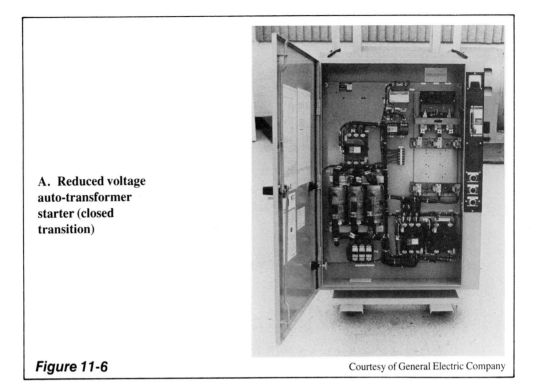

A. Reduced voltage auto-transformer starter (closed transition)

Figure 11-6

Courtesy of General Electric Company

the selected autotransformer tap.

When time set on timer T runs out, the start contactor is deenergized and the run contactor is energized. This connects the motor to full line voltage.

This type of starting has one disadvantage. The motor is momentarily disconnected from the line during the changing from the start contactor to the run contactor. This is known as *open transition.* In some localities or under some conditions open transition may be objectionable. If in doubt, consult the local power company.

The circuit for open transition is shown in Figure 11-6B.

A more practical solution is to use a *closed-transition* autotransformer-type motor starter. The circuit for the closed-transition type is shown in Figure 11-7. With closed transition, the motor is not disconnected from the line when changing from the start condition to the run condition.

11.5.2 Primary Resistor. See Figures 11-8A and 11-8B.

Two contactors, each three pole, are used in the control circuit. One is a *line contactor* (LC). The other is the *accelerating contactor* (AC). Three banks of

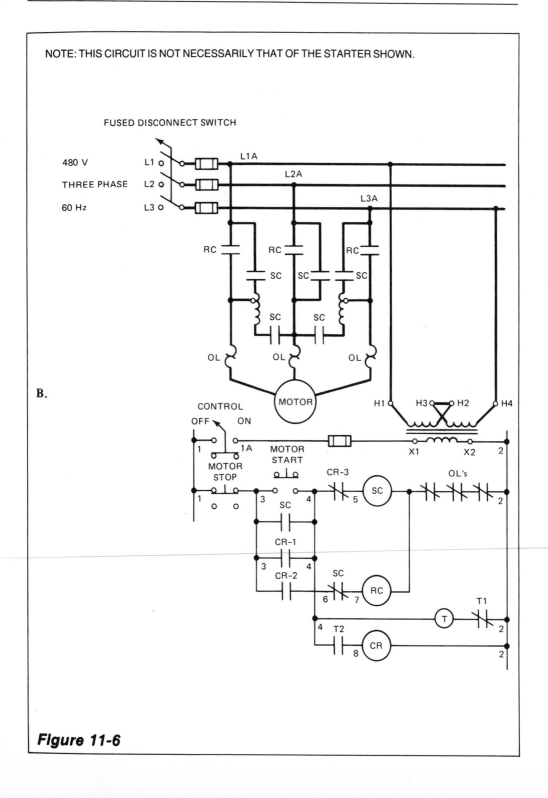

NOTE: THIS CIRCUIT IS NOT NECESSARILY THAT OF THE STARTER SHOWN.

Figure 11-6

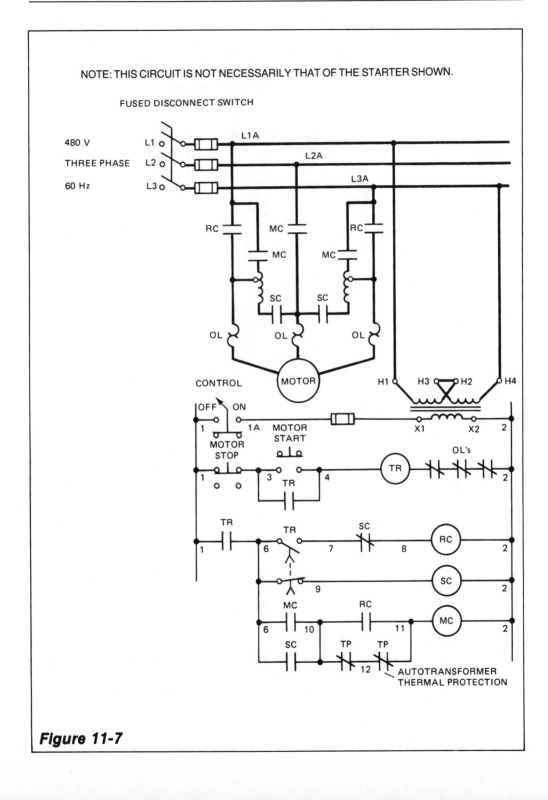

NOTE: THIS CIRCUIT IS NOT NECESSARILY THAT OF THE STARTER SHOWN.

Figure 11-7

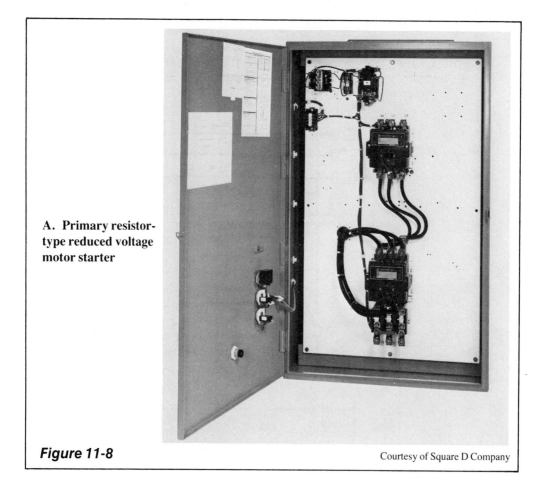

**A. Primary resistor-
type reduced voltage
motor starter**

Figure 11-8

Courtesy of Square D Company

resistors are used. These are shown connected to the line and the motor through
the contacts. The resistors must be of sufficient capacity to handle the horsepower rating of the connected motor. This starter is shown in Figure 11-8A.

On operating the motor START push-button switch, the line contactor is
energized, closing the LC contacts. This connects the motor to the line through
the resistors. After a preset time delay, the AC contacts are closed, connecting
the motor directly to the line.

One advantage of the primary resistor-type starter is the smooth acceleration. As the motor accelerates, the current decreases. Thus, the voltage drop
across the resistors is reduced. This increases the voltage applied to the motor.
The disadvantage is inefficient operation, as the power dissipated in the resistors is
comparatively high. The circuit for this starter is shown in Figure 11-8B.

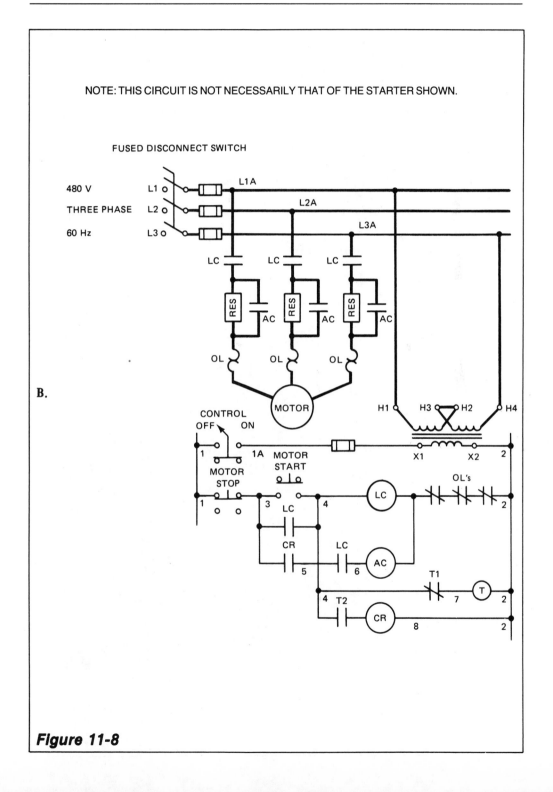

Figure 11-8

11.5.3 Wye (or Star) Delta Starter. See Figures 11-9A, 11-9B and 11-9C.

In Figure 11-9B, it is seen that to use this type of starting, all leads from each three-phase stator winding must be brought out for connecting to the starter. This starter provides 33% normal starting torque.

Three contactors are used. These are the *line contactor* (LC), *start contactor* (SC), and *run contactor* (RC). Operating the motor START push-button switch energizes the line contactor and the start contactor. The start contacts connect the motor in wye or star. This puts approximately 58% of the line voltage across each motor phase. After a preset time delay, the start contacts open and the run contacts close, connecting the motor in delta. This puts full voltage on the motor.

The cost of the control components for this type of starting is less than for the autotransformer and primary resistor types. However, the motor must be supplied with all the leads brought out for this type of starter. Thus, wye-delta

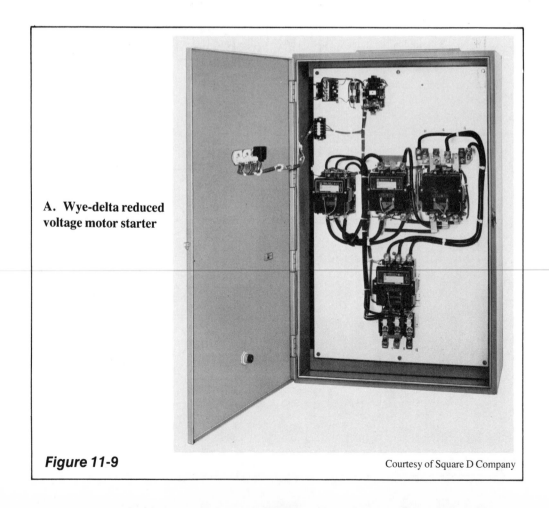

A. Wye-delta reduced voltage motor starter

Figure 11-9   Courtesy of Square D Company

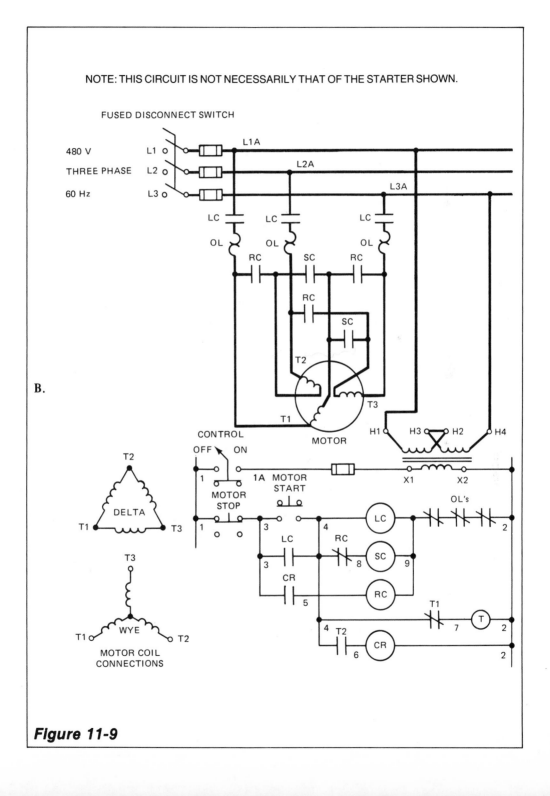

Figure 11-9

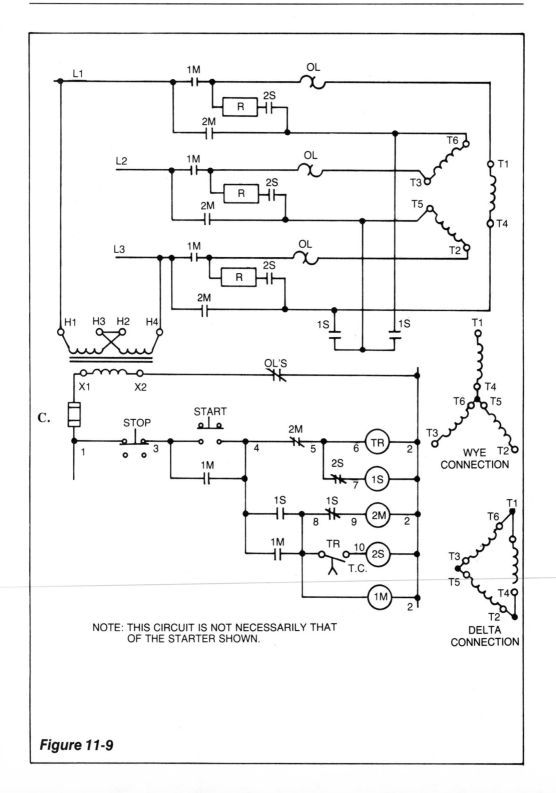

C.

1S   1S

OL'S

X1   X2

START

STOP

1   3   4   2M   5   6   TR   2
                        2S
1M   7   1S

1S   1S
        8   9   2M   2

1M   TR   10   2S
            T.C.

                    1M   2

NOTE: THIS CIRCUIT IS NOT NECESSARILY THAT
      OF THE STARTER SHOWN.

T1
            T4
T6   T5
T3
                T2
WYE
CONNECTION

T1
T6
T3
T5   T4
        T2
DELTA
CONNECTION

Figure 11-9

starters can be used only with wye-delta motors.

Closed transition versions are available which keep the motor windings energized for the few cycles required to transfer the motor windings from a wye connection to a delta connection. Such starters are provided with one additional contactor, plus a resistor bank. Figure 11-9C shows the circuit of a wye-delta closed transition motor starter.

11.5.4 Part Winding. See Figures 11-10A, 11-10B, and 11-10C.

Like the wye/delta starter, the part winding starter depends on the motor having the proper leads brought out for connection to the starter. The part winding starter supplies 48% normal starting torque. Thus, while part winding starting is not truly a reduced voltage means, it is usually so classified because of the reduced current and torque resulting.

If the starter windings of the motor are divided into two or more equal parts, with all terminals for each section available for external connection, then the motor may lend itself to what is known as part winding reduced-voltage starting.

Part winding starters are designed for use with squirrel-cage induction motors generally having two or three windings. A photo of a typical part winding starter is shown in Figure 11-10A.

The circuit for the *two-step part winding starter* is shown in Figure 11-10B. This starter is designed so that when the control circuit is energized, one winding of the motor is connected directly to the line. The winding draws about 60% of normal locked rotor current. It develops approximately 45% of normal motor torque. After about four seconds, the second winding is connected in parallel with the first. The motor is then fully across the line and develops its normal torque.

The *three-step part winding starter* has resistance in series with the motor on the first step. The circuit for this starter is shown in Figure 11-10C. To start, one winding is connected to the line in series with the resistor. After a short interval, approximately two seconds, the resistor is shorted out. The first winding is connected directly to the line. About two seconds later the second winding is connected in parallel with the first winding. Part winding starters provide closed transition starting.

11.5.5 Solid-state Starter

Many applications require the lower starting torque and smooth acceleration offered by solid-state systems. Typical of such applications are conveyor systems, pumps and compressors, etc.

The starter shown in Figure 11-11A is a three-phase, SCR-controlled device specially designed to provide smooth, stepless, reduced voltage control of ac squirrel cage induction motors. The solid-state components featured in this solid-

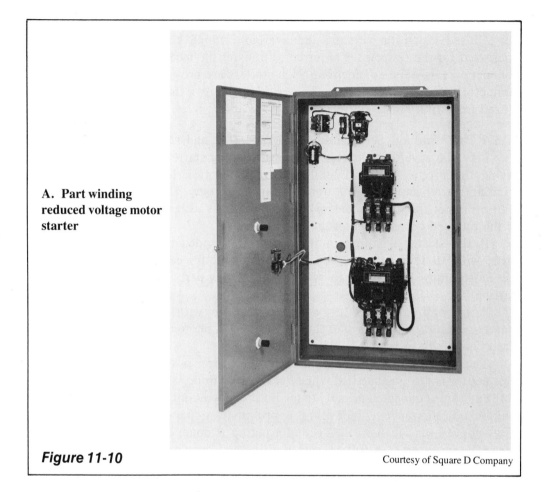

**A. Part winding
reduced voltage motor
starter**

Figure 11-10   Courtesy of Square D Company

state starter offer maximum reliability and flexibility. No contacts mean longer life and less maintenance when compared to conventional electromechanical reduced voltage starting methods. Closed transition is standard on solid-state starters.

The circuit for a typical solid-state reduced-voltage starter may employ two SCRs per phase. They are connected in a back-to-back or reverse-parallel arrangement. They are mounted on a heat sink that makes up a power pole. Each power pole contains the SCR gate firing circuits. There is also a thermal sensor that will deenergize the starter if an overcurrent condition exists. The effective voltage that is applied to the motor is controlled by changing the amount of time that the SCRs are conducting.

The firing circuitry on each power pole is controlled by modules on the logic rack. These modules check for correct start-up and running conditions. An LED (light-emitting diode) provides a visual indication of the starter status.

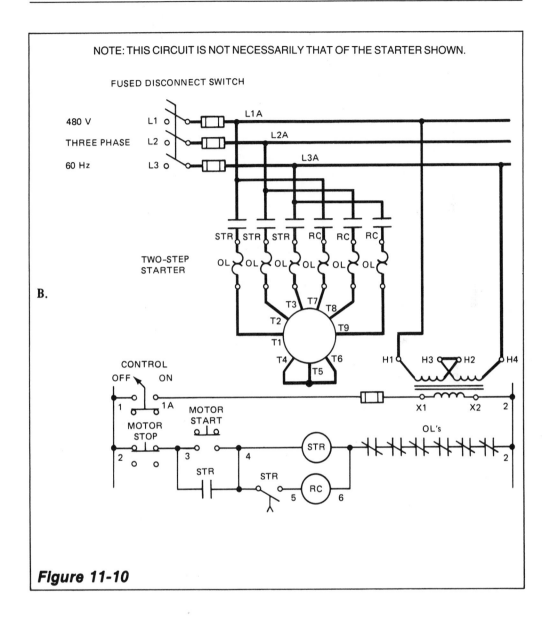

NOTE: THIS CIRCUIT IS NOT NECESSARILY THAT OF THE STARTER SHOWN.

Figure 11-10

Figure 11-11B shows the relationship between various types of reduced-voltage starters for the line current drawn from zero to full-load speed. It is also compared to full-voltage starting. Note that the entire range of most reduced-voltage starting is covered by solid-state reduced-voltage starters. This gives the solid-state type more flexibility over other types of reduced-voltage starting.

Solid-state reduced-voltage starters are available in a voltage rating of 200, 230, 460 and 575–60 Hertz. The maximum horsepower ranges are approximately

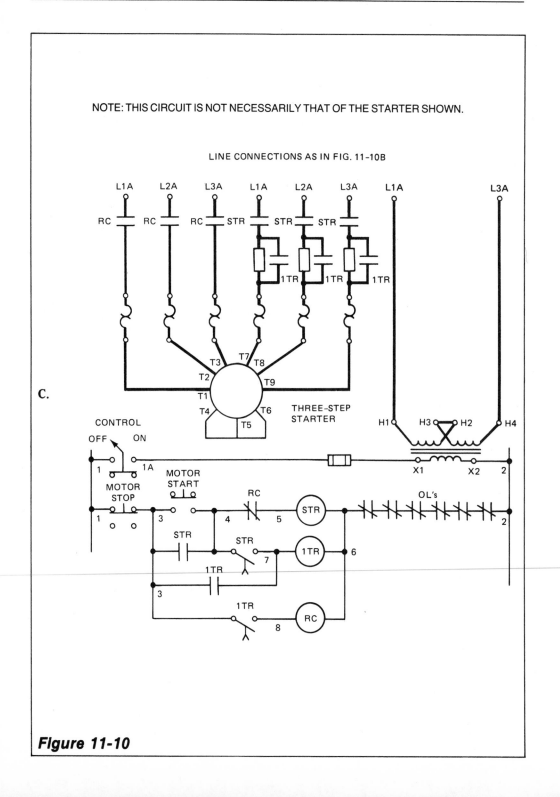

NOTE: THIS CIRCUIT IS NOT NECESSARILY THAT OF THE STARTER SHOWN.

LINE CONNECTIONS AS IN FIG. 11–10B

THREE–STEP STARTER

C.

CONTROL

OFF   ON

MOTOR STOP

MOTOR START

Figure 11-10

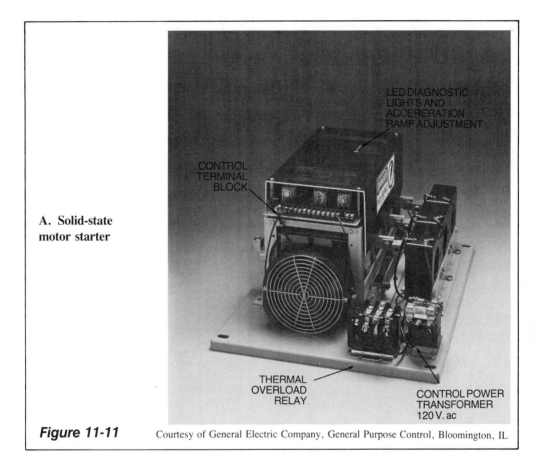

A. Solid-state motor starter

Figure 11-11   Courtesy of General Electric Company, General Purpose Control, Bloomington, IL

10–1000 at 460 or 575 volts. The maximum range for 200, 230 volts drops to approximately 10–300 HP.

11.6 STARTING SEQUENCE

A circuit showing the sequencing of two main drive motors using magnetic reduced-voltage starters is shown in Figure 11-12. The reason for the sequence starting is to reduce the line inrush current.

Associated with the two main drive motors are two auxiliary motors. They are used for pilot pressure and cooling or lubrication. These motors must be started first and must drop out (deenergize) the main motors in case either auxiliary motor deenergizes through an overload.

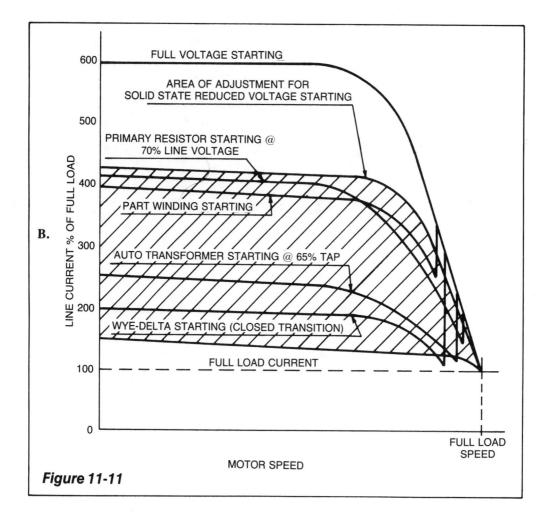

B.

Figure 11-11

When the auxiliary motor starters are energized, the #1 and #2 auxiliary contacts close. The sequencing of the two main drive motor starters proceeds as follows:

1. Operate the motor's START push-button switch.
   a. The energizing of the control section of the #1 motor starter is now initiated.
      1a. Relay coil #1-1CR is energized.
      1b. Timing relay coil #1-1TR is energized.
      1c. Timing relay #1 times out, closing #1 TR contact.
      1d. Relay coil #1-2CR is energized.
      1e. Relay contact #1-2CR-1 closes, energizing the green pilot light.

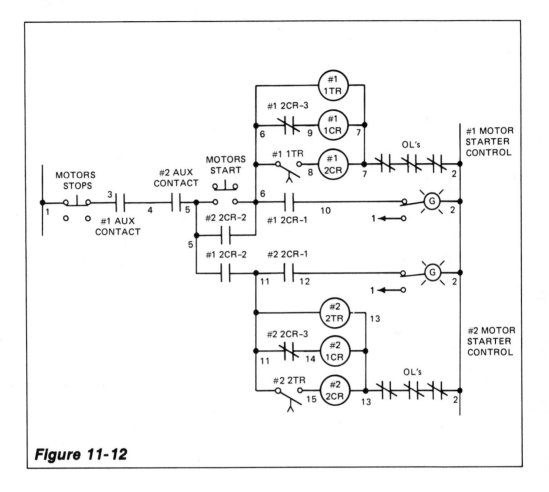

Figure 11-12

1f. Relay contact #1-2CR-3 opens, deenergizing relay coil #1-1CR.

2. The sequence for energizing the control section for #1 motor starter is now complete.

3. Relay contact #1-2CR-2 closes.
    a. The energizing of the control section for #2 motor starter is now initiated.
        3a. Relay coil #2-1CR is energized.
        3b. Timing relay coil #2-2TR is energized.
        3c. Timing relay #2 times out, closing #2-2TR contact.
        3d. Relay coil #2-2CR is energized.
        3e. Relay contact #2-2CR-1 closes, energizing the green pilot light.
        3f. Relay contact #2-2CR-3 opens, deenergizing relay coil #2-1CR.

4. Relay contact #2-2CR-2 closes, interlocking around the motor's START push-button switch.

5. The sequencing for the energizing of #2 motor starter is now complete.
6. Both motor starters are energized.
7. Both motor starters are deenergized at any time by the operating of the motor's STOP push-button switch or the opening of any of the overload relay contacts.

ACHIEVEMENT REVIEW

1. Why are overload relays used in motor starters?

2. Explain how the normally closed relay contact in the overload relay is used to deenergize the motor starter coil. A basic circuit may be drawn.

3. How does a high ambient temperature in the location of the overload relay affect its operation?

4. Why should the overload relay be trip free?

5. Draw the basic power and control circuit for a magnetic full-voltage reversing motor starter.

6. Explain how the reversing of an ac induction motor is accomplished. Why are mechanical and electrical interlocks used?

7. A reduced-voltage motor starter is set to apply 65% of full voltage to the motor for starting. What torque, in percent of full voltage, will be developed in the motor under these starting conditions?

8. Explain how the resistor-type, reduced-voltage motor starter reduces the voltage to the motor on starting. A basic power and control circuit may be drawn.

9. What is the difference between "no-voltage or low-voltage release" and "no-voltage or low-voltage protection"?

10. What is meant by *jogging* in motor starter control?

11. What protection is provided in the solid-state reduced-voltage starter to prevent a high temperature condition due to overcurrent?
    a. Special fuses
    b. Current meter
    c. Thermal sensor

12. The effective voltage applied to a motor using a solid-state reduced-voltage starter is controlled by
    a. a timer.
    b. changing the amount of time that the SCRs are conducting.
    c. the current drawn by the motor.

13. What is the difference between open transition and closed transition in a reduced-voltage motor starter?

CHAPTER 12

APPLICATION OF ELECTRIC HEAT

OBJECTIVES

After studying this chapter, the student will be able to:

- Calculate the heat requirements for a large steel object.

- List three sources of heat loss that may be present in applying electrical heating elements.

- Describe how thermostats can be used in various ways to control heat.

- Calculate the current in both single-phase and three-phase heating element circuits.

- Describe the results obtained in heating when the line voltage is raised or lowered.

- Discuss the advantages and methods of switching heating elements for a low heat holding pattern.

12.1 CALCULATING HEAT REQUIREMENTS

This chapter deals primarily with applications on and in the following:

- Relatively large metal objects, such as extruder barrels and platens for plastic molding machines

- Fluids used with hydraulic power and control systems

For the basic design there are two important questions to answer:

1. How many watts (W) or kilowatts (kW; 1 kW = 1000 W) of electrical power will be required to bring the application to the required temperature in a given period of time?
2. How many watts or kilowatts of electrical power will be required to maintain the required temperature?

Under controlling conditions, the time required to bring the heated item from ambient temperature to the preset temperature, known as the *warm-up period*, varies depending on the amount of power applied. Unless the warm-up period of time is critical, it is generally more economical and practical to try to match the power applied to the load. The load requirements vary depending on the type of material to be heated, the percent of time that the heat is required at the preset level, and heat losses. In practice, the power will generally be on 80% of the time to maintain the required temperature.

From a practical viewpoint, the required temperature is generally within a range. For example, in the plastic molding industry most materials are now processed within a 300°F–700°F range. Therefore, if calculations are made for the top of the range (700°F), then the power will be on a lower percentage of the time and the warm-up period will be shorter for the bottom of the range (300°F).

It can be difficult to make the actual calculation of the watts or kilowatts of electrical power required to heat an object to a desired temperature in a definite time period. The main problem lies in determining heat losses.

Losses include the following:

1. Radiation from the heated object (affected by contour, condition of the surface exposed, and temperature difference between the heated item and the ambient)
2. Material being processed (affected by type, amount, and rate of processing)
3. Conduction to other parts of the machine (affected by temperature differential, and type and placement of an insulating medium)

Tables and charts showing various types of heat losses are available from manufacturers of resistance heating elements. These tables and charts are a great help in approximating losses.

If the heating system could be assumed to be 100% efficient (no losses), and assuming that the object to be heated is steel, the information on kilowatt-hours (kW·h) required can be easily calculated.

$$kW \cdot h = \frac{\begin{array}{c} \text{Weight of object in pounds} \times \text{specific heat of steel (0.12)} \\ \times \text{ final temperature} - \text{initial temperature in degrees Fahrenheit} \end{array}}{3412 \times \text{warm-up period in hours}}$$

For example, given a 300-pound steel object, to raise its temperature from 70°F to 670°F the kilowatts required to accomplish this in 1 hour are

$$\frac{300 \times 0.12 \times 600}{3412 \times 1} = 6.3 \text{ kW}$$

If the warm-up period requirement is cut to one-half hour, then the kilowatt requirement doubles to 12.6 kW.

With cylindrical barrels such as are used on plastic injection molding machines, working in the 300°F–700°F range, the designer will generally start with as many kilowatts for losses as were calculated for heating the barrel. The losses are based on the maximum operating temperature of 700°F. Operation is in an ambient temperature of 70°F.

12.2 SELECTION AND APPLICATION OF HEATING ELEMENTS

Basically, a heating element is resistance wire in a magnesium-oxide filler with a sheath around it. The sheath material may be steel, stainless steel, or copper. The material may be tubular or flat rectangular in cross-sectional shape.

After the total kilowatt requirements are determined, the quantity and wattage of the individual heating elements must be determined. Heating elements with the exact wattage calculated are not usually commercially available. The practical approach is to select the next larger heating element. For example, if requirements are for 975 W per unit, and this size is not available, select a 1000-W heating element.

The number of heating elements is generally dictated by the available mounting space. Coverage of the area available should be as near to 100% as is practical. The most important factor in selecting and installing heating elements is the transfer of heat. This is affected by mating surfaces and how securely the heating elements are held in place. Whether the application is a flat plate, groove, hole, or curved surface, it must be clean and smooth. Efficiency of any heating system is reduced as oxides or foreign materials enter between the heating element and the heated surfaces, thus acting as an insulator.

In the case of heating elements with a high watt density inserted in a groove or hole, the clearances should be held to 0.005 inch to 0.010 inch. (*Watt density* is the number of watts per unit area of surface, for example, watts per square inch [W/in^2]). Two types of heating elements used in industry are shown in Figures 12-1 and 12-2.

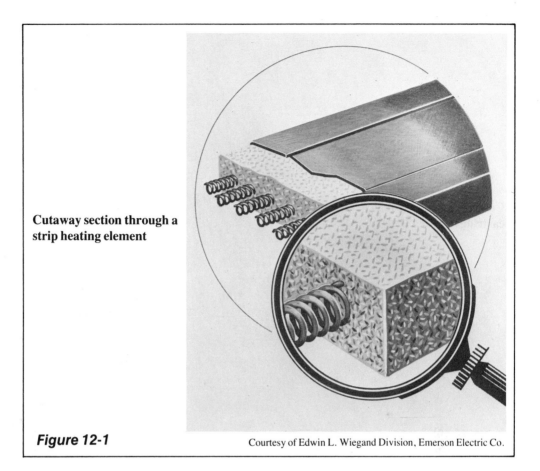

**Cutaway section through a
strip heating element**

Figure 12-1   Courtesy of Edwin L. Wiegand Division, Emerson Electric Co.

12.3 HEATING ELEMENT CONTROL CIRCUITS

Contactors for the transfer of electrical power are often used with heating
elements. In some cases, the heating elements are thermostatically controlled.
A circuit with this arrangement is shown in Figure 12-3. If possible, the load
should be arranged for a balanced three-phase circuit. This arrangement minimizes
the conductor size and may decrease sizes of the contactors and protective devices,
as compared to single-phase connection. This topic is discussed further in a later
section of this chapter.

Referring to Figure 12-3, the sequence of this heating element circuit is as
follows:

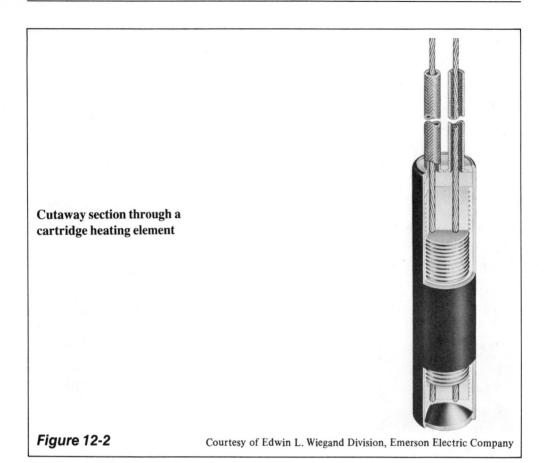

**Cutaway section through a
cartridge heating element**

Figure 12-2   Courtesy of Edwin L. Wiegand Division, Emerson Electric Company

1. Set 1TS at 70 degrees Fahrenheit.
2. Set the control OFF-ON switch to ON.
3. Set the heat OFF-ON switch to ON.
4. Contactor 1CR is energized (with the temperature below 70 degrees Fahrenheit).
    a. Power contacts 1CR-1, 1CR-2, 1CR-3 close, energizing the heating elements.
    b. Auxiliary contact 1-CR Aux. closes, energizing the green pilot light.
5. When the temperature of the part or substance being heated exceeds 70 degrees Fahrenheit, temperature switch 1TS opens.
6. The coil of contactor 1CR is deenergized.
7. Power contacts 1CR-1, 1CR-2, 1CR-3 open, deenergizing the heating elements.
8. Auxiliary contact 1-CR Aux. opens, deenergizing the green pilot light.

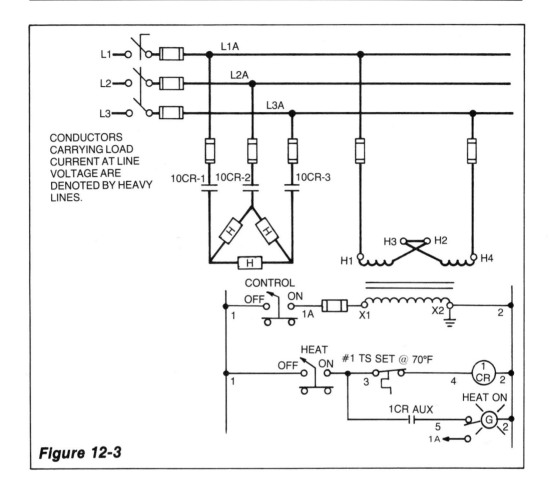

Figure 12-3

From this description, it can be seen that as long as the temperature of the heated part or substance is below 70 degrees Fahrenheit, the heating elements remain energized. When the temperature exceeds 70 degrees Fahrenheit, the heating elements are deenergized.

There are applications where it is required to combine temperature and motor control. For example, the temperature of bearings, operating fluid (oil) and material being processed must be monitored and controlled for the motor to be energized and remain energized during operation.

Figure 12-4 shows such a combination of temperature and motor control. Temperature switches 1TS, 2TS, 3TS, 4TS are in the machine oil reservoir. Temperature switches 5TS, 6TS, 7TS, 8TS are located respectively on four important separate bearings on the machine.

Referring to Figure 12-4, the sequence of this circuit operation is as follows:

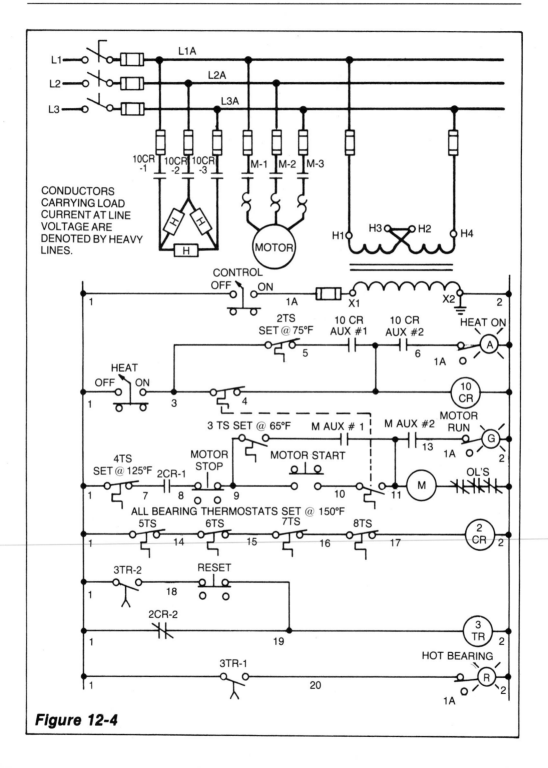

Figure 12-4

1. Set the control ON-OFF switch to ON.
2. Set the heat ON-OFF switch to ON.

With the oil temperature below 70°F, 1TS NC contact is closed energizing contactor 10CR.

3. The power contacts on 10CR (10CR-1, 10CR-2, 10CR-3) close, energizing the heating elements.
    a. Auxiliary contact 10CR Aux. #1 closes, interlocking through 75°F TS #2 and around the 70°F TS #1.
    b. Auxiliary contact 10CR Aux. #2 closes, energizing the heat ON pilot light.

With the oil temperature below 125°F, normally closed #4TS is closed. With the four bearings in normal condition (temperature below 150°F), thermostat contacts #5TS, #6TS, #7TS, and #8TS are all closed.

4. Relay 2CR is energized.
    a. Normally open relay contact 2CR-1 closes.
    b. Normally closed relay contact 2CR-2 opens.

There are now two conditions under which the motor can be started.

-   The oil temperature is below 70°F. It will be necessary to wait until the heating elements raise the oil temperature to above 70°F, operating #1TS.
-   The oil temperature is above 70°F. The motor can be started immediately.

5. Operate the motor start push-button switch.
6. The motor starter contacts M1, M2, M3 close, energizing the motor.
    a. Since the oil temperature is above 70°F, #3TS is closed.
    b. M Aux. #1 contact closes, interlocking through #3TS and around the motor start push-button switch and 1TS normally open contact.
    c. M Aux. #2 contact closes, energizing the motor run light.
7. The motor will remain energized until one or more of four events occur.
    a. The motor stop push-button is operated.
    b. One or more of the four bearing temperatures exceed 150°F. This de-energizes relay 2CR, opening relay contact 2CR-1.
    c. The oil temperature exceeds 125°F, opening #4TS.
    d. The oil temperature drops below 65°F, opening the interlock circuit around the motor start push-button switch.
8. Note that the heat control system keeps the oil temperature between 70°F and 75°F. Unless this system fails and the temperature drops below 65°F, thermostat #3TS will not open.

9. If one or more of the bearings exceed 150°F,
   a. Relay 2CR will be deenergized. Normally closed relay contact 2CR-2 closes, energizing time delay relay 3TR.
   b. Timing contact 3TR (on delay) 3TR-1 closes after a few seconds, energizing the hot bearing light.
   c. After the same few seconds delay, 3TR timing contact (on delay) 3TR-2 closes, interlocking around the normally closed relay contact 2CR-2.
   d. Even if the bearing cools down and the associated thermostat re-closes, 3TR will remain energized for inspection and troubleshooting.
   e. With the bearing problem corrected, the light can be deenergized by operating the reset push-button switch.

12.4 HEATING ELEMENT CONNECTION DIAGRAMS

The connection diagrams for electrical power heating circuits shown in this section are among the most useful found in industry.

In each case the heating elements are shown connected to a three-phase power line, represented by three parallel lines. Where the load is connected single phase, any two of the three lines are used. The connecting lines to the heating element load show the fuses used for protection and the contacts from a contactor. The contactor coil is in a heat control circuit (which is not shown).

These circuits are important in that they show the various arrangements in the connection of heating elements. For each connection, the current in each line feeding the load is calculated.

When either dc or ac is supplied to a resistance heating load, the current (I) resulting can be calculated from the applied voltage (V) and the rated wattage (W) of the heating element.

$$\text{Current in Amperes} = \frac{\text{Watt rating}}{\text{Voltage applied}}$$

The first circuit, shown in Figure 12-5, is connected single phase. The individual heating elements are rated at 240 V. Therefore, they are connected in parallel so that an applied voltage of 230 V is applied to each element. Note that the heating element industry rates the elements at 240 V even though they may be applied to a 230-V line.

In the circuit shown in Figure 12-6, heating elements rated at 240 V are again used. However, they are connected single phase to a 460-V power line. Since the individual elements are only rated at 240 V, two units must be connected in series. It is important when making this connection that both elements are

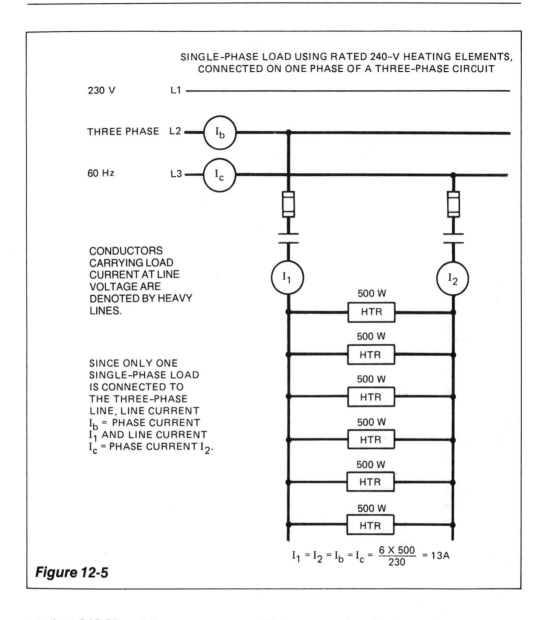

SINGLE-PHASE LOAD USING RATED 240-V HEATING ELEMENTS,
CONNECTED ON ONE PHASE OF A THREE-PHASE CIRCUIT

230 V L1

THREE PHASE L2 I_b

60 Hz L3 I_c

CONDUCTORS CARRYING LOAD CURRENT AT LINE VOLTAGE ARE DENOTED BY HEAVY LINES.

I_1 I_2

SINCE ONLY ONE SINGLE-PHASE LOAD IS CONNECTED TO THE THREE-PHASE LINE, LINE CURRENT I_b = PHASE CURRENT I_1 AND LINE CURRENT I_c = PHASE CURRENT I_2.

500 W HTR

500 W HTR

500 W HTR

500 W HTR

500 W HTR

500 W HTR

$$I_1 = I_2 = I_b = I_c = \frac{6 \times 500}{230} = 13A$$

Figure 12-5

rated at 240 V and the same wattage. This ensures that the line voltage is divided equally across the two elements. Note that since the resulting line current is equal to the wattage rating divided by the applied voltage, the resulting line current will be one-half that of the parallel connection on 230 V.

Rather than having all six heating elements connected across one phase of a three-phase power line, they are now grouped in three groups of equal numbers. (There are two heating elements in each group.) These groups are con-

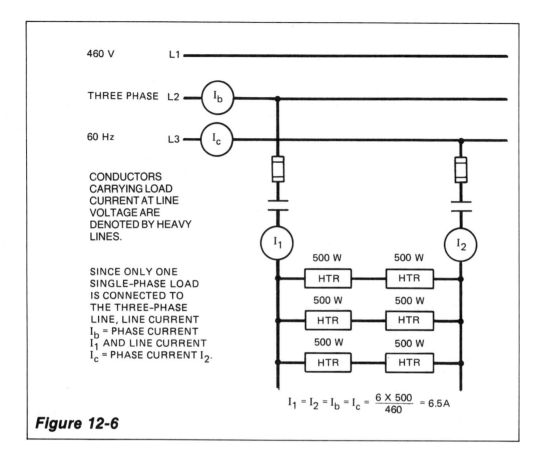

Figure 12-6

nected across the three separate phases of a three-phase power line (L1-L2, L2-L3, L3-L1).

Where a three-phase power line is available, this type of connection has the advantage of reducing the current in the individual phase groups and in the three-phase line current. This condition often permits the use of smaller conductors, contactors, and fuses, resulting in cost savings. The line current in a three-phase delta connection is reduced by a factor of 1.73. Figure 12-7 shows the relationship of current (I) and voltage (E) in three-phase circuits for both delta and wye connections.

Figure 12-8 shows the same heating elements connected to a 460-V, three-phase power line.

Figure 12-9 shows a circuit using 240-V rated heating elements on a 230-V, three-phase line.

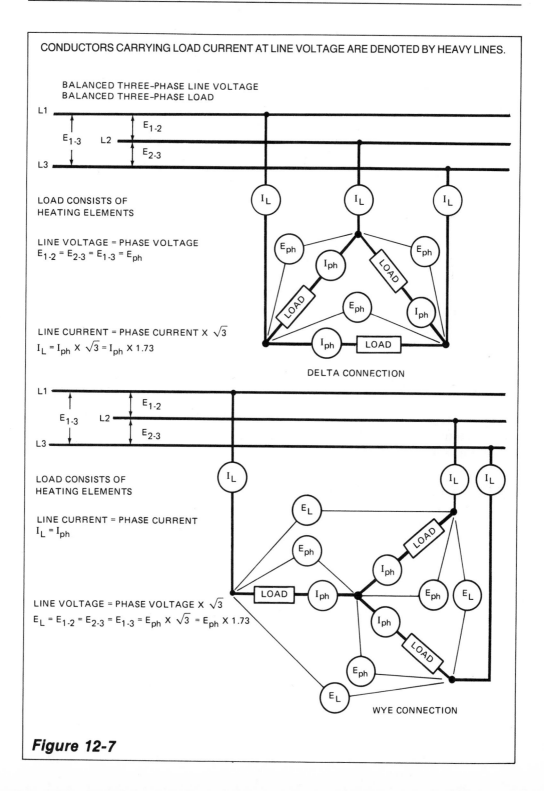

Figure 12-7

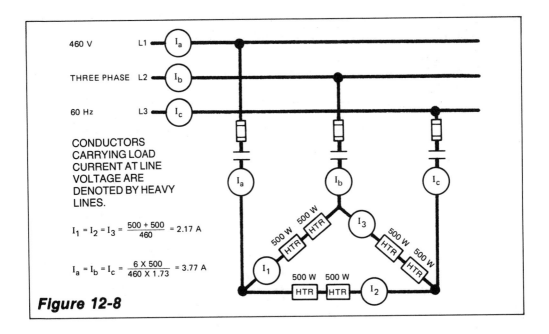

Figure 12-8

Sometimes it is desirable to reduce the heat to a "warm" level. This might be on an installation for overnight or over a weekend.

Heat produced in a resistance heating element is proportional to the square of the voltage applied. Therefore, any reduction in the voltage applied will greatly reduce the heat. For example, if the voltage on a 240-V rated heating element is reduced from 230 V to 115 V, the heat reduction will be

$$\left(\frac{115}{230}\right)^2 = \left(\frac{1}{2}\right)^2 = \frac{1}{4}$$

If the heating elements could be arranged to first be applied to 230 V, and then by switching be applied to 115 V, the resulting heat will be one-fourth the heat at 230 V.

Figure 12-10 shows a single-phase circuit using either a selector switch or drum switch. The switching changes the connections from parallel to series.

Figure 12-11 shows the same type of series-parallel switching applied to a

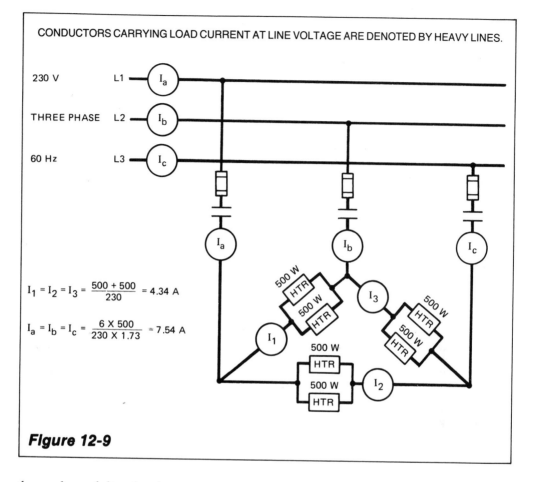

Figure 12-9

three-phase delta circuit.

Figure 12-12 shows a three-phase connection where the full line voltage is applied to each heating element connected in a delta arrangement. By switching the heating elements into a wye connection, the voltage across each element is reduced. The heat is reduced to 33% of that when the elements were connected in delta.

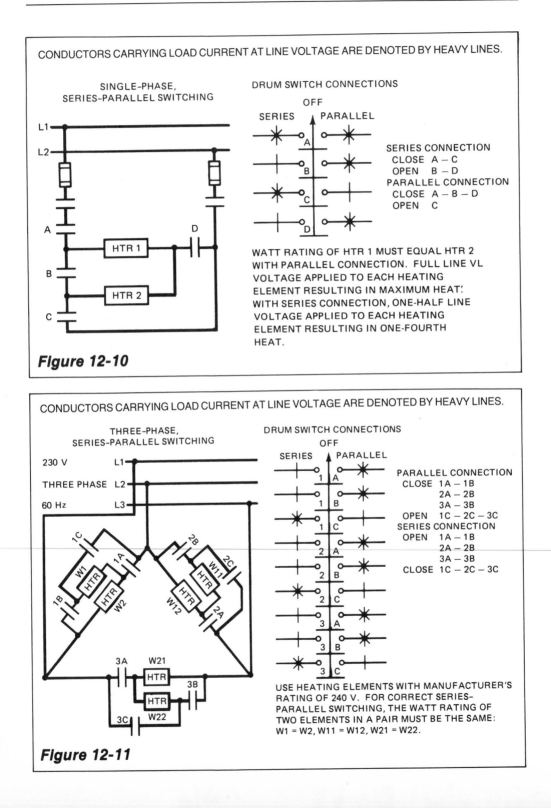

CONDUCTORS CARRYING LOAD CURRENT AT LINE VOLTAGE ARE DENOTED BY HEAVY LINES.

SINGLE-PHASE,
SERIES-PARALLEL SWITCHING

DRUM SWITCH CONNECTIONS

OFF

SERIES ▲ PARALLEL

SERIES CONNECTION
CLOSE A – C
OPEN   B – D
PARALLEL CONNECTION
CLOSE A – B – D
OPEN   C

WATT RATING OF HTR 1 MUST EQUAL HTR 2
WITH PARALLEL CONNECTION. FULL LINE VL
VOLTAGE APPLIED TO EACH HEATING
ELEMENT RESULTING IN MAXIMUM HEAT.
WITH SERIES CONNECTION, ONE-HALF LINE
VOLTAGE APPLIED TO EACH HEATING
ELEMENT RESULTING IN ONE-FOURTH
HEAT.

Figure 12-10

CONDUCTORS CARRYING LOAD CURRENT AT LINE VOLTAGE ARE DENOTED BY HEAVY LINES.

THREE-PHASE,
SERIES-PARALLEL SWITCHING

DRUM SWITCH CONNECTIONS

OFF

SERIES ▲ PARALLEL

230 V   L1

THREE PHASE L2

60 Hz   L3

PARALLEL CONNECTION
CLOSE 1A – 1B
       2A – 2B
       3A – 3B
OPEN   1C – 2C – 3C
SERIES CONNECTION
OPEN   1A – 1B
       2A – 2B
       3A – 3B
CLOSE 1C – 2C – 3C

USE HEATING ELEMENTS WITH MANUFACTURER'S
RATING OF 240 V. FOR CORRECT SERIES-
PARALLEL SWITCHING, THE WATT RATING OF
TWO ELEMENTS IN A PAIR MUST BE THE SAME:
W1 = W2, W11 = W12, W21 = W22.

Figure 12-11

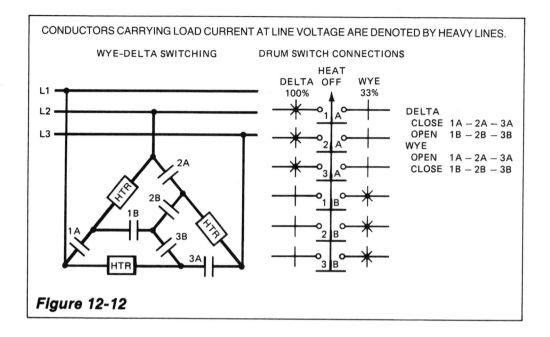

CONDUCTORS CARRYING LOAD CURRENT AT LINE VOLTAGE ARE DENOTED BY HEAVY LINES.

Figure 12-12

ACHIEVEMENT REVIEW

1. Given a block of steel weighing 500 pounds, calculate the kilowatts of electric power required to raise the temperature of the block from a temperature of 80°F to 400°F in 2 hours. Use 0.12 as the specific heat of steel.

2. What is one factor to consider in applying electrical heating elements that may effect their performance efficiency?

3. What are some of the areas that contribute to losses in a heated material or system?

4. You have a motor driving a pump in a fluid power system. It is required that the motor not be energized until the fluid is at 70°F. There are heating elements in the fluid tank that are used to heat the fluid. These heating elements are to be controlled so that the fluid remains between 80°F and 90°F. Draw a schematic diagram of a circuit that will accomplish these results.

5. In a heating element power circuit, there are nine heating elements, each rated at 600 watts, 240 volts. They are connected in parallel across a single-phase, 230-volt power line. Calculate the line current drawn by this load.

6. You believe you can save some money in your installation of the heating elements in Question 5 by connecting them in a balanced load across a

three-phase power line. Calculate the individual phase current and the line current resulting from this type of connection. The power line is 230 V, three phase, 60 Hz.

7. In your plant, you find that at certain periods of time the line voltage drops from the normal 230 V to 200 V. What heat gain or loss in an electrical heating system will you experience under these conditions?

8. A tank of fluid is being heated by two heating elements. You wish to maintain a low reduced heat over the weekend. Show how you can connect these heating elements, using a selector or drum switch to accomplish this reduced-heat condition.

9. When installing heating elements with a high watt density in a groove or hole, the clearances should be held to
   a. 0.1–0.5 inch.
   b. 0.01–0.05 inch.
   c. 0.005–0.010 inch.

10. In Figure 12-5, the rated 240-volt heating elements are now connected in series across one phase of a 460-volt line. The resulting line current (I_b) is
   a. higher.
   b. the same.
   c. lower.

CHAPTER 13

CONTROL CIRCUITS

After studying this chapter, the student will be able to:

- Explain how all complete control circuits progress through three basic areas: information or input, decision or logic, and output or work.
- Draw a bar chart sequence for an electrical control circuit.
- Demonstrate how various electrical components are used to gather information from a machine or system.
- Demonstrate how electrical components are used to make a decision as to how the gathered information is to be used.
- Demonstrate how the decision affects the output or work the machine is to accomplish.
- Draw simple electrical control circuits from a given set of requirements.
- Describe how an interlock circuit is used.
- Explain the function of the timer sequence symbol above each timer contact.

13.1 PLACEMENT OF COMPONENTS IN A CONTROL CIRCUIT

Most of the components used in electrical control of machines are discussed in Chapters 1 through 12 of this text. It will be helpful to the reader to place these components in positions in a control circuit to accomplish several different types of control.

INFORMATION OR INPUT	DECISION OR LOGIC	OUTPUT OR WORK
Manual 　Push-button switches 　Selector switches 　Drum Switches	Relays	Lights
Position 　Limit Switches		Solenoids
Pressure 　Pressure switches		
Temperature 　Pyrometers 　Thermostats	Contactors	Heating 　Elements
Time 　Timers	Motor Starters	Motors

Figure 13-1

Figure 13-1 shows an outline of components and their usage.

There are many examples that show the procedure as the control moves from the *input* (information) to the *output* (work). Three such examples follow.

1. A motor START push-button switch is operated. This energizes a motor starter coil. The motor starter contacts then close, energizing the motor. See Figure 13-2.
2. A normally open thermostat contact closes, energizing the coil of a contactor. The contactor contacts then close, energizing the heating elements. See Figure 13-3.
3. A normally open limit switch contact may be held closed (operated) at the start of a cycle. This limit switch contact could be connected to energize the coil of a relay. A normally open relay contact then closes to energize a solenoid. See Figure 13-4.

In the eight circuit examples which follow in this chapter, some or all of the input components are used in a control circuit to accomplish a given output.

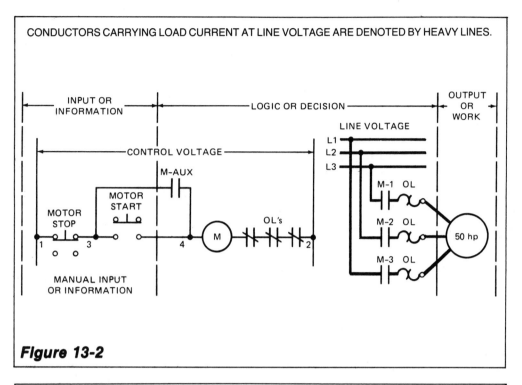

Figure 13-2

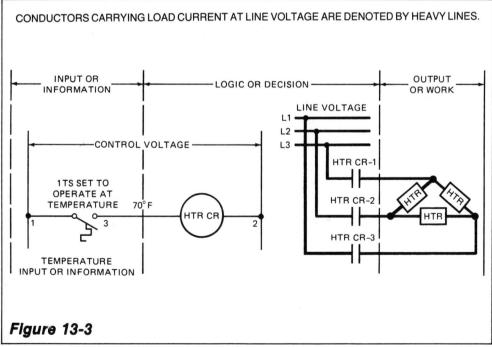

Figure 13-3

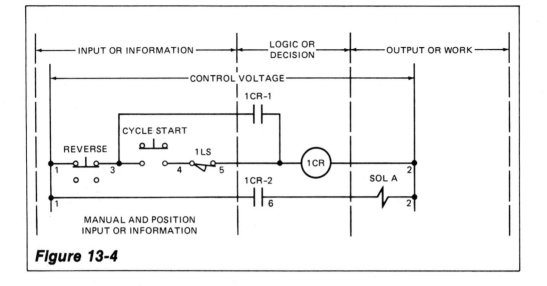

Figure 13-4

For each example, a *bar chart* sequence or analysis of the circuit is shown. The bar chart is the most efficient and simple method of explaining the functioning of the control components from input to output.

The bar chart is arranged with the electrical components to be energized in a vertical column at the left-hand side. The components to be energized are the relays, time-delay relays, contactors, timers, and motor starters.

The events that occur in sequence during a cycle of the machine are shown in a horizontal line at the bottom. These events are due to manual, position, pressure, temperature, or time inputs.

Each time a component is energized during a machine cycle from one event to a following event, a solid bar is shown on a horizontal line opposite to the

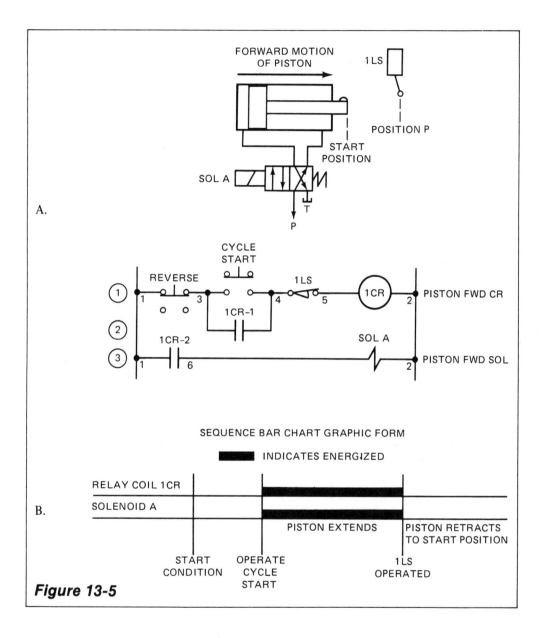

Figure 13-5

component being energized. Thus, at any time during a machine cycle, the energized or deenergized condition of the component is observed. This is done by checking along a vertical line corresponding to the event that has occurred.

Figure 13-5A shows a typical control circuit. A pictorial drawing of the machine or equipment being used is included. In this circuit, at the position that limit switch 1LS is operated, both the coil of relay 1CR and solenoid A are deenergized. Figure 13-5B is a bar chart of this activity.

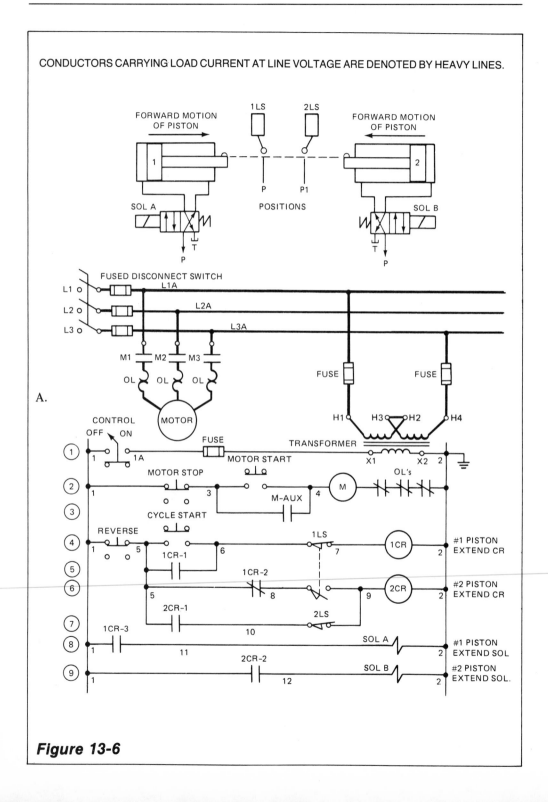

CONDUCTORS CARRYING LOAD CURRENT AT LINE VOLTAGE ARE DENOTED BY HEAVY LINES.

Figure 13-6

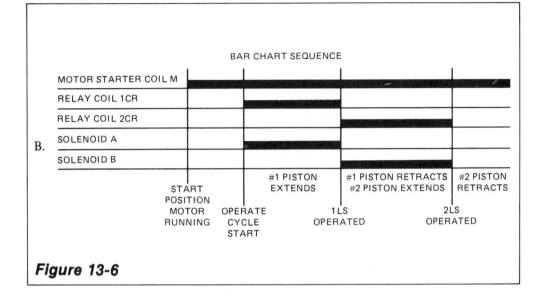

Figure 13-6

Example #1 — Figures 13-6A and 13-6B

A three-phase motor is used to drive a fluid power pump that provides fluid power pressure. A fused disconnect switch provides a disconnecting means and short-circuit protection. The control transformer has fuses in the primary winding circuit. The transformer secondary is grounded on the common side. The other side of the transformer secondary is fused. A selector switch is used for control disconnecting means. The operating cycle for the machine proceeds as follows.

The #1 piston moves from a start position to the right as shown. When the piston reaches a predetermined position P, it stops and reverses to its start position. At the same time that #1 piston reverses, #2 piston moves from a start position and moves to the left as shown. When #2 piston reaches a predetermined position P1, it stops and returns to its start position. Pressure can be available on the rod end of both pistons at the start positions. Provisions must be made to reverse either piston at any time in their forward stroke.

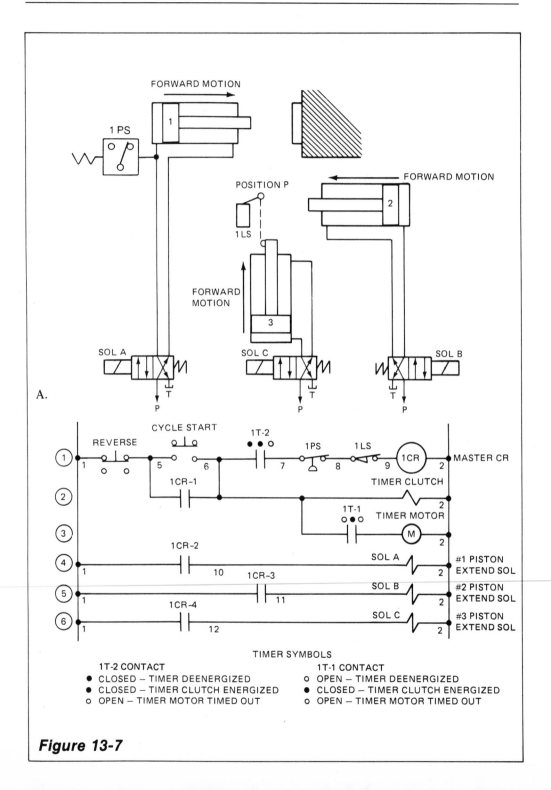

Figure 13-7

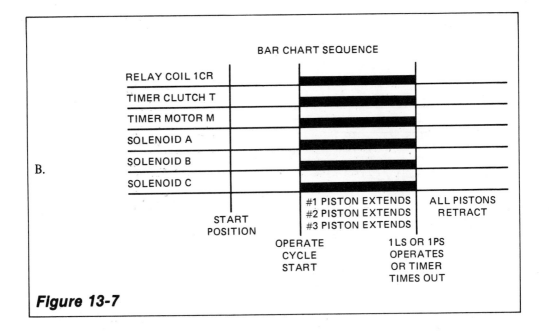

Figure 13-7

Example #2 — Figures 13-7A and 13-7B

The same power and control circuit used for the motor and control transformer in Example #1 is used in this example. Therefore, in Example #2 only the ladder-type schematic circuit is shown.

As shown, the #1 piston moves from a start position to the right. The #2 piston moves from a start position to the left. The #3 piston moves from a start position up.

All three pistons will return to their start positions when either

- piston #1 has engaged the work and built pressure to a predetermined setting; or

- a predetermined time setting on #2 piston has been reached; or

- piston #3 has reached a predetermined position P.

Pressure can be available on the rod end of the pistons in their start positions. Provisions must be made to reverse all pistons at any time in their forward stroke.

One practical advantage of this circuit is that any one of the three inputs can be adjusted independently to control all three pistons. For example:

1. Adjust the pressure switch.
2. Change the timing.
3. Move the limit switch.

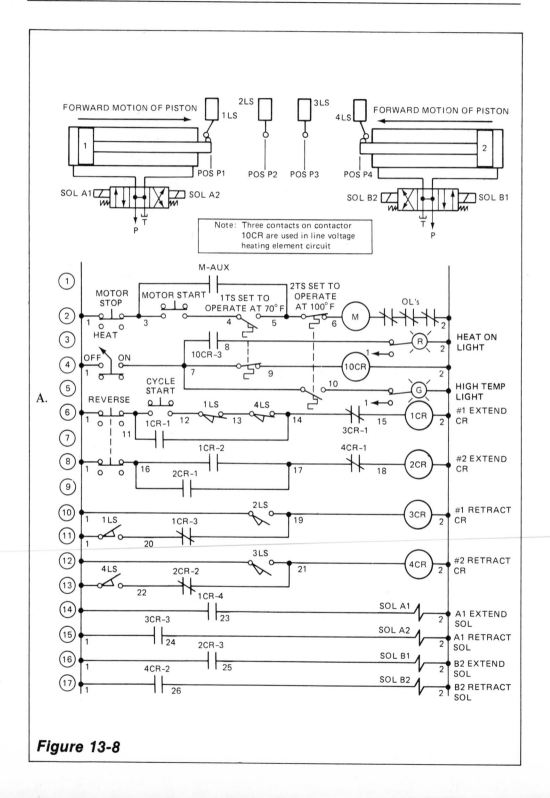

Figure 13-8

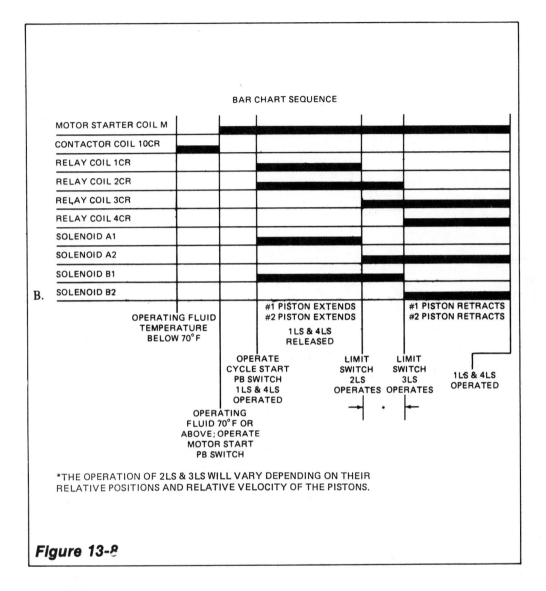

Figure 13-8

Example #3 — Figures 13-8A and 13-8B

In this example the goal is to control the temperature of the operating fluid in the fluid power system. The motor should not be able to start unless the temperature of the operating fluid is more than 70°F and less than 100°F. The same power and control transformer circuit used in Example #1 is used here.

The #1 piston must be at position P1 for start condition. The #2 piston must be at position P4 for start condition. Piston #1 moves to the right as shown. At the same time, #2 piston moves to the left. When #1 piston reaches a predetermined position P2, it stops and reverses to position P1. When #2 piston reaches

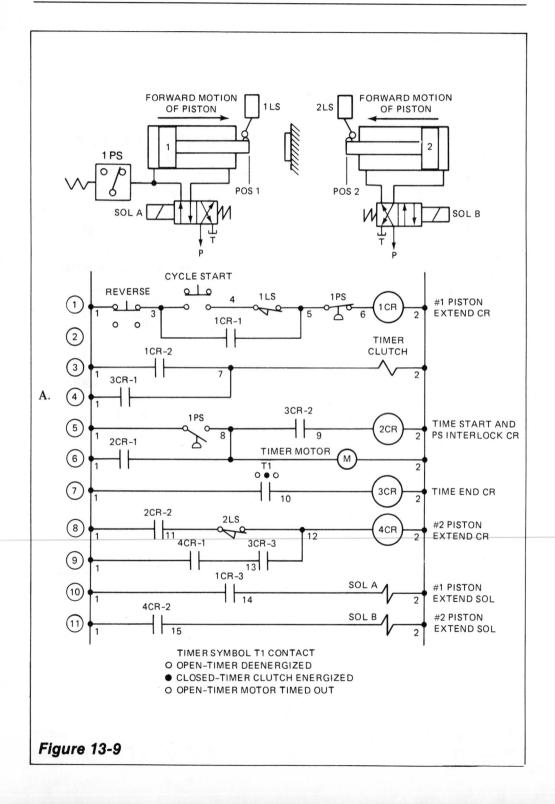

Figure 13-9

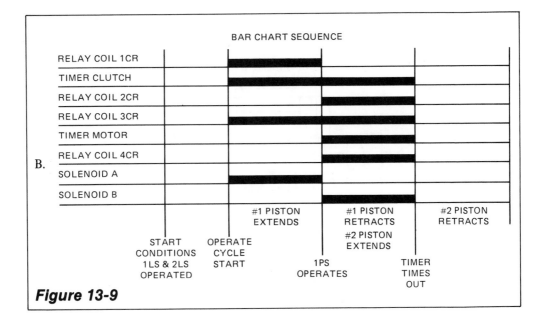

Figure 13-9

a predetermined position P3, it stops and reverses to position P4. No pressure can be on the pistons at the start positions. Provisions must be made to reverse both pistons at any time in their forward stroke.

Example #4 — Figures 13-9A and 13-9B

The same motor and control transformer circuit used in Example #1 is used here. Therefore, only the ladder-type schematic circuit is shown. Two cylinder-piston assemblies are used, #1 and #2. Piston #1 must be at position 1 and #2 piston must be at position 2 for start conditions.

When the cycle START push-button switch is operated, #1 piston moves to the right and engages a work piece. A preset pressure is built on the work piece. When the preset pressure is reached, #2 piston advances to the left. At the same time, piston #1 returns to position 1. Piston #2 advances for a preset period of time. When the preset time expires, #2 piston returns to position 2. A reset-type timer is used in this circuit.

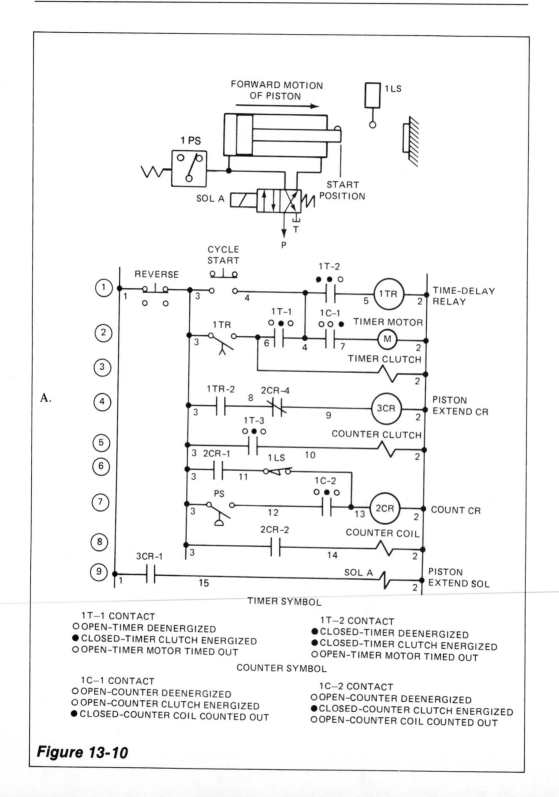

A.

Figure 13-10

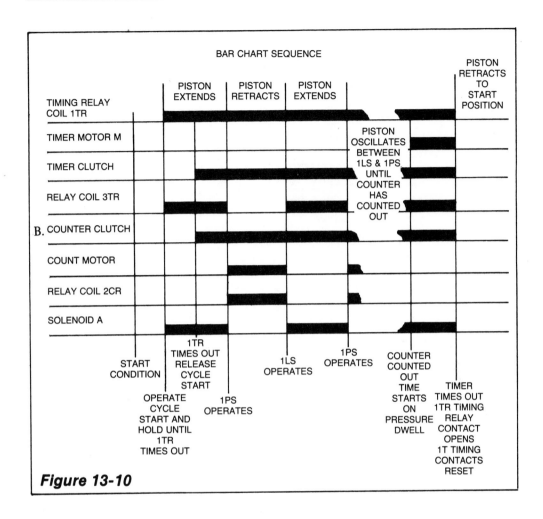

Figure 13-10

Example #5 — Figures 13-10A and 13-10B

The same motor and control transformer control circuit used in Example #1 is used here. Therefore, only the ladder-type schematic circuit is shown.

One piston is used. The piston extends and contacts work. When the system reaches preset pressure, the piston reverses and operates limit switch 1LS. The cycle is repeated until the counter counts out. On the last forward stroke and after a preset time, the piston returns to the start position and stops.

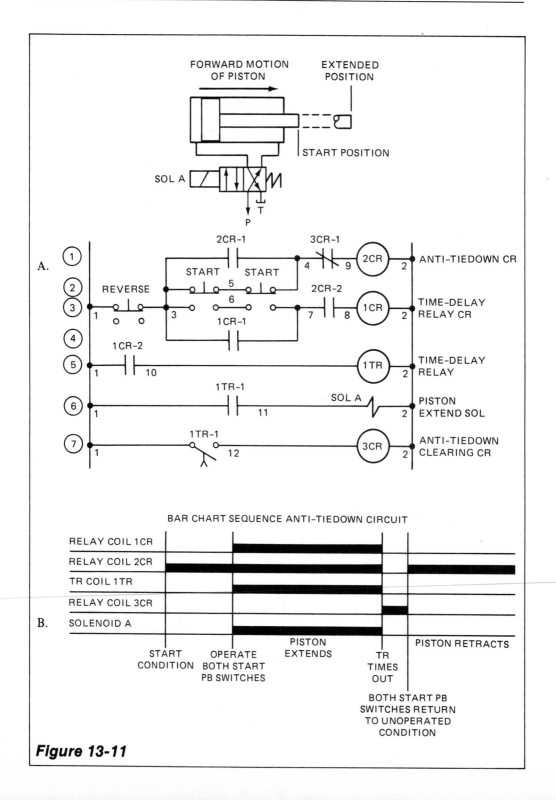

Figure 13-11

Example #6 — Figures 13-11A and 13-11B

The same motor and control transformer circuit used in Example #1 is used here. Therefore, only the ladder-type schematic circuit is shown. This circuit is known as an *anti-tiedown*. That is, it is required that both START push-button switches be operated to start a cycle. Both switches must be released to the unoperated condition before the next cycle can be started.

One piston is used. The piston extends and stops in the extended position until a preset time has elapsed. The piston then returns to the start position.

Example #7 — Figures 13-12A and 13-12B

This circuit shows only a timed starting sequence circuit for two motor starters. The #1 motor starter coil is energized when the motor START push-button switch is operated. This energizes motor #1. After a preset time delay, #2 motor starter coil is energized. This energizes motor #2. Indicating lights are energized, showing that both motor starter coils are energized. The motor START push-button switches must be held operated until both motor starter coils are energized.

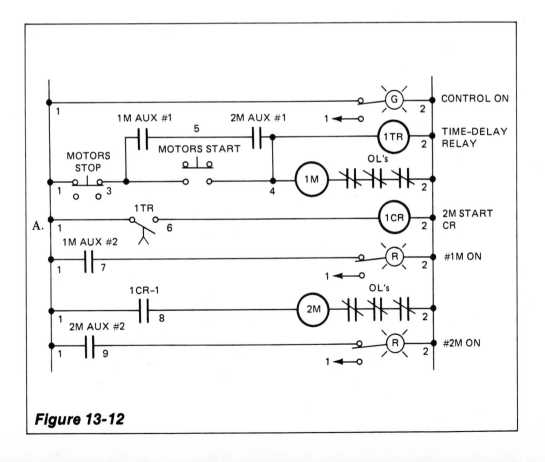

Figure 13-12

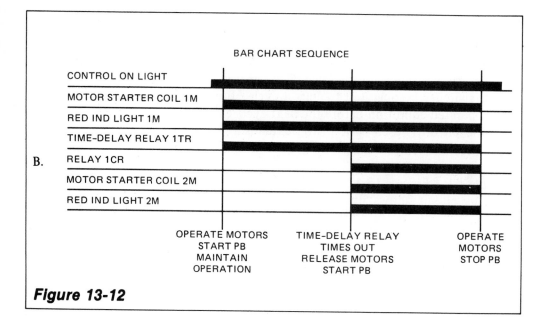

B.

Figure 13-12

Example #8 — Figures 13-13A and 13-13B

In this example, two separate three-phase motors are used. Each motor drives a fluid power pump. Fluid power pressure from #1 pump supplies fluid power pressure to the operating valve, using solenoids A1 and A2. Fluid power from #2 pump supplies fluid power pressure to the operating valve, using solenoids B1 and B2.

The motor start circuit uses only one push-button switch for starting. The #2 motor starter coil is energized through the closing of a normally open interlock contact on #1 motor starter. The operation of a single motor STOP push-button switch deenergizes both motor starter coils. An auxiliary contact on #2 motor starter is used to interlock the motor START push-button switch.

The #1 piston must be at position P1 for start condition. The #2 piston must be at position P4 for start condition. Piston #1 moves to the right as shown. When this piston reaches position P2, it stops and reverses to its start position P1. At a predetermined midpoint P5 in the travel of #1 piston on its return stroke, #2 piston moves to the left. When #2 piston reaches position P3, it stops and reverses to position P4. No pressure can be on the pistons in their start positions. Provisions must be made to reverse both pistons at any time in their forward stroke.

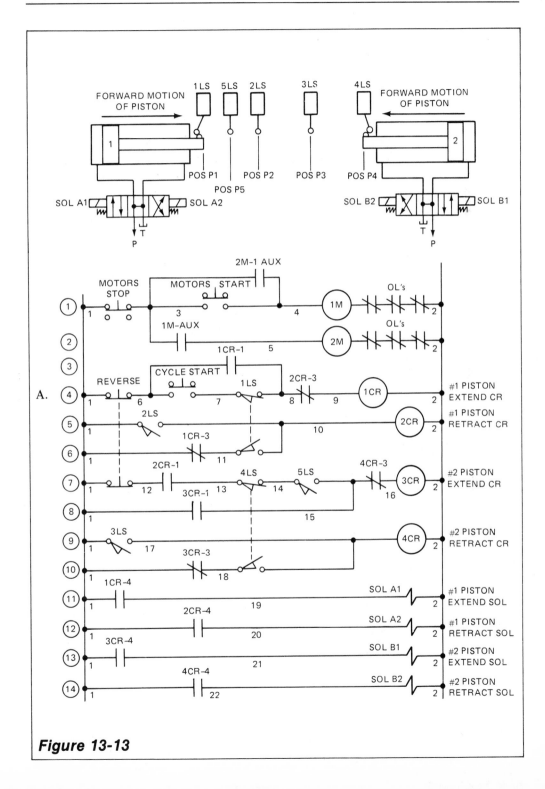

Figure 13-13

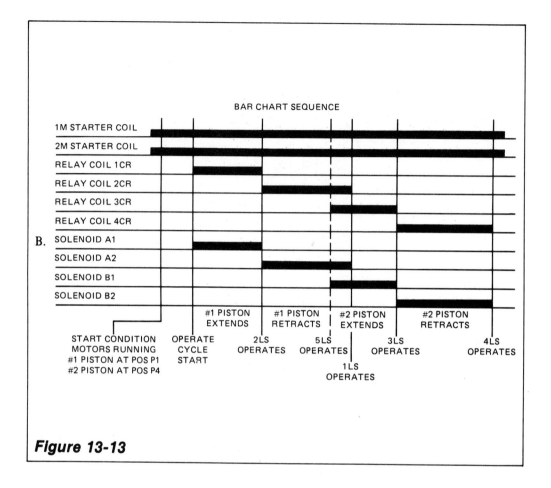

BAR CHART SEQUENCE

Figure 13-13

ACHIEVEMENT REVIEW

1. List five components that provide input (information) to an electrical control circuit.

2. Using electrical components, show how input, logic, and output progress through an electrical control circuit.

3. What electrical components should be shown on the left-hand vertical column of a bar chart?

4. In Example #1 (Figures 13-6A and 13-6B), what use is made of the fused disconnect switch?

5. Redesign the circuit shown in Figure 13-7 to allow a short time delay after operating the START push-button switch for the energizing of solenoid C. A time delay relay may be used.

6. In the circuit shown in Figure 13-8, the operating fluid temperature being sensed by temperature switch 1TS is at 80°F. The heat OFF-ON switch has been turned on. What is the condition now of contactor coil 10CR?
    a. Energized
    b. Deenergized

7. In the circuit shown in Figure 13-9, what is the function of the normally open (held closed) limit switch contact 2LS?

8. In the circuit shown in Figure 13-10, when is the counter clutch energized? Explain the sequence of operations that follow the operation of the cycle START push-button switch.

9. In the circuit shown in Figure 13-11, what is the function of relay contact 2CR-2?

10. In the circuit shown in Figure 13-12, 2M motor has experienced an overload. This has opened the overload relay contacts on the 2M motor starter. Following this, what is the condition of 1M motor starter?
    a. Deenergized
    b. Remains energized until the motor STOP push-button switch is operated.

CHAPTER

USE OF ELECTRICAL CODES AND STANDARDS 14

OBJECTIVES

After studying this chapter, the student will be able to:

- Name the two major goals that codes and standards are designed to accomplish.

- Explain some of the items that help in reading an elementary circuit diagram.

- Explain the major reason for using a disconnecting device on electrical circuits.

- List three types of protection covered in electrical standards.

- List three considerations that should be observed in locating and mounting control equipment.

- Locate the sizes and insulation specifications for conductors.

- Discuss some of the items that are generally covered under "Good Workmanship."

- Name the type of gasket you would use in an electrical control cabinet and describe how it should be used.

- List some of the factors you should consider in installing a motor on a machine.

- Show two types or methods of grounding a control circuit and list what equipment should be grounded.

14.1 MAJOR GOALS

All electrical codes and standards are written with at least two major goals in mind: maximum safety in operation, and long, trouble-free operating life. These goals are best accomplished through good basic engineering design.

It is hard to write a comprehensive single code or standard because of the problems involved in applying electrical control. There are many different types of machines and user requirements. From the small machine shop to the large production-line manufacturer, conditions and requirements differ. Even within a given user's plant, conditions change from section to section and from time to time within the same section. Also, the needs or requirements of today will not necessarily be those of tomorrow.

During the past years, many groups have coordinated their efforts in writing codes and standards. In most cases they are written from the experience and knowledge of that group. Therefore, codes and standards reflect the concepts of good design as held by the group that write them. Differences, additions, or omissions among groups do not necessarily rank one group above another. Rather, it is an indication that they are written for a specific application in a certain locality or field. This, of course, presents a problem to the manufacturer who supplies electrically controlled machines to every group.

One of the manufacturing problems for the builder is economics. (This is not to imply that good design is in direct proportion to cost. Often a poor design costs more.) The problems vary from one builder to another. The problems arise in the fields of engineering, marketing, and production.

Problems are usually resolved by close cooperation between the builder and user on each specific machine or group of machines. Both the builder and user must be concerned with the need for codes and standards. From a practical sense they must weigh the relative merits of economy against complete adherence to safety requirements. Safety must be the top consideration, however.

The builder and user should be acquainted with the NFPA 79 Electrical Standards for Industrial Machinery. The 1985 edition reflects the incorporation of the appropriate material from the JIC Electrical Standards (EMP-1-67 and EGP-1-67) not previously covered.

This applies to Metal Working Machine Tools and Plastics Machinery. The NFPA electrical standard and others are reviewed, revised and re-issued on a regular basis. In addition, codes of local municipalities and states should be followed where applicable.

The codes and standards referred to in this text are those of application. They cover, for example, how the circuit is to be presented, and how, where, and what components are to be installed on a machine or in a control cabinet.

The electrical standards found in NFPA 79-1985 Electrical Standards for Industrial Machinery break the information down into eighteen chapters. These follow in sequence from 1.1 in Chapter 1 to 18.1 in Chapter 18. The purpose of this text is not to cover every detail but to explain some of the reasons for writing standards. For the reader's convenience, a cross reference is made between the divisions of Chapter 15 in this text and the material in each chapter of the standards.

For any specific industrial application, close attention should be given to the NFPA 79-1985 Electrical Standard for Industrial Machinery.

14.2 GENERAL

In Chapter 1 of the standard, the purpose, scope and definitions are covered. For example, quoting from the standard, "The purpose of these Electrical Standards is to provide detailed information for the application of electrical/electronic equipment, apparatus or systems supplied as part of industrial machinery which will promote safety to life and property."

In the section on scope, the types and sizes of machines and the applicable industry is covered.

To fix the areas covered specifically, machine tools, plastic machinery and mass production industrial equipment are defined.

Ref: NFPA 79-1985

Chapter 1

- 1.1 Purpose
- 1.2 Scope
- 1.3 Definitions
- 1.4 Other Standards
- 1.5 Nominal Voltages

14.3 DIAGRAMS, INSTRUCTIONS, AND NAMEPLATES

It is difficult to explain a point to someone if you are not speaking the same language. Therefore, all symbols used in preparing elementary electrical diagrams should be in accordance with a common standard. In the United States, the accepted standard is the *Graphical Symbols for Electrical and Electronic Diagrams*, ANSI Y32.2. Examples of these symbols, as used in this textbook, are shown in the Appendix.

The form used in drawing the diagrams, cross-indexing of coils and their contacts, and the proper marking of each conductor are important. This aids in establishing a standard form and helps in reading the diagram.

On elementary diagrams for large machines, it is sometimes difficult to locate all the components required. A layout drawing showing the relative location of components and their circuit numbers is a great help. This drawing does not always have to be to scale. A further help is to properly identify the components on the machine with nameplates. Nameplates should carry the same description that is on the elementary diagram.

Also, a description of the components used and operating instructions are necessary.

A very important item has been added in these latest standards. This is WARNING MARKINGS. Quoting from the standards, "A warning marking shall be provided adjacent to the disconnect operating handle(s) where the disconnect(s) that is interlocked with the enclosure door does not deenergize all exposed live parts when the disconnect(s) is in the open (OFF) position."

"When the attachment plug is used as the disconnecting means, a warning marking shall be attached to the control enclosure door or cover, indicating that power shall be disconnected from the equipment before the enclosure is opened."

Ref: NFPA 79-1985

Chapter 2

2.1 Diagrams
2.2 Instructions
2.3 Markings
2.4 Warning Markings
2.5 Machine Marking
2.6 Machine Nameplate Data
2.7 Equipment Marking and Identification
2.8 Function Identification

14.4 GENERAL OPERATING CONDITIONS

Chapter 3 in the standards describes general requirements and conditions for the operation of electrical equipment on the machine. Conditions cover Ambient Operating Temperatures, Altitude, and Relative Humidity.

The electrical equipment shall operate satisfactorily under given conditions of Voltage, Frequency, Harmonic Distortion, Radio Frequency Voltage, Impulse Voltage, Voltage Drop and Micro-Interruption.

Ref: NFPA 79-1985

Chapter 3 {
3.1 General
3.2 Electrical Components and Devices
3.3 Ambient Operating Temperature
3.4 Altitude
3.5 Relative Humidity
3.6 Transportation and Storage
3.7 Installation and Operating Conditions
3.8 Supply Voltage

14.5 SAFEGUARDING OF PERSONNEL

The most important consideration of electrical standards is the safety of the worker. Therefore a section has been added in these standards covering this area. In general, it says the electrical equipment shall provide safeguarding of persons against electrical shock both in normal service and in case of fault.

Proper insulation of live parts, enclosure interlocking and grounding are a must.

In the area of extra low voltages (30 volts), consideration shall be given to isolation, insulation and grounding.

There is also the problem of residual voltages that are retained after the equipment is switched off. This voltage shall be reduced automatically to 50 volts, one minute after being disconnected.

Ref: NFPA 79-1985

Chapter 4 {
4.1 General
4.2 Safeguarding Against Electric Shock in Normal Service
4.3 Safeguarding Against Electric Shock by Machine Extra Low Voltage (MELV)
4.4 Safeguarding Against Electric Shock from Residual Voltages

14.6 SUPPLY CIRCUIT DISCONNECTING DEVICES

When work is to be done on the machine, particularly in the electrical circuit, it is a "must" for convenience and safety to have one point where all electrical power feeding the machine can be disconnected.

It cannot always be determined when the disconnecting device may have to be opened, such as in an emergency. Therefore, the disconnect device should be capable of interrupting the sum of the locked-rotor current of the largest motor plus the full load current of all other connected operating equipment.

The ampacity of the disconnecting means shall not be less than 115 percent of the sum of the full-load current required for all equipment which may be in operation at the same time under normal conditions of use.

Interlocking of the disconnecting means is important. See Figure 14-1. It shall be electrically and/or mechanically interlocked with the control enclosure door. This accomplishes two functions.

1. Prevents the closing of disconnecting means while the enclosure door is open, unless an interlock is operated by deliberate action.
2. Prevents closing of disconnecting means while the door is in its latch position or until the door hardware is fully engaged.

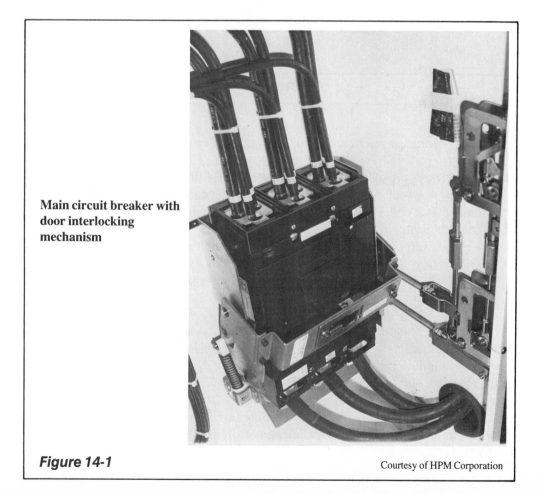

Main circuit breaker with door interlocking mechanism

Figure 14-1   Courtesy of HPM Corporation

For safety, the disconnect device should be located at the top of the control panel, interlocked with the door. The operating handle should be located not more than 6½ feet (ft) above the floor. See Figure 14-2. Note that the operator is wearing safety glasses with side shields. This is an additional precaution when working around electrical equipment.

Ref: NFPA 79-1985

Chapter 5

5.1 General Requirements
5.2 Type
5.3 Rating
5.4 Position Indication
5.5 Supply Conductors to be Disconnected
5.6 Connections to Supply Lines
5.7 Exposed Live Parts
5.8 Mounting
5.9 Interlocking
5.10 Operating Handle
5.11 Attachment Plug and Receptacle

14.7 PROTECTION

Here, *protection* refers to the following three important items:

1. *Short-circuit protection.* For conductors, this protection should be given to both power and control conductors. Reference should be made to specific standards or codes for allowable maximum ratings.
2. *Overload protection* is necessary, particularly in the case of motors. Thermal overloads responsive to motor current are used here.
3. *Undervoltage protection* is used to protect an operator and/or the machine, should full power return after an undervoltage condition and cause the machine to restart.

Reference should be made to the several tables listed in the Appendix, for practical applications. A few of these tables are as follows:

The maximum rating or setting of motor branch circuit short-circuit ground-fault protective devices is shown in Table 6-5(a). See Figure 14-3 for an installation of branch-circuit protection.

There is a relationship between conductor size and maximum rating or setting of short-circuit protective devices for power circuits. These are shown in Table 6-5(b).

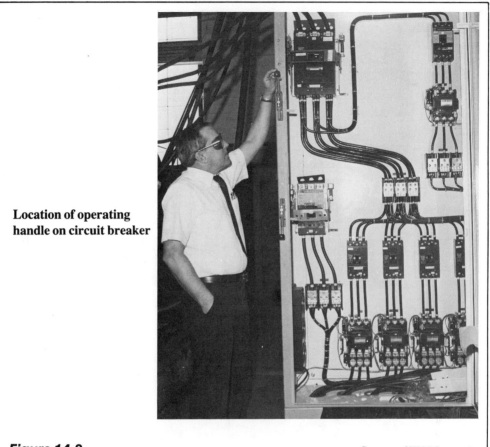

Location of operating handle on circuit breaker

Figure 14-2

Courtesy of HPM Corporation

Installation of branch circuit protection

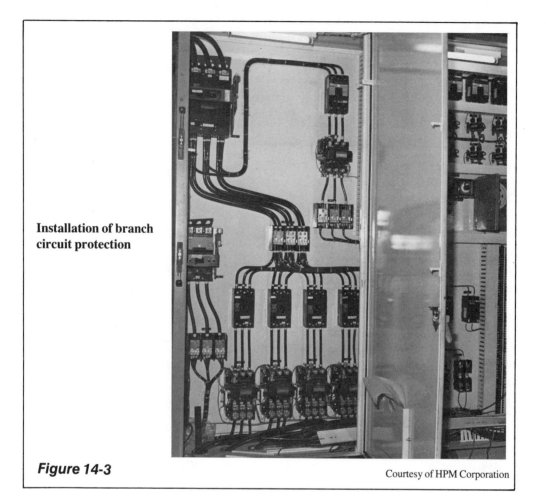

Figure 14-3

Courtesy of HPM Corporation

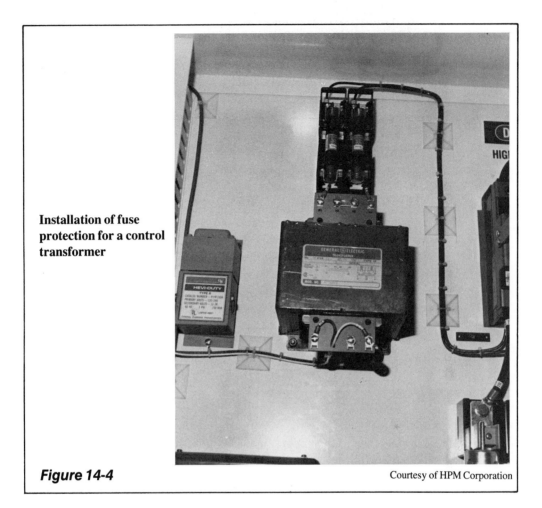

Installation of fuse protection for a control transformer

Figure 14-4

Courtesy of HPM Corporation

Tables are available in the Appendix [13-1(f)] giving the maximum conductor size for a given controller size.

Conductors in control circuits must be protected against overcurrent. The relationship between conductor size and ampacity is shown in Table 13-1(c). While there are exceptions listed, in general conductors of No. 18, 16, and 14 shall be considered protected by an overcurrent device of not more than 20 amperes.

Protection for control transformers is shown in Table 6-12. Figure 14-4 shows the installation of fuse protection for the control transformer primary.

Ref: NFPA 79-1985

Chapter 6 {

6.1   Machine Circuits
6.2   Supply Conductor and Machine
      Overcurrent Protection
6.3   Additional Overcurrent Protection
6.4   Location of Protective Devices
6.5   Motor Branch Circuits
6.6   Motor Overload
6.7   Motor Overload, Special Duty
6.8   Resistance Heating Branch Circuits
6.9   Control Circuit Conductors
6.10 Lighting Branch Circuits
6.11 Power Transformers
6.12 Control Circuit Transformers
6.13 Common Overcurrent Devices
6.14 Undervoltage Protection
6.15 Adjustable Speed Drive Systems
6.16 Motor Field or Tachometer Loss

14.8 CONTROL CIRCUITS

In general, the source of supply for all control circuits shall be taken from the load side of the main disconnect means. With the present programmable control now used extensively in industry, the memory elements may require power at all times to maintain their storage of memory. In this case, the power may be taken from the line side of the disconnecting means.

In most cases the control voltage is 115–120 V. This keeps the potential at a relatively low level and ensures standard component design. Lower voltages may be used for electronic, static, or precision devices. The control voltage is generally obtained from the secondary of an isolated secondary transformer.

Two types of control circuits are commonly used, grounded and ungrounded. Examples of these are shown in Figures 15-2 and 15-3 (Chapter 15).

In most cases the operating coil of an electromechanical magnetic device and transformer primary of a pilot lamp are connected to the same side of the control circuit. However, there are several exceptions listed in the standards. These should be examined.

Stop functions shall be initiated through deenergizing rather than energizing of control devices.

Several important control functions requiring proper interlocking are covered in this chapter.

Ref: NFPA 79-1985

Chapter 7

{
7.1   Source of Control Power
7.2   Control of Circuit Voltages
7.3   Grounding of Control Circuits
7.4   Connection of Control Devices
7.5   Stop Circuits
7.6   Cycle Start
7.7   Jog Circuits
7.8   Mode Selection
7.9   Sequence Control by Pressure Switches
7.10 Feed Interlocked with Spindle Drive
7.11 Machinery Door Interlocking
7.12 Motor Contactors and Starters
7.13 Relays and Solenoids
}

14.9 CONTROL EQUIPMENT

Certain restrictions are placed on the components used. This guards against inferior products and provides the user with standard available equipment. It is important to properly identify the components. They should be marked with the manufacturer's name or trademark, the part number, and a description of their electrical characteristics (voltage, frequency, current, etc.).

Control devices such as relays, timers, and transformers are generally mounted on a panel. Terminal blocks should be mounted on the panel to provide a means of making conductor connections.

Ref: NFPA 79-1985

Chapter 8

{
8.1 Connections
8.2 Subpanels
8.3 Manual and Electro-Mechanical Motor Controllers
8.4 Marking on Motor Controllers
}

14.10 CONTROL ENCLOSURES AND COMPARTMENTS

Housing electrical components in poorly designed or constructed enclosures has made it necessary to provide many detailed instructions. After making a good circuit design and securing top-quality components, it is foolish to house them inadequately or in a location subjected to contamination.

Doors on enclosures should be hinged to swing about a vertical axis. They should be gasketed with an oil-resistant material securely fastened to the door or enclosure. Most control enclosures are constructed of sheet metal. It should be a minimum of 14 gauge.

For safety, the door of the enclosure must be interlocked with the enclosure. **Caution:** If voltages above 50 V are used, it is dangerous to allow personnel to have access to the panel without first turning off the power.

An example of a well-designed control enclosure is shown in Figure 14-5.

In most cases, enclosures and compartments shall be non-ventilated and constructed to exclude such materials as dust, flyings, oil and coolant. However, where equipment requires ventilation, it can be housed in a separate ventilated portion of the enclosure or compartment or in a ventilated enclosure or compartment. Figure 14-6 shows a filtered, ventilating system for a control enclosure.

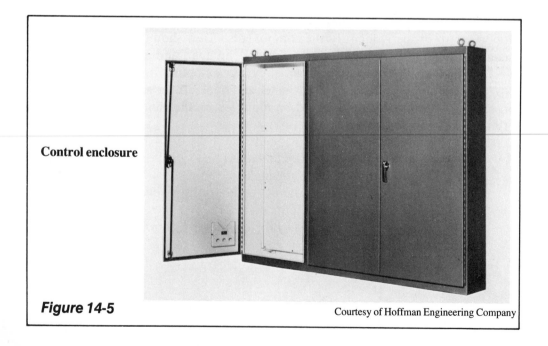

Control enclosure

Figure 14-5   Courtesy of Hoffman Engineering Company

Filtered ventilating system for control enclosure

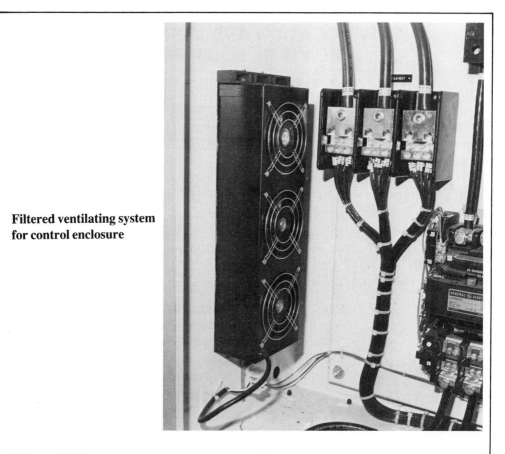

Figure 14-6

Courtesy of HPM Corporation

Ref: NFPA 79-1985

Chapter 9

{
9.1 Type
9.2 Nonmetallic Enclosures
9.3 Compartment Location
9.4 Wall Thickness
9.5 Dimensions
9.6 Doors
9.7 Gaskets
9.8 Interlocks
9.9 Interior Finish
}

14.11 LOCATION AND MOUNTING OF CONTROL EQUIPMENT

More detail is given here for the use of enclosures such as that shown in Figure 14-6. Basically, control components should be located in one enclosure. They should be placed in a position which will aid maintenance, promote safety, and guard against oil, dirt, coolant, and dust.

Components carrying line voltage above the level of the normal machine control voltage of 115–120 V should be grouped together at the top or to the side of the components carrying control voltage.

On machine-mounted components, safety and maintenance are again of prime consideration. They should be rigidly mounted and accessible in a clean, dry location.

Figure 14-7 shows the mounting of control and line voltage equipment.

Ref: NFPA 79-1985

Chapter 10

{
  10.1 General Requirements
  10.2 Control Panel
  10.3 Control Panel Enclosure
  10.4 Clearance in Enclosure
  10.5 Machine Mounted Control Equipment
  10.6 Rotary Control Devices
}

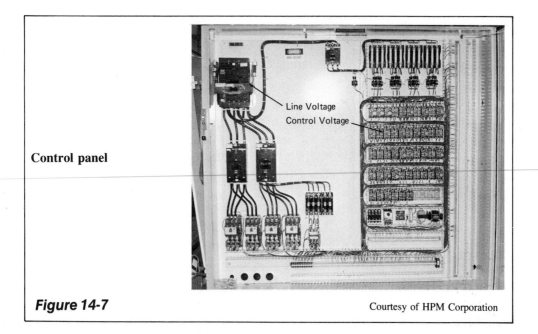

Control panel

Line Voltage
Control Voltage

Figure 14-7   Courtesy of HPM Corporation

14.12 OPERATOR'S CONTROL STATION AND EQUIPMENT

Push buttons, selector switches, and indicating lights should be oil-tight. An exception is made where they are identified for the environment. For safety and convenience of operating, colors have been assigned; for example, Start = Green, Stop = Red. A complete listing of color coding is given in the Appendix. Proper identification is a must. Assigned relative locations help the operator in safe and efficient operation. See Figure 14-8.

The design, location, and legend plates are important for control stations having push buttons, selector switches, and pilot lights. All machines shall incorporate one or more emergency stop controls which stop all machine motions upon momentary operation, and which do not create other hazards when activated.

Operator's control station

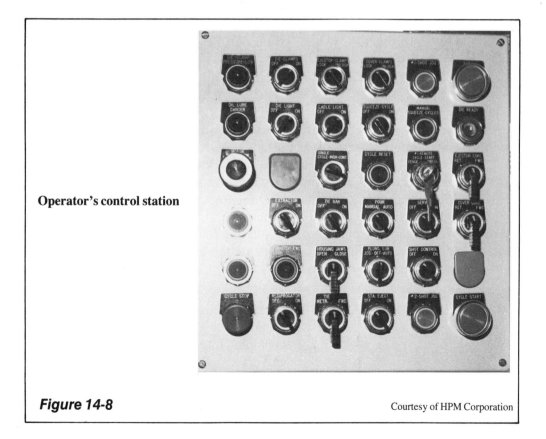

Figure 14-8

Courtesy of HPM Corporation

All operator control station enclosures shall be dust, moisture and oil tight. An exception is made where the enclosure is suitable for the environment.

Control enclosures should be located where they are reasonably free of all outside contaminates. They should be within normal reach of the operator and so located that it is not required to reach past any moving parts.

Ref: NFPA 79-1985

Chapter 11

{
11.1 Pushbuttons, Selector Switches, Indicating Lights
11.2 Emergency Stop Controls
11.3 Two-Hand Controls
11.4 Foot Operated Switches
11.5 Control Station Enclosure
11.6 Arrangement of Control Station Components
11.7 Legends
11.8 Location of Control Stations
11.9 Pendent Stations

14.13 ACCESSORIES AND LIGHTING

This section covers plugs, receptacles, control panels, instruments, and machine lights. Safety is the most important consideration.

Plugs should be of the locking type. They should be effectively protected against oil or moisture. If a grounding connection is provided, it should be made before the line connection and broken after the line disconnects. See Figure 14-9.

The lighting circuit should not be energized in excess of 150 V. The sources of energy can either be from a separate isolating transformer, a control circuit, or a plant lighting circuit. The size and type of the conductor used for stationary lights are important. The specifications are covered in the standards.

Ref: NFPA 79-1985

Chapter 12

{
12.1 Attachment Plugs and Receptacles External to the Control Enclosure
12.2 Receptacles Internal to the Control Enclosure
12.3 Control Panel, Instrument and Machine Work Lights

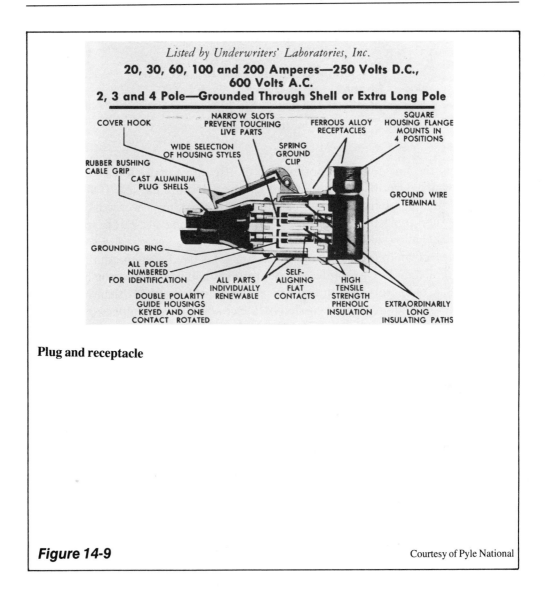

Listed by Underwriters' Laboratories, Inc.

20, 30, 60, 100 and 200 Amperes—250 Volts D.C., 600 Volts A.C.

2, 3 and 4 Pole—Grounded Through Shell or Extra Long Pole

Plug and receptacle

Figure 14-9   Courtesy of Pyle National

14.14 CONDUCTORS

Much of this section presents tables and charts specifying sizes and insulation for conductors. The following tables are found in the Appendix of this text.

13-1(a) Single Conductor Construction Type-MTW
13-1(c) Conductor Ampacity
13-1(f) Maximum Conductor Size for a Given
         Motor Controller Size
13-3(a) Multi-Conductor Cable Stranding
         (Constant Flexing Service)

Probably the single most important factor affecting conductors is the current-carrying capacity. Overloading conductors can be a definite hazard. It also creates problems for the maintenance personnel. In most control or power work, stranded copper with thermoplastic insulation is specified. In some cases of extreme heat, a stranded nickel alloy with high temperature insulation is specified.

Where electronic, precision and static control is used, in general the conductor size is reduced. For example, conductors in raceways shall not be smaller than No. 18. Conductors within enclosures shall not be smaller than No. 26. In any case, the continuous current carried by the conductors shall not exceed values given in Table 13-1(c).

Ref: NFPA 79-1985   ⎧ 13.1 Power and Control
                         ⎪ 13.2 Electronic, Precision and Static
         Chapter 13   ⎨   Control
                         ⎪ 13.3 Machine Wire Type MTW
                         ⎩

14.15 WIRING METHODS AND PRACTICES

This covers the more practical problem of the actual installing of components and their connections. Many of these could come under the general heading of "Common Sense and Good Workmanship." Electrical service and maintenance personnel should exercise great interest and concern in this topic. Such items as conductor color coding, taping connections, and marking terminal blocks are noted. Refer to Figure 14-10 and note that each conductor connection is marked.

For panel wiring, wiring channel is permitted if the channel is made of a flame-retardant insulating material. Under machine wiring, the proper application of conduit and conductors on the machine is covered. The use of liquidtight flexible conduit for small or infrequent movements and stranded multiconductor cable for continuous moving parts is specified.

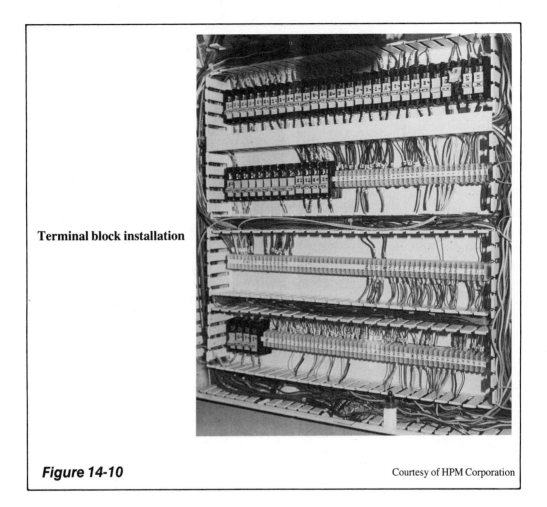

Terminal block installation

Figure 14-10   Courtesy of HPM Corporation

An item that has come into more prominence in the past few years is that of soldering. The practice of soldering takes into consideration such items as the use of rosin as a flux, proper tinning of parts to be soldered and consideration for insulation and parts that might be damaged by the heat of soldering.

Ref: NFPA 79-1985   14.1 General Requirements

Chapter 14   14.2 Panel Wiring

14.3 Machine Wiring

14.4 Wire Connectors and Connections

14.16 RACEWAYS AND JUNCTION BOXES

An item that should be covered by good workmanship but which is often overlooked is the warning against sharp and rough edges and burrs in raceways, conduit, and fittings. See Figure 14-11. Conduit on the right has not been deburred, leaving a sharp edge. Conduit on the left has been deburred, resulting in a smooth edge.

Information is supplied on the enclosures that are to carry the conductors. This covers rigid metal conduit, flexible metal conduit, raceways, wireways, and compartments.

Overloading conduits may result in overheating or insulation damage in pulling the conductors. Care must be taken to not place too many conductors in an enclosure. For example, the combined cross-sectional area of all conductors and cables shall not exceed 50% of the interior cross section of the raceway. Bending too sharp a radius in rigid metal conduit often results in a "bottleneck," as the conduit tends to flatten out. See Figure 14-12. The conduit at the top has been improperly bent, resulting in a flattened area.

Table 16-3(g) in the Appendix covers the minimum radius of conduit bends.

The use of bushings, oil-tight features provided by gaskets, and good construction of compartments are important in a good installation.

Concern for unused knockouts or openings in junction and pull boxes is covered, as is the connections that should be allowed in motor terminal boxes.

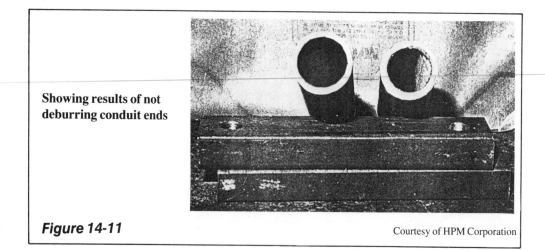

Showing results of not deburring conduit ends

Figure 14-11

Courtesy of HPM Corporation

Results of improper conduit bending

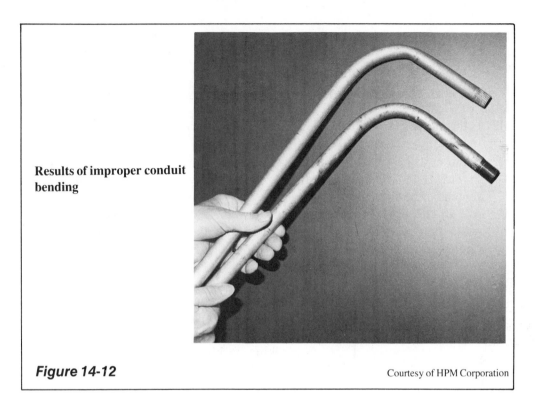

Figure 14-12

Courtesy of HPM Corporation

14.17 MOTORS AND MOTOR COMPARTMENTS

The main considerations for motors can generally be summed up as follows:

1. Is the motor easily accessible for maintenance? See Figure 14-13 for a motor installation easily accessible for maintenance.
2. Is the motor properly mounted and protected against oil, water, chips, and so on? If it is in a compartment, is the compartment clean and dry and vented to the exterior of the machine?
3. Are all the operating characteristics available on a nameplate and easy to read? Is it properly marked in accordance with the latest **National Electrical Code**?

Ref: NFPA 79-1985 $\left\{ \begin{array}{l} \text{16.1 Location on Mounting} \\ \text{16.2 Mounting Arrangements} \\ \text{16.3 Direction Arrow} \\ \text{16.4 Motor Compartments} \\ \text{16.5 Marking on Motors} \end{array} \right.$

Chapter 16

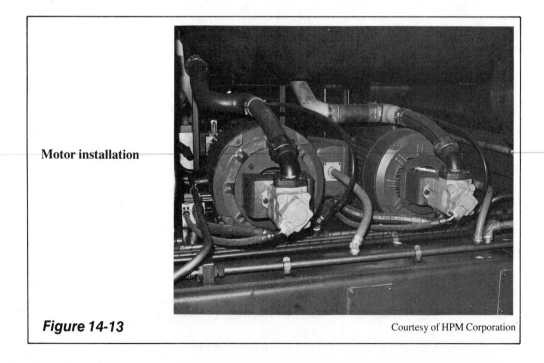

Motor installation

Figure 14-13   Courtesy of HPM Corporation

14.18 GROUNDING CIRCUITS AND EQUIPMENT GROUNDING

The grounding of control circuits varies with the user's conditions and practice. In some cases it is required and desirable to ground one side and fuse the other side. In other cases both sides are fused, and ground detector lights are used to indicate the presence of an undesirable ground or fault in the circuit.

Portable, pendent, and resilient mounted equipment shall be grounded. It shall be bonded by separate conductors. The machine and all exposed noncurrent-carrying metal parts and equipment shall be effectively grounded. A table listing the sizes of grounding conductors is in the Appendix, Table 18-2(c).

It is important that the continuity of the grounding circuit be ensured by effective connection through conductors or structural members.

Machine lighting circuits should be grounded. If the lighting circuit is taken from an isolation transformer, the grounding shall occur at the transformer.

Ref: NFPA 79-1985

Chapter 17

17.1 General
17.2 Grounding Conductors
17.3 Equipment Grounding
17.4 Exclusion of Switching Devices
17.5 Grounding Terminal
17.6 Continuity of Grounding Circuits
17.7 Control Circuits
17.8 Lighting Circuits

14.19 REFERENCED PUBLICATIONS

Ref: NFPA 79-1985
Chapter 18
Referenced Publications
18-1   The following documents or portions thereof are referenced within this standard and shall be considered part of the requirements of this document. The edition indicated for each reference is current as of the date of the NFPA issuance of this document. These references are listed separately to facilitate updating to the latest edition by the user.

NFPA 70-1984   NATIONAL ELECTRIC CODE
ANSI C80-1-1977 SPECIFICATIONS for RIGID STEEL CONDUIT, ZINC COATED
NEMA FB 1-1977 (Rev. Dec. 1980), FITTINGS AND SUPPORTS for CONDUIT and CABLE ASSEMBLIES

ACHIEVEMENT REVIEW

1. List two major goals that Electrical Codes and Standards attempt to achieve.

2. What aids can you list that help to establish a standard form and assists in reading the diagram?

3. List three items that are a help to the safety of the worker in the electrical field.

4. The ampacity of the disconnecting means in percent of the sum of the full-load current required for all equipment which may be in operation at the same time is:
   a. 100%
   b. 115%
   c. 150%

5. The operating handle for the disconnect device located at the top of the control panel and interlocked with the door shall be located not more than: _____ feet above the floor.
   a. 5
   b. 6½
   c. 7

6. Name three important items of protection on a machine in reference to installing electrical equipment.

7. What important information should be given on all electrical components for identification?

8. Make a sketch showing the proper mounting of line voltage and control voltage equipment.

9. What can you say about the practice of overloading raceways and conduit with too many conductors?

10. What grounding consideration should be given to pendent, portable and resilient mounted equipment?

CHAPTER 15

TROUBLESHOOTING

OBJECTIVES

After studying this chapter, the student will be able to:

- List five areas that should be considered when starting a troubleshooting job.

- Describe two methods that can be used in troubleshooting a job.

- Explain a procedure for checking fuses.

- Discuss why a loose connection in an electrical power circuit can be a major problem.

- Explain why contacts on electrical operating equipment should not be filed.

- List several problems resulting from low line voltage.

- Discuss the merits of good housekeeping.

- Show how an open in a common line can result in a faulty circuit operation.

- Explain one method for checking an electrical control circuit.

- List four steps of procedure in locating problems resulting from momentary faults.

15.1 ANALYZING THE PROBLEM

Effective troubleshooting starts with an analysis of the problem. Too frequently, troubleshooting is approached in a "hit or miss" fashion. This generally creates more expense and wastes time.

15.1.1   To analyze any problem, it helps to break it down into types or sections. Breaking down a problem limits the size of the job. In troubleshooting, the following causal areas might be considered:

- Electrical

- Mechanical

- Fluid power

- Pneumatic

- Personnel

In many cases, the problem may be a combination of two or more of these areas.

The following examples illustrate how breaking down a problem into types of causes can simplify the troubleshooting procedure.

A. From advance information supplied by a machine operator or early examination by an electrician, what first appears to be an electrical problem may turn out to be a mechanical one.

B. The failure of a limit switch to function properly may be caused by problems in the electrical contacts or the mechanical operator. The result of a cycle failing to complete is the same, however, regardless of which of these two items caused the problem.

C. As long as people operate machines, problems will arise that do not respond to the usual form of troubleshooting. Such problems may be intentional or unintentional. They may stem from misunderstanding, lack of cooperation, or lack of knowledge of the machine.

   Whatever the cause of this type of problem, it will be recognized quickly by the troubleshooter who takes the approach outlined here. The problem should be handled carefully and diplomatically so that the machine can be returned quickly to its intended job.

15.1.2   Problems can be further separated into physical location or type of operation.

For example, in a large machine where the cycle of operation moves from one section of the machine to another, the trouble may be localized in one section. This may immediately eliminate 75% of the total machine as a possible trouble source. A practical application here could be trouble developing in loading or unloading equipment on a press. If the proper clearance signal has been given to the loader or unloader from the press control, trouble developing after this cycle starts can generally be localized in the loading or unloading control.

Success in troubleshooting is the ability to segregate the problem area from other unrelated circuitry.

Before getting into the "mechanics" of troubleshooting, let us examine some of the problem spots.

15.2 MAJOR TROUBLE SPOTS

It would be impractical, if not impossible, to list all potential trouble spots. However, there are a few areas that contribute to a large percentage of troubles. These areas are discussed in sections 15.2.1 through 15.2.11.

15.2.1 Fuses The checking of fuses is generally a good place to start. Too often this is overlooked. The details for checking fuses will be discussed in a later section.

The replacement of an open (defective) fuse can be an important safety factor. As discussed in Chapter 2, there are three different types of fuses. Within each type there are different voltage and current ratings. Too often just any type of fuse is used as a replacement. Unless changes have been made in the machine circuit and components, the replacement fuse should be exactly the same type, voltage and current rating as the fuse removed.

The policing of fuses can be a problem. However, the replacement policy given here must be rigidly maintained if safety to personnel and the machine is to be realized.

One area that is extremely important is when a machine is connected to a power source that has a high short-circuit current available. In such cases, it may be advisable to use current limiting fuses with high interrupting capacity.

15.2.2 Loose Connections On today's machines, there may be hundreds of connections. Each of these spots may be a source of trouble. Many advancements have been made in terminal block and component connectors to improve this condition. The use of stranded conductors in place of solid conductors has in general improved the connection problem.

The problem starts when the machine is built and continues throughout the life of the machine.

The problem may be of greater importance in power circuits, as the current handled is of greater magnitude. A loose connection in a power circuit can generate local heat. This spreads to other parts of the same component, other components, or conductors. An example of where direct trouble arises is with thermally sensitive elements. These can be overload relays or thermally operated circuit breakers.

For the correction of loose connections, the best advice is to follow a good program of preventive maintenance in which connections are periodically checked and tightened.

15.2.3 Faulty Contacts   This applies to such components as motor starters, contactors, relays, push buttons, and switches.

A problem that appears quite often and one of the most difficult to locate is with the normally closed contact. Observation indicates that the contact is closed but does not reveal if it is conducting current.

Any contact that has had an overload through it should be checked for welding.

Such conditions as weak contact pressure and dirt or an oxide film on the contact will prevent it from conducting. Many times contacts can be cleaned by drawing a piece of rough paper between the contacts. **Caution:** Use only a fine abrasive to clean contacts. Do not file contacts. Most contacts have a silver plate over the copper. If this is destroyed by filing, the contact will have a short life. If contacts are worn or pitted so badly that a fine abrasive will not clean them, it is better to change the contacts.

Another problem that may occur with a double-pole, double-break contact is cross firing. That is, one contact of the double break travels across to the opposite contact, but the other remains in its original position. If both the NO and NC contacts are being used in the circuit, a malfunction of control may occur.

15.2.4 Incorrect Wiremarkers   This problem usually appears on the builder's assembly floor or in reassembly in the user's plant. The error can be difficult to locate, as a cable may have many conductors running some distance to various parts of the machine.

One common problem is the transposition of numbers. For example, a conductor may have a 69 marked on one end and a 96 on the other end. Another problem that may occur is in connecting conductors into a terminal block. With a long block and many conductors, it is a common error to connect a conductor either one block above or below the proper position.

15.2.5 Combination Problems   Reference has been made to combination problems, but its importance should be emphasized. The following are typical types of combination problems:

Electrical-mechanical
Electrical-pressure (fluid power or pneumatic)
Electrical-temperature

The greatest problem is that the observed or reported trouble is not always indicative of which aspect is at fault. It may be both.

It is usually faster to check the electrical circuit first. However, both systems must be checked as both may contribute to the problem.

As an example, very few solenoid coils burn out due to a defect in the coil. Probably over 90% of all solenoid trouble on valves develops from a faulty mechanical or pressure condition which prevents the solenoid plunger from seating properly, and thus draws excessive current. The result is an overload or a burned out solenoid coil.

15.2.6 Low Voltage If no immediate indication of trouble is apparent, one of the first checks to make is the line and control voltage. Due to inadequate power supply or conductor size, low voltage can be a problem.

The problem generally shows up more on starting or energizing a component, such as a motor starter or solenoid. However, it can cause trouble at other spots in the cycle.

A common practice in small shops is to add more machines without properly checking the power supply (line transformers) or the line conductors. The source and line become so heavily loaded that when they are called on for a normal temporary machine overload, the voltage drops off rapidly. This may result in magnetic devices such as starters and relays dropping off the line (opening their contacts) through undervoltage or overload protective devices.

Heat is one result of low voltage that may not be noticed immediately in the functioning of a machine. As the voltage drops, the current to a given load increases. This produces heat in the coils of the components (motor starters, relays, solenoids), which not only shortens the life of the components but may cause malfunctioning. For example, where there are relative moving metal parts with close tolerances, heat can cause these parts to expand to a point of sticking. In cases where electrical heating is used, the heat is reduced by the square of the voltage. For example, if the voltage is dropped to one-half of the heating element's rated voltage, the heat output will be reduced to one-fourth.

15.2.7.A Grounds—Typical Locations There are many locations on a machine where a grounded condition can occur. However, there are a few spots in which grounds occur most often. These are discussed in the following section.

There are many locations on a machine where a grounded condition can occur. However, there are a few spots in which grounds occur most often. These are discussed in the following section.

A. **Connection points in solenoid valves, limit switches, and pressure switches**
Due to the design of many components, the space allowed for conductor entrance and connection is limited. As a result, a part of a bare conductor may be against the side of an uninsulated component case. Where bolt

and nut connections are made, the insulating tape may not be wrapped securely, or it may be of such quality that age destroys its insulating properties. This may occur in the field (user's plant), where due to the urgency for a quick change, sufficient care is not taken when handling the wiring and connections on a replaced unit.

B. Pulling conductors

In pulling conductors through conduit where there are several bends and 90° fittings, or into pull boxes and cabinets, the conductor insulation may be scraped or cut. If care is not taken to eliminate sharp edges or burrs on freshly cut or machined parts, cuts and abrasions occur. This is one good reason for the use of the insulation required for machine tool wiring.

C. Loose strands

The use of stranded wire has greatly reduced many problems in machine wiring. However, care must be taken when placing a stranded conductor into a connector. All strands must be used. One or two strands unconnected can touch the case or a normally grounded conductor, creating an unwanted ground. Even if the ground condition does not appear, the current-carrying capacity of the conductor is reduced.

15.2.7.B Grounds—Means of Detection For the best operation of an electrical system, some means of detecting the presence of grounds should be available. There are two methods shown here. Each has merits.

Using the grounded method as shown in Figure 15-1, the circuit is deenergized by the opening of the fuse. This means that the machine will be down until the ground is located and removed. In the small shop this inconvenience is usually not too serious. The ungrounded method shown in Figure 15-2 is usually used in large production shops. This method is an advantage because production can continue with one ground until there is time to locate the ground and remove it.

In the first method, all coils are tied solidly to a common line, and this line is grounded. The opposite side of the control power source is protected by a fuse or circuit breaker.

If a ground condition should appear in the circuit shown in Figure 15-1 on circuit point 8 between contact 1CR-2 and solenoid A, the load (coil) is bypassed. This is because there is a direct path to ground on both sides (relay 1CR energized — 1CR-2 closed). This puts a direct short on the control power source, opening the protective device and removing the control power from the circuit.

When the ground is located and removed, the fuse can be replaced or the circuit breaker reset. Thus, the control circuit is again ready for operation. Note that the ground condition must be removed first.

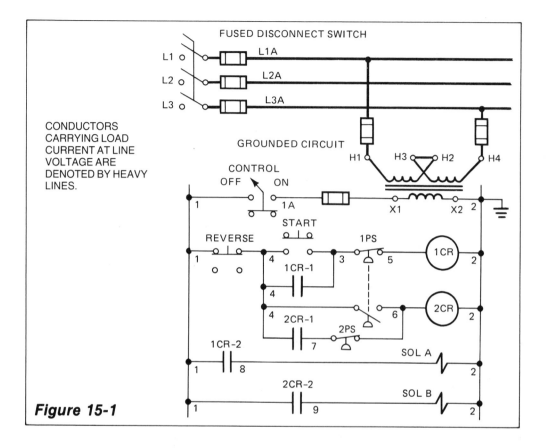

Figure 15-1

In Figure 15-2, the common side is not grounded. A set of two ground detector lights (standard 120-V indicating lights) are connected in series across the power source. A solid ground is then placed between the two lights.

With the system showing no grounds, both lights will glow at half brilliance, since two 120-V bulbs are connected in series across 120 V (60 V on each).

If a ground appears in the system, one of the lights will go out. The other will go to full brilliance. The determining factor of how these two lights perform depends on which side of the load the ground appears. For example, if a ground should appear on circuit line 16 between the fuse and solenoid A, lamp A will go out when contact 3CR-1 closes. Lamp B will glow at full brilliance.

An advantage of this method is that in some cases a ground can be present but not cause any immediate trouble. This gives maintenance time to check for the ground without immediately taking the machine out of production.

The important point is that some system should be used to detect the presence of a ground.

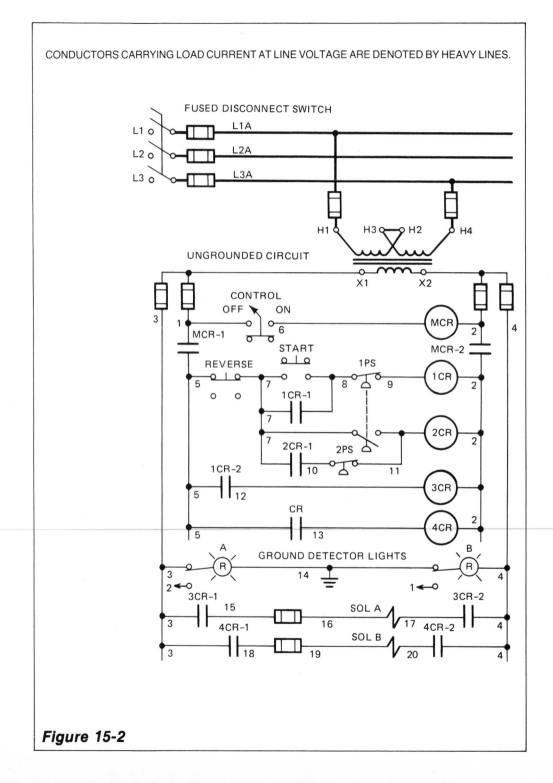

CONDUCTORS CARRYING LOAD CURRENT AT LINE VOLTAGE ARE DENOTED BY HEAVY LINES.

Figure 15-2

15.2.8 Poor Housekeeping Poor "housekeeping" leads to more work for the troubleshooter. There is an overall economy in having a clean machine and a well-organized and well-executed preventive maintenance program.

Dust, dirt, and grease should be removed periodically from electrical parts. Their presence only causes mechanical failure and forms paths between points of different potential, causing a short circuit.

Moving mechanical parts should be checked, particularly in large motor starters. Such items as loose pins and bolts and wearing parts are sources of trouble.

Overheated parts generally indicate trouble. Without proper instruments, it is difficult to determine the temperature of a part or how high a temperature can be sustained. Certainly any signs of smoke or baking of insulation are cause for immediate concern.

Manufacturers of components have done considerable work in their design to prevent dust, dirt, and fluids from entering.

When it is necessary to remove a cover or open a door for troubleshooting, immediately replace it after the trouble is corrected.

Many users have gone to great lengths to develop and rigidly enforce a good electrical maintenance program. Records are kept of each reported trouble and the work is done to correct the problem. These records are compiled periodically and are available to the supervisor. This not only leads to faster troubleshooting in the future, but it also gives the supervisor an indication of why the production in a given department may be down.

15.2.9 Trouble Patterns As troubleshooting work progresses over a period of time with a particular machine or group of machines, a pattern of trouble may develop. For example, it may be necessary to increase the production rate with a particular machine. Certain areas of this machine may not have been originally designed to handle this increased rate of operation. Unless the machine is redesigned and rebuilt, a pattern of trouble may develop.

Another example is a machine that is relocated in another section of the user's plant. Here the environment may be different. A change in the atmosphere and a presence of dust, dirt, or metal chips may create a pattern of trouble for a machine if it is not designed to operate under these conditions.

15.2.10 Opens in a Common Line Many circuits have two or more connection points that are common for multiple connections. For example, from the elementary diagram shown in Figure 15-3, line 1 has two points of connection. Line 2 has five points of connection. Line 4 has six points of connection. As it is not good wiring practice to place more than two conductors under any one terminal, connections are generally jumpered on the panel components or brought back to

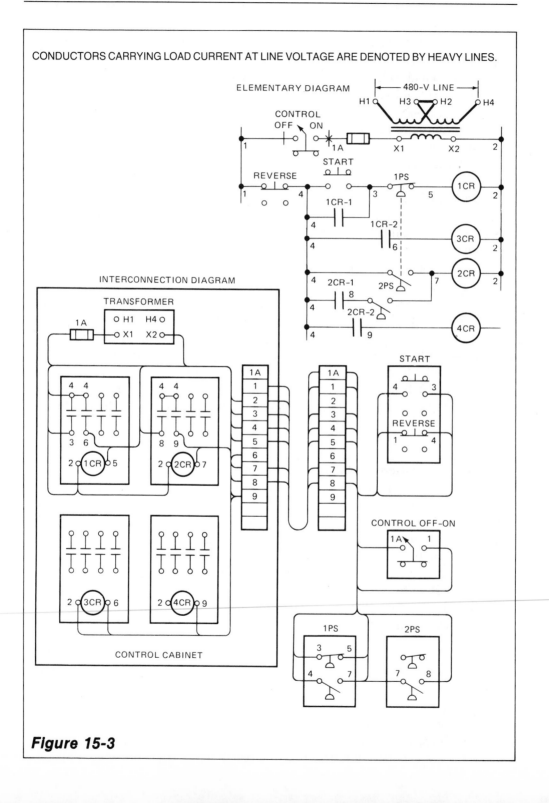

CONDUCTORS CARRYING LOAD CURRENT AT LINE VOLTAGE ARE DENOTED BY HEAVY LINES.

Figure 15-3

the terminal block. It is not unusual for a condition to arise where a jumper is omitted. In Figure 15-3, a jumper from circuit connection #4 is omitted from relay 2CR to circuit connection #4 on either relay 1CR or to connection point #4 on the terminal block. Also, the common coil connection on relay coil 4CR is not jumpered to relay coils 1CR, 2CR, or 3CR, or brought back to the terminal block. The schematic circuit is shown as it would appear with these two jumpers missing. This shows the effect on the condition of the circuit and the resulting faulty operation.

In this rather simple circuit, it does not appear that this error would be committed. However, there are cases where there may be many of these common connections. In the larger and more complicated circuit diagrams, the probability of jumpers being missed is much greater.

15.2.11 Wiring the Wrong Contact Many components such as relays, limit switches, and pressure and temperature switches have NO and NC contacts available for use.

A wiring error is often made, particularly where only one of the two available contacts is used. The error consists of wiring the wrong side of the contact. That is, the NO contact may be put into the circuit where the NC should be used. The reverse of this may also be true.

Unless the person who does the wiring is completely familiar with the component and doublechecks the work, this error will occur frequently.

The routine checks discussed in the following section will reveal this error quickly.

15.2.12 Momentary Faults The momentary fault is one of the most difficult problems to troubleshoot. With this type of fault, the machine or control can be under close observation for hours with no failures. The fault may occur at any time, however, and if the observer's attention is even briefly diverted, any direct evidence that might have been seen is lost.

There is no direct solution to this problem. The best approach is a well-organized analysis. The following steps might be helpful:

1. Attempt to localize within the total cycle. If the fault always occurs at the same place in the cycle, generally only the control associated with that part of the cycle is involved. If the fault occurs at random spots through the cycle, then the spots to examine are those that are common through the entire cycle. Examples are a drum or selector switch used to isolate an entire section of control.

2. Examine for loose connections, particularly in the area of the fault. Attention should be paid to areas where mechanical action may have damaged conductors, pulled them loose from connectors, or cut or broken them. The

complete break of a stranded conductor within the insulation is rare.

3. Localize attention to components. Many times casual observation will not disclose the trouble. In these cases the complete replacement of the component(s) in question is the quickest and best solution. Here, the plug-in components have a distinct advantage in returning a machine to operating condition in a minimum of time.

4. Examine the circuit for unusual conditions. This type of trouble rarely occurs in the user's plant. However, there are cases where, either through an oversight on the part of the circuit designer or by a change of operating conditions on the machine, a circuit change is indicated as a solution to the problem.

15.3 THE EQUIPMENT FOR TROUBLESHOOTING

15.3.1 Safety is number one and always the most important factor in troubleshooting. The person involved should be well acquainted with the electrical field and the potential hazards that may be present.

Problems generally arise when units have been abused, carelessly applied or misapplied. Experience is one of the best teachers that a troubleshooter can have.

In addition to well-insulated hand tools, meters are an important item. The multi-range clamp-on ammeter and the volt-ohm-milliammeter (VOM) are almost a necessity.

The clamp-on ammeter should have a range to cover the largest line current of the largest motor that may be checked. A low range (one ampere or less) is also useful to check coil currents.

The sensitivity of the VOM may range from 200 to several million ohms per volt (for example, the vacuum tube voltmeter). The voltage range will generally be from 150–600.

Several ohm ranges should be available so that accurate resistance checks can be made.

One troubleshooting instrument that is widely used in industry is a voltage tester, commonly called a Wiggy®. Figure 15-4 shows a solenoid type with polarity indication and slotted receptacle for holding prods. The voltage range covers 120–240–600 volts ac—50/60 Hertz. Also 120–240–600 volts dc.

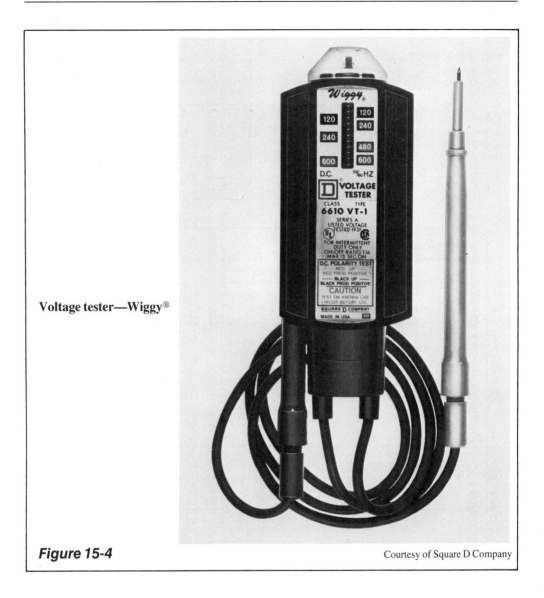

Voltage tester—Wiggy®

Figure 15-4

Courtesy of Square D Company

15.4 PROCEDURES USED IN TROUBLESHOOTING

15.4.1 In Figure 15-5 a power and control circuit is shown. The power source is 480 V—3 PH—60 Hertz. Two motors are used with magnetic full-voltage starters. The control voltage is taken from the secondary of an isolated secondary transformer and is 120 volts.

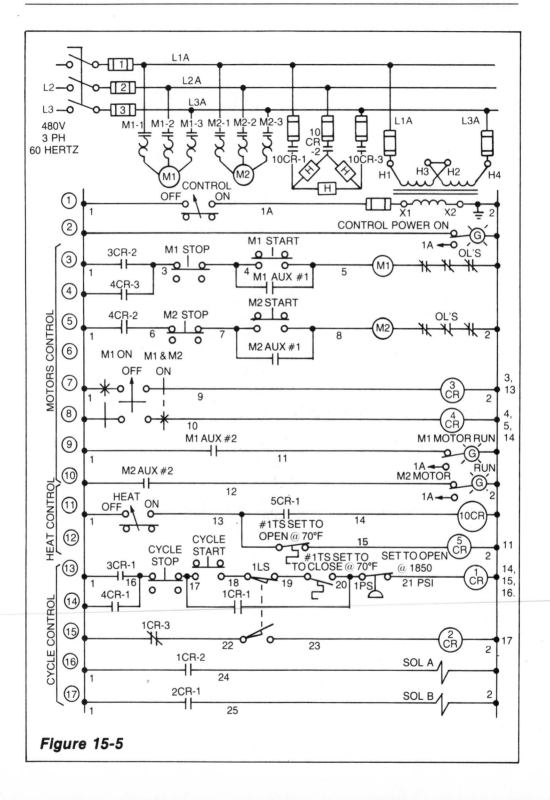

Figure 15-5

Depending on the machine speed that is desired, one or two motors can be used. Heating elements are used in the machine operating fluid to bring the fluid temperature up to 70°F required by the machine. The machine must be in a given position for start condition and reverses by pressure.

The approach to effective troubleshooting, as noted earlier, is to first segregate the section where the trouble is observed to be occurring. In the circuit shown in Figure 15-5, there are three sections indicated, motors, heat and cycle control. In a particular problem you have observed that the motors and heat control are OK. The problem is that solenoid A is not energizing at the proper time. Thus you should be concerned only with the cycle control section.

However, let us first take a look at this entire circuit. Always keep in mind the possible causes of trouble as outlined earlier in the chapter.

15.4.2 Motors and Motor Control

1a. Using the ac voltmeter (600-V range), check the voltage from L1-L2, L2-L3, L3-L1. With approximately 480 volts showing on all three checks, the plant line voltage is established.

   b. Close the line fused disconnect switch. Check the voltage from L1A-L2A, L2A-L3A, L3A-L1A. If any of the three checks show no voltage, check the fuses in the disconnect switch. For example,
   a. No voltage from L1A-L2A.
   b. No voltage from L2A-L3A.
   c. There is voltage from L1A-L3A.
   You now know that #2 fuse is open. Open the disconnect switch and replace the fuse. Note that the same general checking procedure can be used to determine which of the three fuses is open. With no voltage at any of the three checks, then two of the fuses or possibly all fuses are open.

   c. Turn the control OFF-ON selector switch to ON. The control power on light should be illuminated. If not, using the 150 V scale on the ac voltmeter, check the voltage from X1-X2 on the transformer secondary. This should read approximately 120. If there is voltage here, check from X2-1A. With no voltage here, the control fuse is probably open. Open the line disconnect switch and replace the control fuse. Reclose the disconnect switch and recheck voltage from 1–2. With no voltage here, the selector switch may be wired incorrectly. Again, open the disconnect switch and check the wiring on the selector switch. Always remember in all checking that there is a possibility of an open conductor at a component terminal or at a terminal block. With voltage established at 1–2, and the control power light is not illuminated, depress the push-to-test pilot light to determine if the bulb is burned out. When you originally checked for voltage at X1-X2 you found no voltage. The fuses in the transformer primary should be checked. See Figure 15-6. When the open fuse is located, open the fused disconnect switch and replace the fuse or fuses.

d. Operate the M1-M2 motor starter selector switch to M1 ON. Relay 3CR should be energized, closing relay contact 3CR-2. If not, using the 150 V range on the ac voltmeter, check from circuit number 2–9. With no voltage here, either the selector switch is wired incorrectly or there is an open conductor. Turn the control OFF-ON selector to OFF and check the wiring on the selector switch and for an open conductor.

e. Operate the M1 motor start push-button switch. M1 motor starter coil should be energized. If not, start checking the circuit line from 2–1, again using the 150 V range ac voltmeter. From this point on, there can be two methods used.

    a. **Series checking.** With this method, one prod is placed on point 2 and the other progressively moves from 5–4–3–1. At the point where voltage is indicated, the open circuit is in the first component to the right. For example, with the motor start push-button switch held operated, there is voltage at 3 but not at 4; the open, then is in the M1 stop push-button switch.

    b. **Half-Split checking.** With this method, one prod is placed on point 2

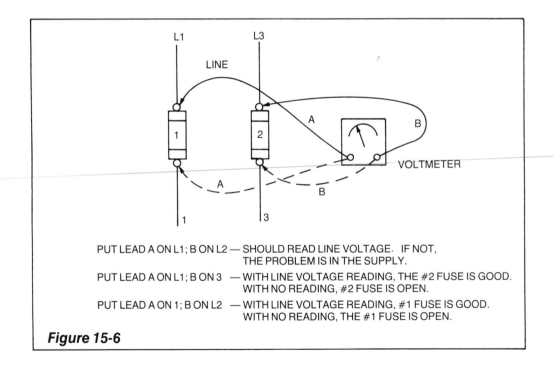

PUT LEAD A ON L1; B ON L2 — SHOULD READ LINE VOLTAGE. IF NOT, THE PROBLEM IS IN THE SUPPLY.

PUT LEAD A ON L1; B ON 3 — WITH LINE VOLTAGE READING, THE #2 FUSE IS GOOD. WITH NO READING, #2 FUSE IS OPEN.

PUT LEAD A ON 1; B ON L2 — WITH LINE VOLTAGE READING, #1 FUSE IS GOOD. WITH NO READING, THE #1 FUSE IS OPEN.

Figure 15-6

and the other directly to 4. Thus, in the problem indicated above, only one additional move with the second prod is required to locate the problem. While in this particular case it does not seem to have an advantage, there are cases where there are many components in series. In these cases, considerable time can be saved with the half-split method.

 f. Turn the M1-OFF-M1&M2 selector switch to M1 & M2. Relay 4CR should be energized, closing relay contacts 4CR-2 and 4CR-3. Operate the M1 motor start push-button switch and the M2 motor start push-button switch. M1 motor starter coil and M2 motor starter coil should be energized. With one or both of the motor starter coils not energized, follow the same general procedure in checking as indicated for M1 only.

  Apart from the possibility of an open motor starter coil about the only other problem that may come up in the motor starter circuits is the possibility that one of the overload relay contacts in series with the motor starter coil may be defective and open or a jumper may be missing on the starter.

2a. **Heat Control.** Operate the heat OFF-ON selector switch to ON. With the temperature of the machine operating fluid above 70°F, neither relay coil 5CR or contactor coil 10CR will be energized.

  With the temperature below 70°F, temperature switch NC contact will be closed. This allows relay coil 5CR to be energized. Relay contact 5CR-1 closes, energizing contactor coil 10CR. Contactor contacts 10CR-1, 10CR-2, and 10CR-3 close, energizing the heating elements.

  If the contactor 10CR is not energized:

a. Check to see if relay coil 5CR is energized.

b. Check the voltage from 2–15. If voltage here, the coil is open or there is an open conductor.

c. No voltage at 2–15 but voltage at 2–13. Check the temperature switch. It may be defective or incorrectly wired.

d. No voltage at 2–13, the heat OFF-ON switch may be incorrectly wired or there may be an open conductor.

3a. **Cycle Control.** The same general pattern is followed in checking the cycle control as was followed with the motor and heat control. For example, with the M1 motor starter control energized, relay 3CR is energized, closing the NO 3CR-2 contact. Operate the cycle start push-button switch.

a. Control relay 1CR does not energize.

  Using the series checking method explained, place one voltmeter prod on circuit point 2. Then, using the other prod, proceed consecutively through points 21, 20, 19, 18, 17 and 16. At the point that voltage is detected, the open circuit is in the component immediately to the right of this point. Remember that the cycle start push-button switch must be held operated during these checks.

An example of a problem might be if voltage was noted at point 18 but not at 19. There can be three possibilities:

a. Operating cam not holding the switch in the operated condition.

b. Limit switch incorrectly wired.

c. Open conductor.

Using the half-split method, you would probably start with the second prod at point 18. In this case it would have required only one move to have located the problem area. This cuts down on the time to troubleshoot.

The problem of open conductors has been mentioned several times. Remember that push-button switches, selector switches, and pilot lights are generally located on an operator's panel. Relays, contactors and motor starters in a control cabinet. Limit switches, pressure switches and temperature switches can be mounted remotely on the machine. This means several terminal block connections. Be sure these connections are kept tight.

15.5 MOTOR PROBLEMS

A few items are added in this chapter on basic motor problems.

15.5.1 Motor problems can generally be attributed to approximately four items: dirt, moisture, vibration and friction.

15.5.1A Dirt

An accumulation of dirt (metallic filings, chemical dust, lint and so on) can blanket the motor windings and prevent proper heat radiation. Periodically the motor should be blown out with dry compressed air at a pressure of 30–50 psi.

15.5.1B Moisture

Moisture combines with dirt to produce a sticky mess. Moisture also absorbs alkali and acid fumes. Leaking oil can do serious damage to a motor. Be careful not to grease bearings more than is required.

15.5.1C Vibration

With excessive vibration, parts can be shaken loose and electrical connections broken. Check frequently for any misalignment. Check for loose bearings. Check for loose mounting bolts.

15.5.1D Friction

Follow the lubrication instructions of the manufacturer and use the right kind of oil for lubrication. With ball bearings, be sure that the inner race is tight enough to ride with the shaft, but not so tight as to cause distortion. Use the right grease and be sure the grease is clean.

ACHIEVEMENT REVIEW

1. Explain two different methods of troubleshooting an electrical circuit.

2. Draw a sketch which explains how to check fuses.

3. What types of contacts may cause trouble and yet appear to be in good condition?

4. Why should you never file a contact?

5. What is a *combination problem*, as applied to machine control?

6. Explain two methods of detecting grounds in an electrical circuit.

7. Explain how an open in a common line can cause trouble.

8. List a few locations where you may be able to find grounds that have been created in wiring a machine.

9. What are some of the electrical problems found on machines which could have been avoided by good housekeeping practices?

10. What condition may lead to a low-voltage problem in an industrial shop?

11. What is the value of a good preventive maintenance program?

12. In the circuit shown in Figure 15-5, the electrician has incorrectly wired the NO and NC limit switch contacts. NO contact is wired to 22-23 and the NC contact is wired to 18-19. With this change, can a cycle be started? If not, why not?

13. In the circuit shown in Figure 15-5, the NC contact on relay 1CR is found in circuit line
   a. 14.
   b. 15.
   c. 16.

14. In the circuit shown in Figure 15-5, with the M1-OFF-M1 and M2 selector switch set to OFF, what motors can be started?
   a. M1
   b. M2
   c. Neither motor can be started.

CHAPTER 16

DESIGNING CONTROL SYSTEMS FOR EASY MAINTENANCE

OBJECTIVES

After studying this chapter, the student will be able to:

- List several important points in the assembly of a machine using electrical control.

- Show how to cross-reference relay contacts with circuit line numbers.

- List several items that can be of help in making better drawings for electrical control.

- Draw a typical control panel showing where components such as relays, motor starters, and disconnecting means should be placed.

- List the advantages and disadvantages of returning all circuit connections to the panel.

- Discuss several concerns with the use of mechanical limit switches.

- Explain the advantages of having all electrical components accessible.

- Name several places where indicating lights can help the troubleshooter.

16.1 DESIGN CONSIDERATIONS

In its final form, a design generally involves some compromise. Considerations such as safety, cost, manufacturing, assembly, and operating conditions enter into the final machine.

Maintenance problems are considered in the design through items such as the following:

Consider economics in building a competitive product.

Recognize the environment where the machine is to be used.

Consider the problem of installing electrical components after other mechanical and fluid power work has been completed.

Recognize vibration, shock, or other mechanical motion that may be present or that can develop to affect the electrical components.

Present adequate wiring diagrams, layouts, and instructions.

Display evidence of experience in the electrical control field.

Avoid sloppy workmanship.

All of these considerations, and others, show that careful planning of design is a prerequisite to ease of maintenance.

In the assembly of a machine, several points are important:

Circuit wiring should be exactly as called for on the electrical prints. In case of an error on the prints, corrections should be made before the machine is checked out.

All conductors and terminals should be properly marked with numbers corresponding to the wiring diagram.

A sufficient number of terminal block checkpoints should be provided.

Doublecheck all terminal connections to be sure the conductors are tight in the connectors.

Care should be taken in pulling conductors through conduit and fittings so that no conductor insulation is cut or scraped.

Conduit fill, as set up in electrical standards, should not be exceeded.

In long or difficult-access conduits, three or four spare conductors should be pulled.

In the field installation, a greatly improved overall job can be obtained by close cooperation between the user and the builder.

The user must know:

Total power requirements with the amount of possible low power factor load (induction motor)

Foundation and assembly with relative location of separate control enclosure, if used

Location of main power connection

Location of terminal disconnect points on the machine (where machine is disassembled for shipment)

The builder must know:

User's power voltage, phase and frequency
Any special conditions or limitations on power supply
Unusual ambient temperature condition
Unusual atmosphere condition
User's specification of manufacturer of components (if any)
User's specification on motor starters, if not full voltage
A complete and accurate sequence of functions the machine is to follow. This item is probably the most difficult to obtain. It is generally due to the following:

- Inexperience, particularly in the case of a new process

- General lack of knowledge

- Problems of communication; This is generally best solved by a cooperative effort on the part of the user and builder.

In all cases, a well-studied, careful analysis and procedure should be followed in selecting, applying and installing the electrical control. This always pays off in reduced maintenance.

In supplying some practical help as a guide to designing for easy maintenance, there are two major areas to examine:

1. Diagrams and layouts
2. Locating, assembly, and installing of components

16.2 DIAGRAMS AND LAYOUTS

In making original elementary diagrams and layouts, there are aids to make them more useful and cut maintenance and troubleshooting time.

16.2.1   The first example deals with the numbering of elementary circuit lines and cross-referencing the relays and their contacts. Normally closed contacts are so indicated by a bar under the line number. See Figure 16-1. Note that the line numbers are enclosed in a geometric figure to prevent mistaking the line numbers for circuit numbers.

16.2.2   All contacts and the conductors connected to them should be properly numbered. Numbering should carry throughout the entire electrical system.

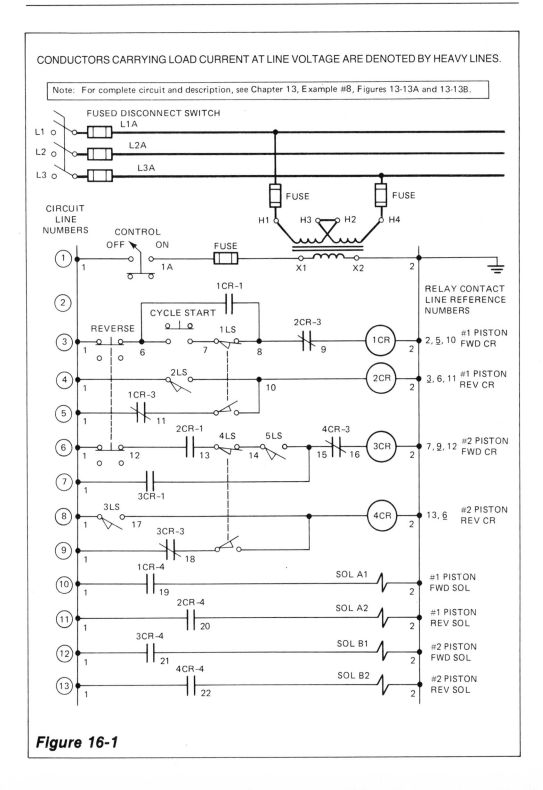

Figure 16-1

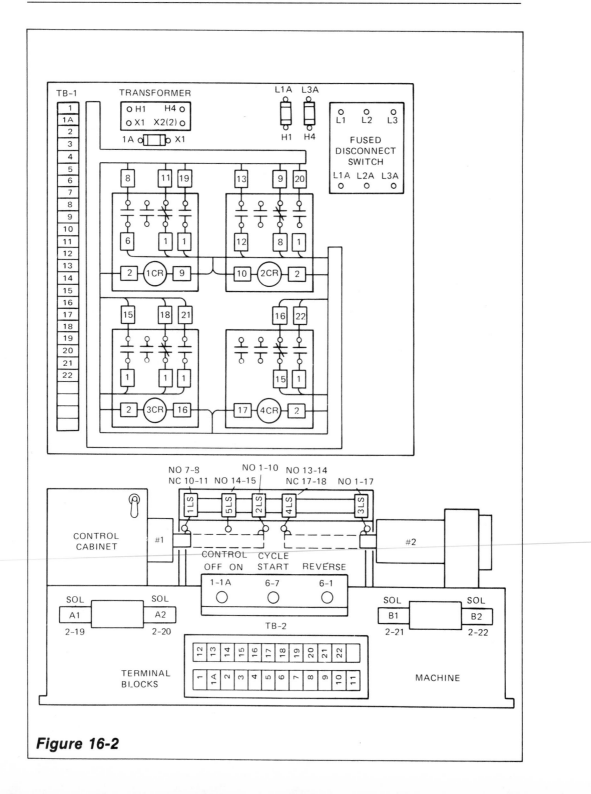

Figure 16-2

This may involve going through one or more terminal blocks. The incoming and outgoing conductors as well as the terminal blocks should carry the proper electrical circuit numbers. If at all possible, connections to all electrical components should be taken back to one common checkpoint.

All electrical elements on a machine should be correctly identified with the same marking as shown on the wiring diagram. For example, if a given solenoid is marked "solenoid A1" on the drawing, the actual solenoid on the machine should carry the same marking, "solenoid A1."

A drawing should be made showing the relative location of each electrical component on the machine. The drawing need not be a scale drawing. However, it should be reasonably accurate in showing the location of parts relative to each other and in relative size.

For example, if solenoid A1 is located on the left-hand end of the machine base, it should be so shown on the machine electrical layout. Figure 16-2 illustrates these points.

Figure 16-3 illustrates the advantages of displaying the identification and approximate location of various control components on a machine. In this specific case all inputs and outputs are indicated.

16.2.3 It is important to list components on the sheet with the elementary diagram and/or layout. They should be cross-indexed in some manner with the components as they appear in the circuit diagram or layout.

The component should be described so that the user can obtain a replacement if necessary. An example is shown in Figure 16-4. Only basic information is used to describe the components and to cross-reference them to the circuit.

16.2.4 Within any organization, aid can be given for maintenance or troubleshooting through standardization of circuit numbers and component numbers of designation.

For example, in a motor starter circuit, the numbers 1, 2, 3, and 4 can always be used in the same location. After some experience, the maintenance worker or troubleshooter can remember that the coil is always 4,1; the interlock and START button is 3,4; and the STOP button is always 2,3.

It is understandable that this arrangement cannot always be used completely. However, with a group of machines that are similar in design, this pattern of standard circuit numbers can be followed to a great degree.

It follows that components can carry a similar standardization. For example, solenoid S1 may be a clamp forward solenoid. A limit switch 1LS may be a clamp forward stop.

In a group of similar machines, it helps the maintenance personnel if the same designations for specific electrical components can be maintained on all the machines. The worker soon becomes acquainted with the functions of specific

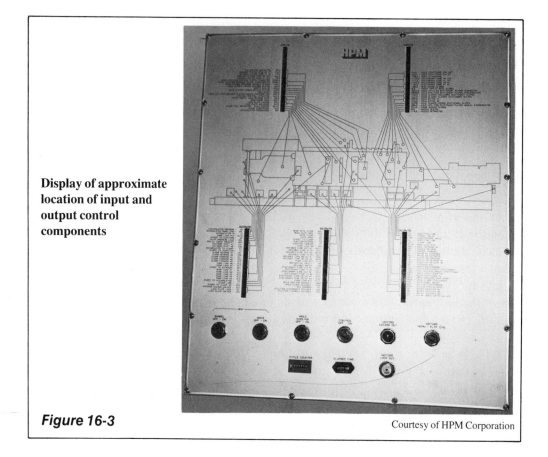

Display of approximate location of input and output control components

Figure 16-3

Courtesy of HPM Corporation

components, such as solenoids and limit switches, regardless of the machine involved.

Another help is to advance circuit numbers for similar components in the ten's place. For example, if more than one motor starter were used, #1 motor starter would carry circuit numbers 1, 2, 3, and 4. The #2 motor starter would carry circuit numbers 21, 22, 23, and 24. The #3 motor starter would carry circuits 31, 32, 33, and 34. The numbers should mean more than just numbers. For example, with the proper numbers in the one's place and the ten's place, the troubleshooter can immediately spot the circuit numbers of 32 and 33 as the STOP button on the #3 starter.

16.2.5   In an elementary circuit diagram carrying more than one solenoid, each solenoid should carry a reasonable description of its function. This will help maintenance or service personnel to quickly locate a possible source of trouble without referring to the fluid power circuit. The description will aid in quickly

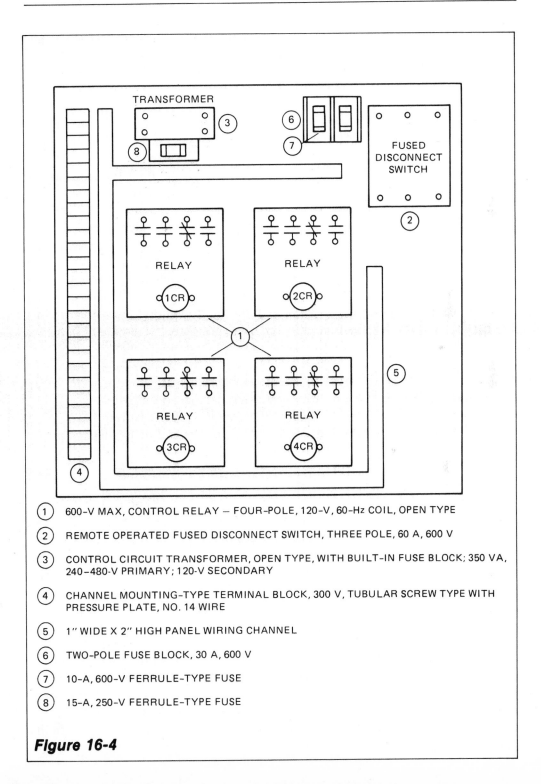

1. 600-V MAX, CONTROL RELAY — FOUR-POLE, 120-V, 60-Hz COIL, OPEN TYPE

2. REMOTE OPERATED FUSED DISCONNECT SWITCH, THREE POLE, 60 A, 600 V

3. CONTROL CIRCUIT TRANSFORMER, OPEN TYPE, WITH BUILT-IN FUSE BLOCK; 350 VA, 240–480-V PRIMARY; 120-V SECONDARY

4. CHANNEL MOUNTING-TYPE TERMINAL BLOCK, 300 V, TUBULAR SCREW TYPE WITH PRESSURE PLATE, NO. 14 WIRE

5. 1" WIDE X 2" HIGH PANEL WIRING CHANNEL

6. TWO-POLE FUSE BLOCK, 30 A, 600 V

7. 10-A, 600-V FERRULE-TYPE FUSE

8. 15-A, 250-V FERRULE-TYPE FUSE

Figure 16-4

locating the solenoid on the machine. This help can also carry to the relays, where a specific relay controls the initial energizing of a solenoid.

Referring again to Figure 16-1, a circuit is shown with several solenoids, indicating not only their function but also picturing their relative location.

16.2.6   An item that is almost a must with complex circuits is some form of sequence of operations. This may be written out in detail, listing component by component each step in the operation of the circuit.

A shorter method that gives nearly the same information is the sequence bar chart. (Several examples are shown in Chapter 13.)

16.2.7   The size of the drawing is important. This can be understood first in the problem of storing the drawings. If every size and shape were allowed, the task of systematic and protective filing of drawings could be tremendous.

Page sizes of 8 1/2″ x 11″ or 9″ x 12″ and multiples thereof are generally accepted. The maximum size is usually 36″ x 48″.

The drawing size can also be a problem for the troubleshooter or maintenance personnel. If the drawing is too large, it is unwieldy to handle at the machine. If it is too small, it is hard to read the schematic.

16.2.8   In the mechanics of drawing the elementary diagram, there are a few suggestions to "clean up" the drawing and make it easier to follow:

    A. Evenly space circuit lines. Approximately one-half inch is satisfactory in most cases.

    B. Space component symbols on the line so they are clear and so that space is left for the circuit numbers.

    C. It is not necessary to carry all circuit lines to their completion. For example, take the case of the line connection on a push-to-test pilot light. An arrow leading from the line connection on the light and indicating the proper circuit number termination of the arrow is sufficient. (See Chapter 13, Example #3, Figure 13-8A.)

    D. When the contacts of limit switches or selector switches are separated in a circuit drawing, there may be a problem for the maintenance personnel reading the drawing. There are three things that can be done to help when the circuit is drawn.

        1. Arrange the operating switches in the circuit so that the contacts are grouped in close proximity to each other. (See Figure 13-13A.)

        2. If the NO and NC limit switch contacts can be located in the circuit drawing so there are no other lines or contacts between, the two symbols can be joined by a broken line. Do not use the broken line

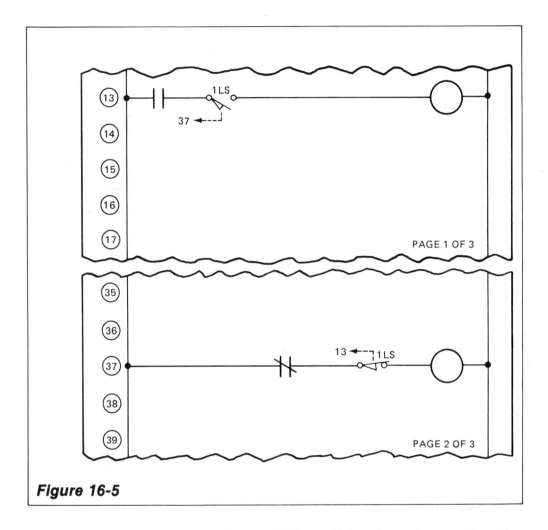

Figure 16-5

and cross several other circuit lines. This only tends to confuse the reader. Figure 16-1 is an example of the proper use of the broken line connection.

3. A cross-reference can be made between limit switch contact symbols by using an arrow and line number to "tie" the two contacts. See Figure 16-5. This arrangement can be used for widely spaced limit switch contact symbols on the same page, or where two contacts of the same limit switch appear on different pages of the circuit diagram.

16.2.9 To increase their utility, many machines are designed with optional features. These may not always be required by the user on the initial job. However, to make it easy to add the option at a later date in the field, provision can be

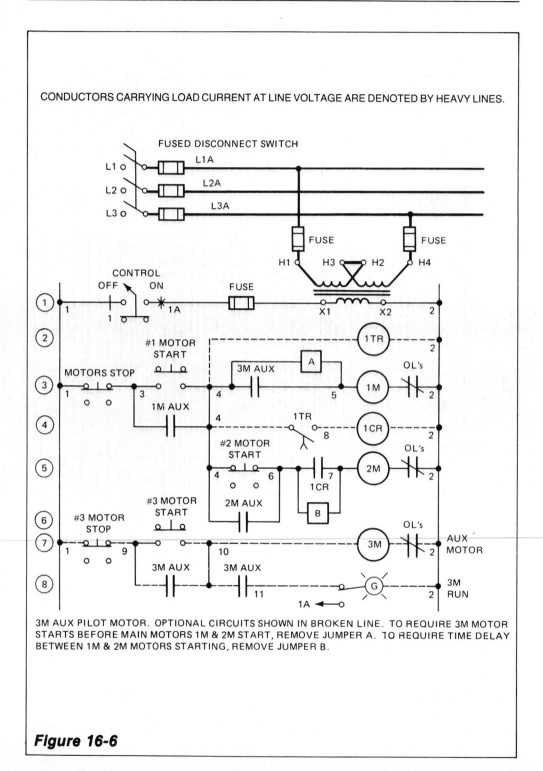

CONDUCTORS CARRYING LOAD CURRENT AT LINE VOLTAGE ARE DENOTED BY HEAVY LINES.

3M AUX PILOT MOTOR. OPTIONAL CIRCUITS SHOWN IN BROKEN LINE. TO REQUIRE 3M MOTOR STARTS BEFORE MAIN MOTORS 1M & 2M START, REMOVE JUMPER A. TO REQUIRE TIME DELAY BETWEEN 1M & 2M MOTORS STARTING, REMOVE JUMPER B.

Figure 16-6

made in the circuit to show its function. The conductors are brought to a terminal block and given circuit numbers. The opening in the circuit that is to receive the optional feature is then jumpered. Appropriate notes instruct the user on adding the option. The auxiliary circuit may be shown in broken line if desired.

This feature is also helpful when the user has equipment to work in conjunction with the machine. For example, this could be a safety device, material feeder, or material removal equipment.

The circuit in Figure 16-6 shows an example of this feature.

16.3 LOCATING, ASSEMBLING, AND INSTALLING COMPONENTS

16.3.1 After the control circuit is designed, one of the first areas to examine is the control panel, Figure 16-7. While more actual physical work is probably done on the machine, the control cabinet is generally the key to locating trouble.

Maintenance work can be reduced by starting with a systematic and standard arrangement of components. (In Chapter 14, Figure 14-5 shows the layout of a typical control panel with explanatory notes on requirements for good design.)

16.3.2 The wiring duct and plastic ties for cabling have come into universal use in recent years. Both save time and expense in the original wiring of the panel and machine. The greatest saving for maintenance is in checking a circuit. It is time consuming to search out and check a particular conductor that may be at the center of a group of 40 or 50 conductors, all tightly laced. After locating the conductor and changing or correcting the trouble, the cable is usually altered.

Wiring duct is available in various widths and heights. It can be obtained with slots or holes in the sides or open slots to the top. Recent designs by relay manufacturers use a top strip only, fitted between the relay and supported by center posts. This arrangement further reduces component space requirements.

Where wiring duct is not applicable, plastic ties can be used. They can be installed either by a tool or by hand. If it is necessary to get into the cable, the plastic ties can be quickly cut and new ones applied later in a short time.

Figure 16-8 shows various types and sizes of wiring duct. Figure 16-9 shows the use of wiring duct and plastic ties.

16.3.3 The use of terminal blocks for electrical control on a machine is important. They can be of tremendous aid in checking circuits, thus reducing maintenance time. They are valuable when machines must be disassembled for shipment

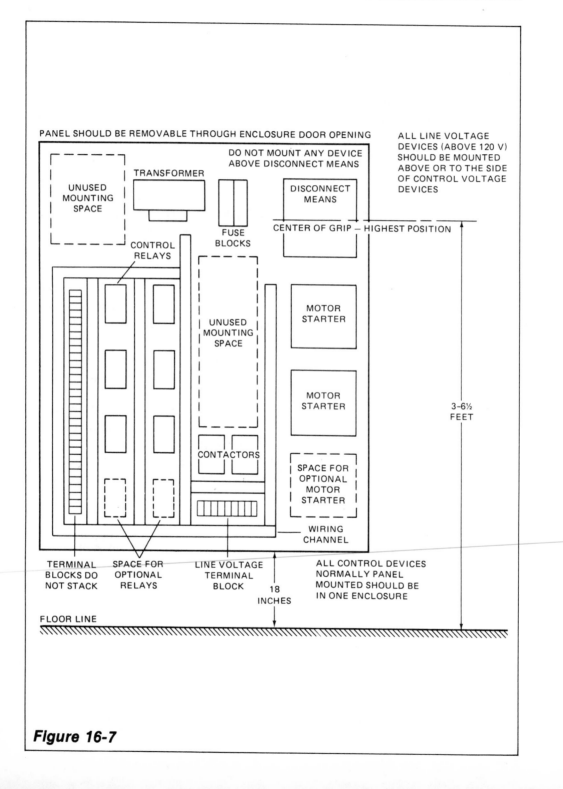

Figure 16-7

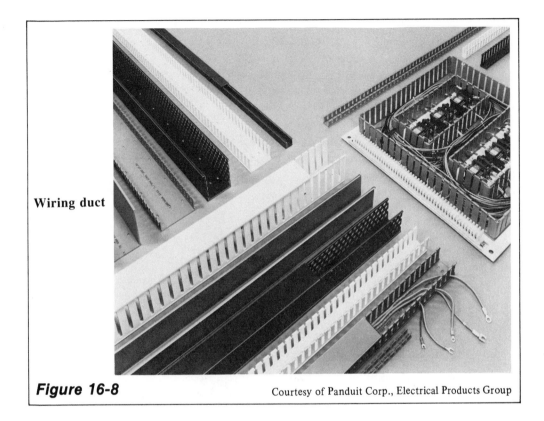

Wiring duct

Figure 16-8   Courtesy of Panduit Corp., Electrical Products Group

and reassembled in the user's plant. Here the terminal block on each part containing electrical equipment is a must.

Figure 16-10 shows a variety of terminal blocks available for industrial use. Figure 16-11 shows the use of terminal blocks on a machine.

16.3.4   Provide a central point for checking by bringing all connections between components back to the control panel. In some cases this may add expense in the initial wiring of the machine. It also increases the amount of conductor used, and the size of conduit increases. Additional terminal blocks and more wiring time are required.

These disadvantages must be carefully weighed against the maintenance time saved in the user's plant. A much faster job of troubleshooting can be done if all connections are available at one point.

For example, look at the circuit in Figure 16-12. Limit switches 1LS, 2LS, and 3LS and solenoid A are probably on the machine. Normally it would only be necessary to bring points 1, 2, and 3 back to the control panel. This is because the source of power and relay 1CR is available here. To provide checkpoints in

Use of wiring duct and plastic ties

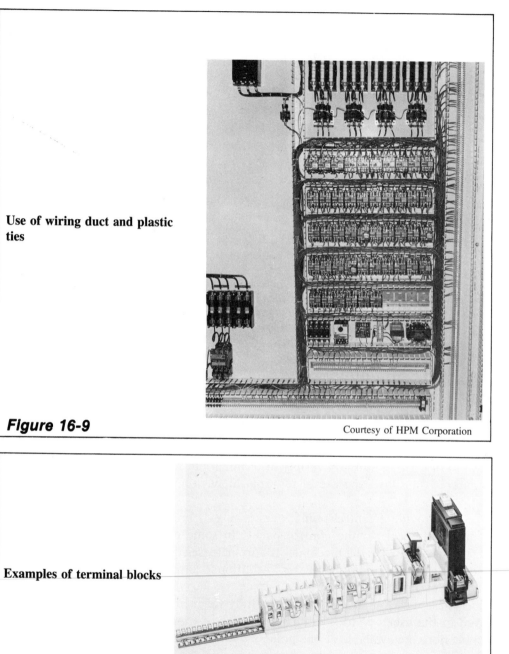

Figure 16-9

Courtesy of HPM Corporation

Examples of terminal blocks

Figure 16-10

Courtesy of Allen-Bradley Co.

Use of terminal blocks

Figure 16-11

Courtesy of HPM Corporation

the control panel for the entire circuit, points 4, 5, and 6 could also be brought back to a terminal strip in the control cabinet.

16.3.5 Many designers find it economical to use either fuses or overload heating elements in the solenoid circuit. They find that it requires less time and is less expensive to replace a fuse or reset an overload than to replace a solenoid coil. In this case the solenoid coil connection would be brought back to the control panel to pick up this protection.

An example of this is similar to Figure 16-12, except that a protective device is added, Figure 16-13. The fuse selected in this case should have a short time lag to override the momentary inrush current but should open on sustained overload.

16.3.6 Limit Switch Problems . Experience has shown that a high percentage of limit switch problems results from their application, not from basic design of the unit.

Much of the material in this section is taken from experience. The section is separated into three divisions: mechanical, vane, and proximity switches.

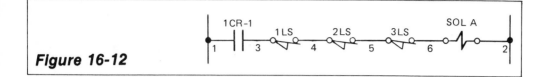

Figure 16-12

A. Mechanical
1. Environment

Limit switches are usually located on the machine. Therefore, they are subjected to the working area environment. Flying liquids, metal chips and debris, corrosive fumes, excessive temperatures, and shock or vibration are some of the problems.

Generally, a temperature limit of approximately 175°F should not be exceeded. In cases of higher temperatures, thermally insulated barriers can be of some help.

If heavy shock or vibration is present in the machine, the electrical circuit may be interrupted by the contacts opening. Under these conditions, a low control mass and high contact pressure are important. The switch should be installed so that the direction of shock forces is in a different plane from the plane in which the contacts operate.

Very few of the housings and/or integral parts of a limit switch will stand up for long in the presence of corrosive fumes. It is helpful to seal the switch at the conduit end, or use a small, light-duty, hermetically sealed switch with a relay.

The problem of flying chips, liquids, or other debris can sometimes be handled by relocation, change of switch operators, or addition of mechanical guards. If the problem is liquids entering the switch through the conduit, change the design of the conduit, or use a multiconductor synthetic rubber cable with a compression seal.

2. Actuation

One of the most common problems is the design of the actuating cam. Such problems as impact forces, direction of travel, speed of the moving part, frequency of repeat operation, and overtravel are always present.

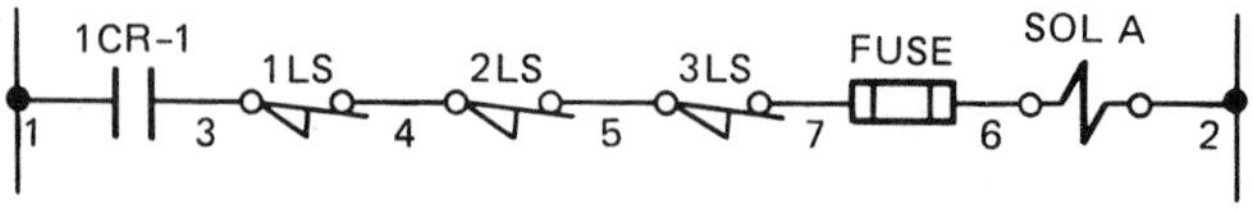

Figure 16-13

Excessive impact from improperly designed actuating systems is the leading cause of premature failure of the electromechanical limit switch. At slow speed, impact is rarely troublesome, but as speed increases, impact applied to the switch becomes critical. In today's higher speed machines, therefore, it is important to give proper consideration to correctly designed actuating systems. The following recommendations are designed to assist you in obtaining greater life from limit switches.

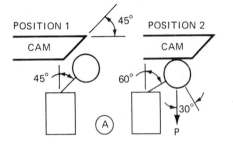

Switch levers should be positioned as nearly parallel with leading edges of cams as possible. Cam A is satisfactory for speeds up to 50 r/min; cam B for speeds up to 200 r/min (nonuniform acceleration of switch lever); cam C for speeds to 400 r/min (uniform or other controlled acceleration).

Cam D is designed with a trailing edge so it can override the switch lever and then return. Cam actuates switch on return also. Lever angle is very important in applications of this type.

The black sector in the roller indicates recommended design limits of angle of pressure P. Pressure applied by actuating mechanism to switch operating lever should approximate direction of lever rotation with a variation not to exceed 30° angle of pressure. Changes drastically with rotation of lever. Cam must be designed for proper pressure angles of all positions of lever travel.

Information courtesy of R.B. Denison Manufacturing Co.

Limit switch cam design

Figure 16-14

Industrial application of a limit switch with circuit conductor termination enclosure

Figure 16-15

Courtesy of HPM Corporation

For example, the operating frequency should not exceed 2 times per second. The impact speed should not exceed 400 feet per minute. The minimum designed overtravel should be approximately one-third of that required to actuate the switch. The maximum designed overtravel should not exceed 10 degrees. The angle of the cam engaging the switch actuator should not produce forces that tend to deflect the actuator.

Figure 16-14 shows examples of various cam designs as they are applied to limit switch operation. Solutions to many of the above problems are illustrated.

Figure 16-15 shows an actual application of limit switches on a machine. A conductor termination enclosure is also shown. This enclosure carries terminal blocks for the various electrical components located in that area.

3. Contacts

The electrical contacts on limit switches are very similar to contacts on other electrical components. They will not hold up under excessive overloads. Even at normal loads, excessive speed of operation tends to create local heating in the contact, which can be damaging. Excessive voltage and high inductive loads will also contribute to contact failure.

A few of the factors that affect sensing and therefore the output switching action are as follows:

1. The speed and direction of the material that is to be sensed
2. The size, shape, and material that is to be sensed
3. Sensitivity adjustment is required to properly use the type and amount of energy received by the sensing head.

In general, the use of the proximity switch requires greater overall knowledge of electrical work. Due to its relatively complex design as compared to the mechanical or vane switches, more complex maintenance problems can be expected. Maintenance problems generally lie with the amplifier and/or switching unit. Unless there is a complete familiarity with these units, it is better practice to change the complete unit and allow the manufacturer to make the repairs. Plug-in elements make this arrangement relatively simple.

Opposite polarities should not be connected to the contacts of a limit switch unless the switch is designed for such service. Likewise, power from different sources should not be connected to the contacts of a single switch. Normally, if good circuit design is followed, this does not become a problem.

In many cases the use of double-pole switches can replace two single-pole switches. This may eliminate a relay that would otherwise be required to provide the extra circuit.

B. Vane Switches

Many of the problems noted for mechanical limit switches can be avoided with the vane limit switch. For example, all integral parts are encapsulated within a cast-aluminum enclosure. This prevents the entrance of dust, dirt, coolants, oil, and lubricants into the switch mechanism. **Caution**: Misuse can lead to maintenance problems.

The following are two factors to consider when using the vane switch.

1. Excessive contact load

The contacts on the vane switch are rated at 0.75 A make and 0.2 A continuous. This is at 115 V, 60 Hz. It can be seen that this is considerably below the contact rating of the heavy-duty mechanical limit switch. The normal procedure is to use a relay in conjunction with the vane switch to increase its output capacity.

2. Vane design

The operation of the vane switch is accomplished by passing a mild steel operating vane through the vane slot. The proper size and shape of the vane, and the direction and accuracy of the vane entrance into the slot become design problems for each specific application to give maximum operating efficiency.

Figure 16-16 shows various designs of vane switch operators.

C. Proximity switches

Most of the information noted for the vane switch can be applied to the proximity switch. In particular, the proximity switch has the advantage of operating in adverse environments.

Under some conditions, the requirements for a separate output unit for switching action may be a disadvantage. One example is supplying space for the amplifier and switching units where many switches are required.

16.3.7 Accessibility of Components   This should include the components in the control cabinets, push-button station, and on the machine. Any maintenance

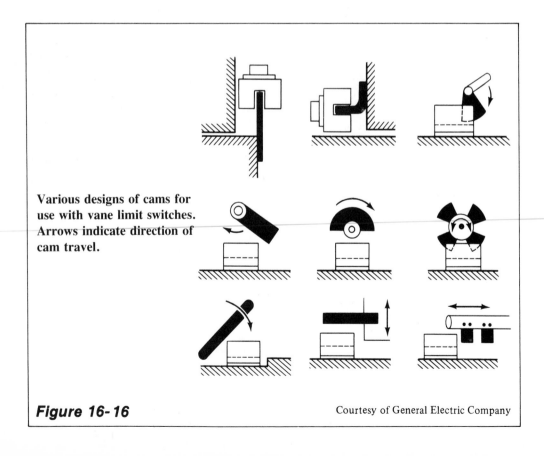

Various designs of cams for use with vane limit switches. Arrows indicate direction of cam travel.

Figure 16-16   Courtesy of General Electric Company

problem can be made easier by accessibility.

Problems with accessibility may arise from inadequate planning in the original design stage, or by lack of cooperation between mechanical and electrical assembly and installation.

For the maintenance worker, the first problem is to locate the trouble. Then the faulty component and/or conductors must be repaired or replaced.

The second problem is time. Downtime is a loss both in production and labor. It therefore follows that the downtime is directly proportional to the time it takes to locate the trouble and repair or replace faulty parts. This is when accessibility of components is important for easy maintenance.

For the components in the three groups mentioned (the control cabinets, push-button station, and machine), a few suggestions follow.

A. Control Cabinets

1. Keep the components within an easy and safe range of working height. This is approximately 2 feet to 5 1/2 feet above the floor.
2. Design the control panels so they can be removed through the cabinet door opening.
3. Do not stack terminal blocks one on top of the other.
4. Space components and terminal blocks so that the terminals are accessible to add or remove conductors.
5. Use relays with such features as building block assembly for time delay, mechanical latch, and additional standard relay contacts. The changing of NO and NC contacts or the universal contact arrangement are useful.

B. Push-button Stations

1. Space the individual units so that the wiring to the contact block is accessible.
2. Where multiple units are used, it may be desirable to hinge the mounting plate to the enclosure so the units are accessible for checking.

C. The Machine

1. Motors should be located so the mounting means are accessible and the motor can be easily removed. The conduit box should be clear so the cover can be removed.
2. In the installation of pressure switches, shut-off valves should be provided in the fluid power lines for the removal of the switch unit.
3. Temperature switches installed in a reservoir for temperature indication or control should be easily removed without draining the reservoir.
4. Solenoids on valves should be clear of connecting tubes, pipes, conduits, or other mechanical devices. The electrical connections to the

solenoids should be accessible for disconnecting. Their connection should be of a bolt and nut type, and insulated with approved tape. The bolts or screws securing the solenoid to the valve should be accessible for complete removal of the solenoid from the valve.

Solenoid valves should not be installed immersed in an oil tank. This does not refer to the oil-immersed solenoids now used on many valves.

5. Planning for conduit and/or wireways and pull boxes within the bed or main structure of a machine always pays off. It clears, from the outside of the machine, a multitude of conduits that might otherwise interfere with component installation. Also, it provides for access to the conduits through pull boxes for easier adding or pulling of conductors within the machine.

16.3.8 Numbering and Identification of Components The use of the layout drawing is of help in locating the area for components installed on a machine. A further help is to identify these units on the machine with a permanent nameplate. The nameplate should carry the identification of the unit exactly as noted on the layout and elementary diagram.

If at all possible, the nameplate should not be attached directly to the component. It should be attached to one side and on a permanent base. The reason is that if it is on the component being removed, the nameplate may be lost or may not be changed to the new unit installed.

16.3.9 Use of Indicating Lights It is quite possible that too many indicating lights could be used on a machine. However, this is generally the exception. More often, the problem is the absence of indicating lights which give valuable information.

The number of lights is usually in direct proportion to the size and complexity of the control system. For example, a machine with ten motors may have ten motor run indicating lights, as compared to one light with one motor.

When a machine goes through a multiple of sequences, it is valuable to know that each preceding sequence has been completed and the next sequence started.

The machine may have several types of operations available. This may be set up through selector or drum switches. In such cases a light should indicate which type of operation is being used.

Safety and indicating lights are closely tied. For example, with multiple operators, each operator should have a visual indication of the action that the other operators have taken. Other examples of where indicating lights are valuable are: oil temperature, over or under safe operating conditions; safety guards that may have been removed; excessive loading; overtravel; and malfunction of a component in a critical part of a cycle.

16.3.10 Use of Plug-in Components In recent years the design of plug-in components has involved most of the component field. Relays, limit switches, solenoids, timers, and pyrometers are examples.

The use of a plug-in component provides many advantages. However, it does not solve all problems. The following points should be considered:

- Does experience indicate a high rate of failure in a particular application? Can a different component give a better life? Can operating conditions be improved or is replacement the best solution?

- Will downtime for changing the component seriously affect production?

- Does the nature of the component (complexity of circuit) indicate a change rather than repair on the machine?

- Can repair work be accomplished at a bench easier and faster than on a machine?

- Is the machine a prototype or custom design that may require many changes before a final design is reached?

- Is the level of maintenance experience at a given user's shop such that disconnecting and reconnecting circuits on a component may become a problem?

- Are changes being considered by the designer or user for an additional or different type of control after a short time or use?

ACHIEVEMENT REVIEW

1. What is the importance of the use of circuit numbers? Do you believe a pattern of numbers as used on circuits between various machines in a plant would help? Explain your answer.

2. List a few items that would help to clean up an electrical circuit diagram and make it easier to read.

3. Sketch a typical control panel showing the location of components, terminal blocks, wiring duct, and the clear unused space. Use your own judgment as to what you think would be good design.

4. List the advantages and disadvantages of bringing all conductors from limit switches, solenoids, and oil-tight units back to a control panel for checking purposes.

5. In using mechanically operated limit switches, what are the limitations on impact speed, overtravel, and contact rating?

6. Are there contact limitations when using vane limit switches? Would you recommend the use of a vane switch in a dusty atmosphere?

7. List five suggestions to follow when mounting and installing electrical components in a control cabinet or push-button station.

8. What are some advantages to be gained through the use of indicating lights?

9. What factors would you consider in deciding on the use of plug-in components?

10. The generally accepted size of drawings is
    a. 8½" x 11".
    b. 10" x 20". } And multiples thereof
    c. 9" x 12".

11. The use of terminal blocks for electrical control on a machine is important because they
    a. aid in checking circuits.
    b. are always required when pilot lights are used.
    c. are used for changing voltage levels.

12. When mounting components on a panel in a control cabinet, they should be located within a range of _____ feet above the floor.
    a. 1–3
    b. 2–5½
    c. 1–6

CHAPTER 17

POWER FACTOR CORRECTION

OBJECTIVES

After studying this chapter, the student will be able to:

- Explain how apparent power is obtained.
- Express the relationship between actual or useful power and apparent power.
- Discuss the differences among kilowatts, kilovolt-amperes, and reactive kilovolt-amperes.
- Draw a power triangle.
- Draw a curve showing the relationship between load current and power factor for an induction motor.
- Discuss how the addition of a capacitor improves power factor.
- Show how a capacitor can be connected to a motor circuit to help in correcting system power factor.
- State a rule of thumb for determining which size of capacitor to use on a motor with a specified horsepower rating.
- List some of the design features found in a capacitor.
- Explain why the power company is interested in high power factor in the user's plant.

17.1 APPARENT POWER AND ACTUAL POWER

Consideration of power factor is essential in manufacturing plants which use relatively large amounts of alternating-current power. This is generally due to the use of a large number of three-phase, squirrel-cage induction motors that are lightly loaded. In this chapter, the problem of power factor is explained.

Power factor is a term used to describe the ratio between apparent power and actual, or useful, power. These two power components will be present when squirrel-cage induction motors are used.

To understand apparent power, assume that a coil is wound on a magnetic core and connected to a source of direct current. A switch is provided to close and open the circuit, Figure 17-1. When switch A is closed, a current flows through the coil. This sets up a magnetic field around the coil, Figure 17-2.

The magnetic field remains as long as the switch is closed and current flows in the coil. If the switch is opened, the magnetic field collapses, causing the lines of force to cut through the winding. This action causes a current to flow through the coil in the direction opposite to the current that flowed into the coil.

Now, assume that in place of using a direct-current source of power, the coil is connected to a source of alternating-current power, Figure 17-3.

With alternating current, the current flows in one direction, dies to zero, and then flows in the opposite direction, Figure 17-4. Each time the current goes to zero, the lines of force collapse and cut through the winding. The current that flows out of the coil builds to a maximum value just as the supply current drops to zero. The current that flows out of the coil is said to be *out of phase* with the supply current.

If there is no loss in the coil due to heating, the current flowing into the coil (supply current) and the current flowing out of the coil due to the collapsing field will have the same value. If an ammeter is placed in the circuit, it will indicate a current. The amount of current will depend upon the design of the coil.

It would appear that the coil (electromagnet) is consuming power. This, however, is not the case. The current is simply flowing back and forth. A rough mechanical analogy can be made by considering a coil spring being compressed and released. It is from this action of the current flowing back and forth that the term *apparent power* is obtained.

Actual, or *useful*, *power* is that power used in a heating element such as an electric iron or light bulb. Actual or useful power is also present in the induction motor when the load is applied on the shaft.

When a squirrel-cage induction motor is connected to a source of alternating current, a current flows in the windings of the stator. This current produces a revolving magnetic field. The lines of force from this magnetic field cut through the squirrel-cage rotor and set up a second magnetic field. The second magnetic field reacts against the field of the stator, forcing the rotor to rotate.

With no load connected to the motor, all the actual or useful power required is that amount which is necessary to overcome the iron and copper losses. There

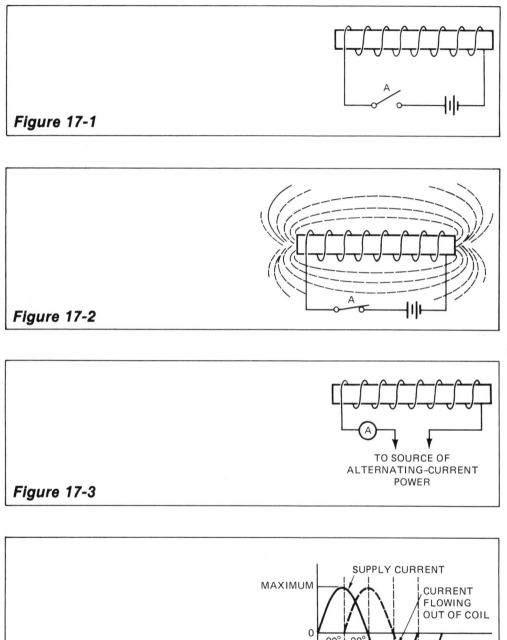

Figure 17-1

Figure 17-2

Figure 17-3

TO SOURCE OF
ALTERNATING-CURRENT
POWER

SUPPLY CURRENT

MAXIMUM

CURRENT
FLOWING
OUT OF COIL

0

90°   90°

MAXIMUM

Figure 17-4

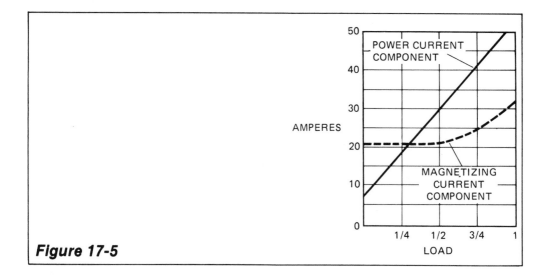

Figure 17-5

are also friction and windage losses caused by the rotor turning. Therefore, most of the power used by a motor running idle is apparent power.

When a load is applied to the motor, such as connecting it to a pump, real power is supplied. This real power increases directly with the increase of load.

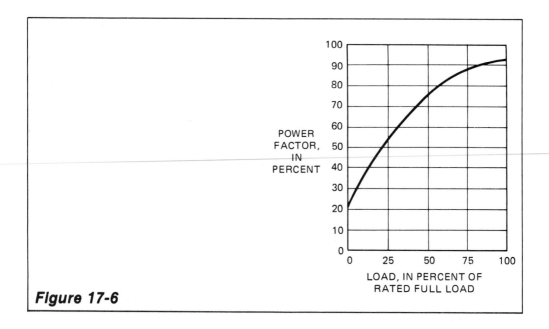

Figure 17-6

Thus, there are two components of current flowing in the motor: *magnetizing current* and *power current,* Figure 17-5.

The ratio between useful power and apparent power can be expressed as

$$\text{Power Factor} = \frac{\text{Useful Power}}{\text{Apparent Power}}$$

A typical curve of a squirrel-cage induction motor is shown in Figure 17-6. This illustration shows that there is a low power factor when the motor is running at light loads. Therefore, the motor requires a large amount of apparent power as compared to the amount of useful power.

17.2 MAGNETIZING CURRENT AND POWER CURRENT

Our explanation of apparent power started by describing current in a coil. The two current components, magnetizing and power, are present in the induction motor. However, they cannot be added directly, because the magnetizing current is 90° out of phase with the power current. The magnetizing current is said to *lag* the power current by 90°.

Magnetizing current and power current can be added by using a triangle, Figure 17-7. By knowing the amount of magnetizing current and power current, a triangle can be drawn to scale. By scaling the drawing, the total current can be determined.

When working with power, it is generally more convenient to work with the

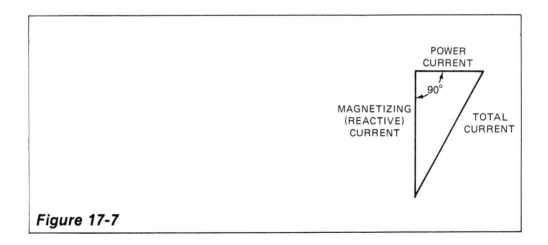

Figure 17-7

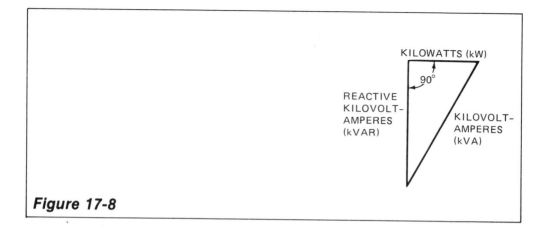

Figure 17-8

units *kilowatts* (kW), *kilovolt-amperes* (kVA), and *reactive kilovolt-amperes* (kVAR). Since the prefix *kilo* means 1000,

- 1 kilowatt = 1000 watts (watts = volts × amperes)

- kilovolt-amperes = $\dfrac{\text{amperes} \times \text{volts}}{1000}$

- reactive kilovolt-amperes = $\dfrac{\text{reactive amperes} \times \text{volts}}{1000}$

The triangle can now be called a *power triangle,* as shown in Figure 17-8.

For example, using values of current in amperes and power in kilowatts, assume the following. A three-phase, 480-V squirrel-cage induction motor is carrying a line current of 121 A in each phase. The kilowatt load, measured with a wattmeter, is 80 kW. The kVA of a three-phase circuit is

$$kVA = \frac{A \times V \times \sqrt{3}*}{1000}$$

*The factor of $\sqrt{3}$, or 1.73, is used because of the three-phase circuit.

Therefore,

$$kVA = \frac{121 \times 480 \times 1.73}{1000}$$
$$kVA = 100$$

The horizontal line of the power triangle is 80 kW. The power triangle is drawn to scale in Figure 17-9.

As shown in Figure 17-9, the extreme points of the horizontal line are marked A and B, and a vertical line is drawn down from A. Using point B, an arc is struck

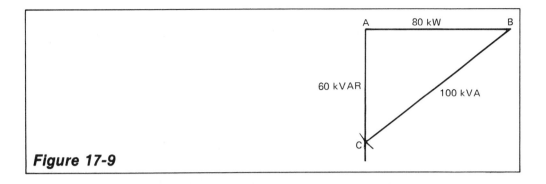

Figure 17-9

with a radius equal to the 100 kVA. The intersection of this arc with the vertical line is marked C. The reactive kVA can now be determined by measuring distance AC. The power factor may be expressed as the ratio of kilowatts, or working power, to the total kilovolt-amperes, or apparent power.

$$\text{Power Factor} = \frac{AB}{BC} = \frac{kW}{kVA} = \frac{80}{100} = 80\%$$

The objective is to improve the power factor. Therefore, the magnetizing current, or reactive kVA, must be reduced in proportion to the power current, or kilowatts. To do this, a capacitor can be connected across the three lines feeding the motor. The capacitor will supply magnetizing current.

Using the example in Figure 17-9, assume that a 15-kVAR capacitor is connected across the motor feed lines. As the capacitor supplies a leading power

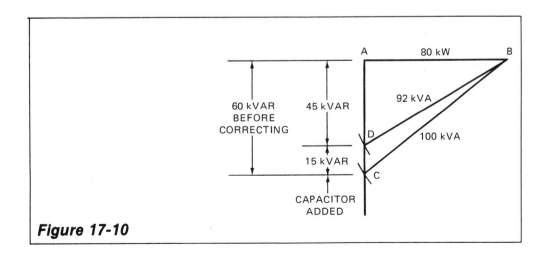

Figure 17-10

factor current, the reactive kVA of the capacitor is subtracted from the reactive kVA now being supplied to the motor, Figure 17-10.

As represented by BD, the new kVA is 92. The power factor (PF) is now 87%, as follows:

$$PF = \frac{kW}{kVA} = \frac{80}{92} = 87\%$$

17.3 DETERMINING THE AMOUNT OF CORRECTION REQUIRED

A reasonable approximation of the capacitor size can be obtained from the following "rule of thumb." For a 1200-r/min motor, use a figure of 1 kVAR for each 5 hp of induction motor load. This figure will vary slightly with motor speed and the amount of power factor correction required.

A more exact method is to use a table such as the one shown in Figure 17-11. This table gives the values of capacitors to correct power factor to approximately 95%.

There are many areas in an electrical system where low power factor can present a problem. The concern can start with the alternator in the power company's station and continue through power distribution lines and transformers (both step up at the power station and step down at the user's plant). The distribution conductors in the user's plant are also involved.

This is of primary concern to the power company, as it must supply the magnetizing current that goes throughout the system. Therefore, it is to the power company's advantage if the power factor can be kept as high as practical (90–95%). As a result, most power companies include a penalty clause in their rate structure for low power factor and/or a discount for high power factor.

A high power factor also provides an advantage for the user. For any given distribution system in a user's plant, increased machine load can be added by improving the power factor, without increasing the size of the power plant feeders and associated equipment.

In the example shown in Figure 17-10, a capacitor is used to correct power factor. This is one of the most simple, direct, and relatively inexpensive methods. This method of correction can be made on the incoming lines or at individual motors. Remember, the correction from the capacitor always occurs toward the source of power.

If correction is desired only for the purpose of gaining rate advantage, one or more capacitors can be added on the incoming lines. A more satisfactory method

HORSEPOWER	MOTOR SYNCHRONOUS SPEED, r/min	CAPACITOR RATING CORRECT POWER FACTOR TO APPROXIMATELY 95%* kVAR
20	3600	3.0
	1800	4.5
	1200	5.0
	900	8.0
30	3600	4.0
	1800	5.0
	1200	7.5
	900	8.5
40	3600	5.0
	1800	5.5
	1200	9.0
	900	14.0
50	3600	6.5
	1800	7.0
	1200	10.0
	900	14.5
60	3600	7.5
	1800	10.0
	1200	12.0
	900	17.5
75	3600	9.5
	1800	10.5
	1200	16.5
	900	18.0
100	3600	13.0
	1800	15.0
	1200	17.5
	900	22.0
150	3600	20.0
	1800	22.5
	1200	36.0
	900	45.0

*When these exact sizes are not commercially available, use the next smaller size.

Figure 17-11

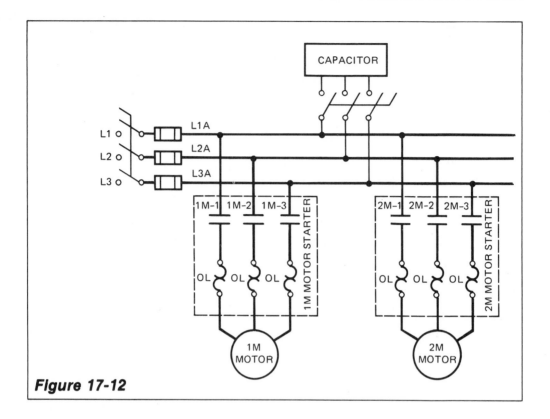

Figure 17-12

for improving the user's load power factor is to connect individual capacitors on the load terminals of each motor starter. The optimum condition in installation economy for any given user's plant, however, may be the use of both methods. It is not unusual to find that savings in the user's power bill pay for the capacitors in one or two years.

Figures 17-12 and 17-13 show two installation circuits using capacitors.

Generally, it is not economical to apply capacitors directly to motors below 20 hp. If a plant has a predominance of small motors (below 20 hp), connecting capacitors on the line bus is practical, Figure 17-12. However, this type of application can become difficult. It is suggested that the local power company be consulted on specific applications.

A check can be made to determine if too large a capacitor has been added on a motor. A voltmeter is attached across any two terminals of the motor (load side of starter). The load is disconnected by deenergizing the motor starter. The resultant rise in voltage as observed on the meter should not exceed 150% of the motor nameplate rated voltage.

The most practical location for the installation of a capacitor is on the outside of the motor starter enclosure. However, with long leads (more than 10

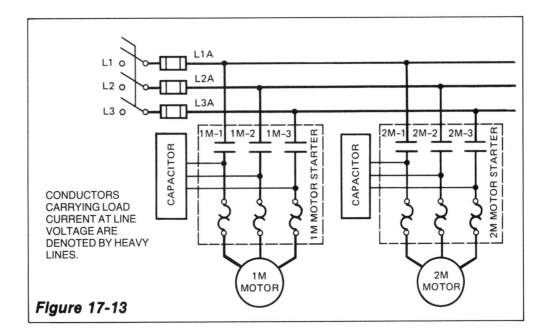

Figure 17-13

feet) from the starter to the motor, the capacitor should be mounted near the motor and the connections made in the motor terminal box.

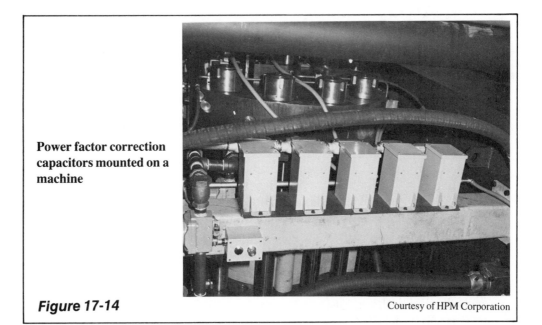

Power factor correction capacitors mounted on a machine

Figure 17-14

Courtesy of HPM Corporation

Figure 17-14 illustrates the use of capacitors mounted on a machine. The connections are made in the motor terminal box.

17.4 TYPICAL CAPACITOR DESIGN FEATURES

Figure 17-15 shows a typical capacitor as used for power factor correction. The following design features are listed in reference to the number designations on the photograph.

1. CASE – The capacitor case shall be made from heavy gauge mild steel with all joints welded and reinforced at point of wear for protection of working element during handling, shipment and service. Capacitor to be either single phase or three phase internally delta connected.

2. GLAND SEALED BUSHING – The gland sealed bushing assembly shall be constructed such that the silicon rubber thick walled gland is to be compressed between the porcelain insulator, copper stud and the top cover of the capacitor. This compression seal shall prevent leakage of the Wemcol fluid even if the bushing assembly is tilted 15 degrees, which occurs during installation of capacitor elements.

3. WORKING ELEMENT – The working element shall consist of individually wound sections of special capacitor grade paper, low loss capacitor grade synthetic film and dead soft aluminum foil. Multiple sections to be arranged to yield the specified KVAC at rated voltage. The grouped sections shall be insulated from the steel case by means of high purity Kraft insulation material.

4. FUSES – Each three phase capacitor unit shall be protected by two type CLN current limiting fuses. Each single phase capacitor shall be protected by one CLN fuse. Fuses to be full range current limiting type with an interrupting capacity of 200,000 amps. The fuse shall have a (1/2 x 13) male thread for customer connection.

5. DUST COVER – A dust cover shall be provided consisting of a bushing enclosure and cover. The bushing enclosure shall be formed from 12 gauge mild steel with a neoprene gasket positioned between the bottom of the enclosure and the top of the capacitor unit. A bolt-on steel cover shall compress a vinyl gasket to assure a dustproof and weatherproof seal between the the top cover and the bushing enclosure. The dust cover shall be bolted to the top of the capacitor unit by means of weld nuts and slotted head studs enclosing the bushings, fuses and connections.

6. DISCHARGE RESISTORS – Individual internal discharge resistors shall be

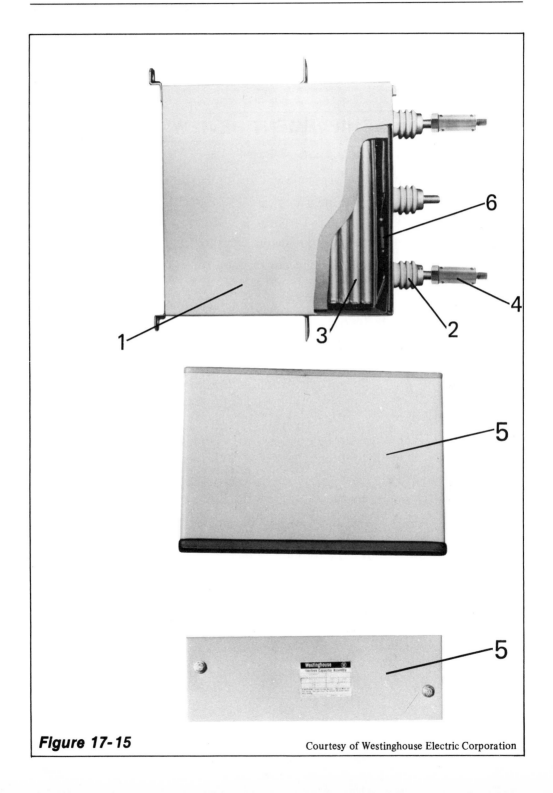

Figure 17-15

Courtesy of Westinghouse Electric Corporation

connected across the terminals to reduce the residual voltage to 50 volts or less within one minute after removal from the energized circuit.

ACHIEVEMENT REVIEW

1. Explain how apparent power is present in a squirrel-cage induction motor.

2. Draw a power triangle using a kilowatt load of 100 and kVA load of 150. Determine the reactive kVA (kVAR).

3. Calculate the power factor of the load in Question #2.

4. If a 40-kVAR capacitor is connected on the load in Question #2, what will be the new power factor?

5. If a 80-kVAR capacitor is connected on the load in Question #2, what will be the new power factor?

6. Why are power companies concerned about their customers' plant power factor?

7. Using a 30-hp, 1200-r/min induction motor, approximately how much reactive kVA would you use to correct the power factor to 95%? You may use a rule of thumb.

8. How would you check to determine if too large a capacitor had been connected to an induction motor?

9. Is there any advantage to the user of electric power in improving power factor, apart from receiving a better rate from the power company? Explain.

10. The major load in a user's plant consists of ten 100-hp and ten 30-hp squirrel-cage induction motors. Where would be the most satisfactory place to connect capacitors for power factor improvement?

11. Magnetizing current and power current can be added
    a. directly.
    b. directly, if the current is multiplied by the voltage.
    c. using a triangular relationship.

12. A practical power factor level for an industrial plant to maintain is
    a. 70–80%.
    b. 80–90%.
    c. 90–95%.

CHAPTER 18

INTRODUCTION TO SOLID-STATE CONTROL

OBJECTIVES

After studying this chapter, the student will be able to:

- Discuss the difference between discrete components and the integrated circuit.
- Explain why logic gates provide a function in a circuit similar to the function provided by switch contacts.
- Draw the symbols for the following logic gates: AND; OR; NOT or INVERTER; NAND; NOR; and EXCLUSIVE-OR.
- Show how the AND gate and the NOT gate are combined to make the NAND gate, and show how the OR gate and the NOT gate are combined to make the NOR gate.
- Using symbols, explain how a NAND gate can be used to obtain the basic AND, OR, and NOT functions.
- Using symbols, explain how a NOR gate can be used to obtain the basic AND, OR, and NOT functions.
- Describe the meaning of the bubble or circle at the output of the NOT or INVERTER symbol.
- Discuss the difference between an OR gate and an EXCLUSIVE-OR gate.
- Draw the truth tables for the basic logic functions.
- List some of the problems that may be encountered when using a low-voltage direct current on standard input components.
- Show how a signal is latched, and explain the reason for latching.
- Follow signals through a solid-state circuit, from the input signal to the output signal.

18.1 INTRODUCTION

In the early chapters of this book we explain the use of relays, contactors, time-delay relays, timers, and motor starters in the decision-making or logic section of machine control. This type of control is generally referred to as *electromechanical control.*

Another form of control, *solid-state control,* does not make the use of electromechanical control obsolete. For example, in many cases involving a small, simple control circuit, electromechanical control performs satisfactorily and is more economical. Also, when using solid-state control, sometimes the output device is too large to be energized directly from the solid state. Heavy solenoids or motor starters are examples of such devices. In these cases, a relay or contactor may be used as an interface between the solid-state control and the work device. Solid-state control, then, is preferable in applications where an adverse environment exists, complex circuits are required, or high-speed operation is used.

This chapter introduces the student to the knowledge and practical use of some basic solid-state elements. It is hoped that this material will interest the student in further study and use of solid-state control.

In the chapters on electromechanical control, the reader is not required to learn the design features of components such as relays, timers, and temperature controllers. In our practical applications, the important concern is that a contact must be open or closed at the proper time within a circuit to accomplish a required output.

The same approach is taken here. We present the practical application of solid-state control. The student is not required to know the design details of the various solid-state modules. The important goal is to obtain the required outputs from the use of sensing element inputs at the proper time and in the proper sequence.

In Chapter 13, Figure 13-1 shows the control broken down into three sections. This explanation is repeated in Figure 18-1 for solid-state control. Our purpose is to emphasize that where control is changed from electromechanical to solid state, only the logic (decision-making) section is affected in most cases. The input (information) section and the output (work) section remain substantially the same. In a few cases contactors and motor starters shift from the logic section to the output section. This generally depends on their size and function.

The decision-making or logic section using solid state is far more versatile than can be obtained with relays.

A *solid-state* device is one which incorporates various components made of solid material. These components exhibit specific electronic characteristics.

INFORMATION OR INPUT	DECISION-MAKING OR LOGIC	WORK OR OUTPUT
Push buttons	Relays	Solenoids
Limit Switches	Reed Relays	Heating Elements
Proximity Switches		
Pressure Switches	Magnetic Amplifiers	Motors
Temperature Switches	Solid-state Elements	Contactors
Tape Readers		Indicating Lights
Pulse Generators		Motor Starters
Photo Cells		

Figure 18-1

Logic gate circuits are made using discrete combinations of diodes, transistors and resistors.

The basic function of a *transistor* is that of a switch which is capable of accepting and combining multiple command signals. The resulting signal is then directed to produce an output or work function. The early use of solid-state elements involved mounting the individual units (such as diodes, transistors, resistors, and capacitors) on a small board. The individual units were known as *discrete components*. The board circuit was then wired or soldered according to a specific circuit design. See Figure 18-2.

More recent developments in industry have introduced the *integrated circuit*. An integrated circuit (IC) is a single monolithic chip of semiconductor material in which the electronic circuit elements are fabricated.

For example, a typical integrated circuit might contain two zener diodes, five diodes, eleven transistors and five resistors combined on a single silicon chip. The size of the chip would only measure about .05 inch. Thus, the use of integrated circuits has given the electronics field a great space savings. Many such integrated circuits would be found in an application shown in Figure 18-3.

Regardless of whether discrete components or integrated circuits are used, several different functions can be developed. Each function can be represented by a symbol. Some of the basic symbols and their use in designing a control circuit are discussed in the following section. These symbols are generally used in the design of solid-state control circuits.

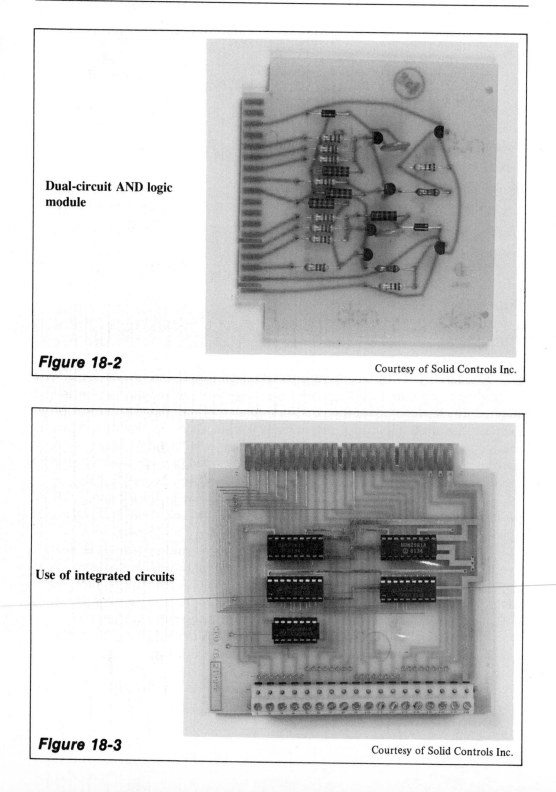

Dual-circuit AND logic module

Figure 18-2

Courtesy of Solid Controls Inc.

Use of integrated circuits

Figure 18-3

Courtesy of Solid Controls Inc.

18.2 SYMBOLS (IEEE STD. 91-1973 — ANSI Y32, 14-1973; DISTINCTIVE SHAPE USED)

Contacts are either open or closed in electromechanical control. They have only two conditions. In solid-state control there also are only two conditions or states. These conditions deal with two voltage levels: HIGH or LOW.

- TRUE or FALSE
- ON or OFF
- CLOSED or OPEN
- 1 or 0.

In the last century, a form of algebra was developed by an Englishman named George Boole. It was used at that time as a way of expressing statements that could be either true or false. This form of reasoning is called Boolean algebra.

With the advent of solid-state control where conditions called for a two-valued logic concept, i.e., TRUE-FALSE, ON-OFF, 1-0, etc., Boolean algebra fit in as an easy way to analyze and express logic statements.

In the material that follows, the system used is known as *positive logic*. In this system, two binary digits (1 or 0) are used to represent the two conditions. The digit 1 represents the higher voltage. The digit 0 represents the lower voltage.

Referring again to electromechanical control, a normally open contact on a relay closes when the relay coil is energized. The relay contact opens when the relay coil is deenergized. That is, the relay contacts have only two conditions; they are either open or closed. In a somewhat similar manner, when solid-state logic functions switch from 1 to 0, or from 0 to 1, they have only two conditions. Therefore, in this way they may be considered as operating in a similar manner to switch contacts. They are known as *logic gates* or *gates*. They also may be referred to as Boolean gates.

The basic operators of Boolean algebra, as they are related to digital logic functions, are the AND, OR and NOT. The AND gate is composed of two or more inputs and one output. Figure 18-4A shows the AND gate logic symbol. (This figure also illustrates the electromechanical equivalent circuit.) The AB notation on the output of the AND symbol means *A and B*, not A times B. There must be a HIGH (1) signal at A and at B in order to obtain a HIGH (1) signal at the output AB.

The AND gate is composed of two or more inputs and one output. Figure 18-4A shows the AND gate logic symbol. (This figure also illustrates the electromechanical equivalent circuit.) The AB notation on the output of the AND symbol means *A and B*, not A times B. There must be a HIGH (1) signal at A and at B in order to obtain a HIGH (1) signal at the output AB.

A *truth table* displays the various conditions under which an output is one or

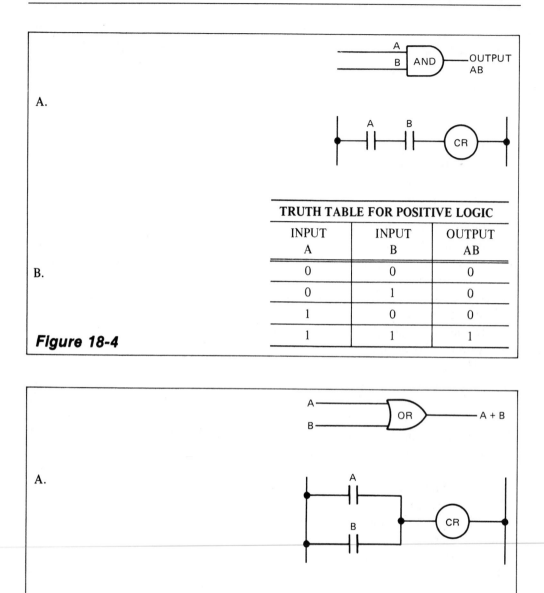

A.

B.

TRUTH TABLE FOR POSITIVE LOGIC

INPUT A	INPUT B	OUTPUT AB
0	0	0
0	1	0
1	0	0
1	1	1

Figure 18-4

A.

B.

TRUTH TABLE FOR POSITIVE LOGIC

INPUT A	INPUT B	OUTPUT A + B
0	0	0
1	0	1
0	1	1
1	1	1

Figure 18-5

the other of two values. The truth table for the AND function is shown in Figure 18-4B. In this truth table it can be seen that there is a 1 output only when there is a 1 input at A and a 1 input at B.

The OR gate logic symbol is shown in Figure 18-5A. (Figure 18-5A also shows the electromechanical equivalent circuit.) The OR gate is composed of two or more inputs and a single output. The A + B notation on the output of the OR gate symbol means *A or B,* not A plus B. There must be a HIGH (1) signal at A or at B in order to obtain a HIGH (1) signal at the output A + B.

The truth table for a two-input OR gate is shown in Figure 18-5B. In this truth table it can be seen that there is a 1 output if there is a 1 input at A or a 1 input at B, or if there is a 1 input at both A and B.

A third symbol is the NOT or INVERTER gate. This gate has a single input and a single output. The *bubble,* or circle, at the output is the standard symbol used to represent inversion. Figure 18-6A shows the NOT or INVERTER gate logic symbol and its equivalent electromechanical circuit. The $\overline{A}$ (read *not A*) at the output indicates that the output is the opposite or complement of the input A. That is,

- If there is a HIGH (1) at the input A, there will be a LOW (0) at the output $\overline{A}$.

- If there is a LOW (0) at the input A, there will be a HIGH (1) at the output $\overline{A}$.

The truth table for the NOT or INVERTER gate symbol is shown in Figure 18-6B.

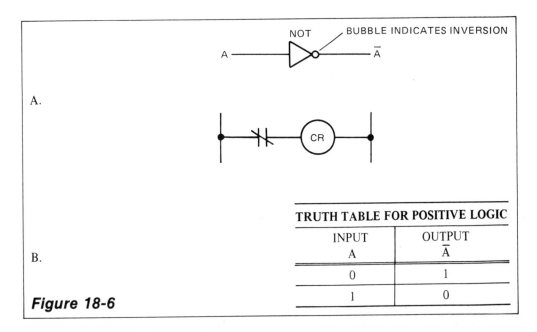

Figure 18-6

TRUTH TABLE FOR POSITIVE LOGIC

INPUT A	OUTPUT $\overline{A}$
0	1
1	0

The relationship between the logic symbol, logical statement and Boolean equation is shown in Figure 18-7.

There are two important gate combinations that are coming into more frequent use. These two combinations are the NAND gate and the NOR gate.

The NAND gate consists of an AND gate followed by a NOT or INVERTER gate. NAND gates may have more than two inputs, but have only a single output. shown in Figure 18-8A.

Figure 18-8B shows the symbol for the NAND gate. The symbol consists of an AND gate symbol followed by a small circle or bubble. Note that, as in the NOT or INVERTER symbol, the circle or bubble indicates inversion (the opposite). The term $\overline{AB}$ reads *not A* and *not B*, or *not AB*.

The truth table for the NAND gate is shown in Figure 18-8C. This table helps to explain the resulting outputs from various inputs. Notice that the output of the NAND gate is 1 until both inputs are 1.

One important feature of the NAND logic gate is that it can be used to produce any of the basic logic functions. Therefore, the NAND logic gate is generally referred to as a *universal gate*.

As shown in Figure 18-9, the AND gate can be obtained by using two NAND logic gates and joining the inputs to the second NAND gate.

In Figure 18-10, the OR gate is obtained by using three NAND gates. The inputs to the first two NAND gates are joined for a single input to each of the gates.

The NOT or INVERTER gate is obtained with a single NAND gate, Figure 18-11. The inputs to the NAND gate are joined for a single input.

The second logic gate combination is the NOR gate. The NOR gate may have more than two inputs, and a single output. The NOR gate consists of an OR gate followed by a NOT or INVERTER gate.

LOGIC SYMBOL	LOGICAL STATEMENT	BOOLEAN EQUATION
A ──┐ B ──┘ ⟩──Y	Y IS 1 IF A AND B ARE 1.	$Y = A \cdot B$ OR $Y = AB$
C ──┐ D ──┘ ⟩──Y	Y IS 1 IF C OR D IS 1.	$Y = C + D$
A ──▷○──Y	Y IS 1 IF A IS 0. Y IS 0 IF A IS 1.	$Y = \overline{A}$

Figure 18-7

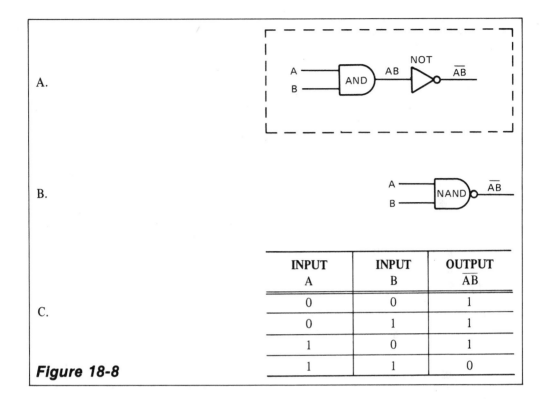

A.

B.

C.

INPUT A	INPUT B	OUTPUT $\overline{AB}$
0	0	1
0	1	1
1	0	1
1	1	0

Figure 18-8

The combination of the OR gate and the NOT or INVERTER gate is shown in Figure 18-12A.

Figure 18-12B shows the logic symbol for the NOR gate. The symbol consists of an OR gate symbol followed by a small circle or bubble. As in the NOT or INVERTER symbol, the circle or bubble indicates inversion (the opposite). The term $\overline{A + B}$ on the output reads *not A or B*.

The truth table for the NOR gate is shown in Figure 18-12C. Note that the output of the NOR gate is 1 only when there is a 0 input at both A and B. If there is a 1 at input A or a 1 at input B, or a 1 at both inputs A and B, the output is 0.

An important feature of the NOR logic gate is that it can be used to produce any of the basic logic functions. Therefore, like the NAND logic gate, the NOR logic gate is generally referred to as a *universal gate.*

Figure 18-13 shows that the function of an AND gate can be obtained by using three NOR gates. Two inputs are joined together so there is only a single input provided on each of two NOR gates (A – B). The outputs from these two NOR gates are connected to the input of a third NOR gate. The output (AB) is an AND function.

It can be seen in Figure 18-14 that the function of an OR gate can be obtained

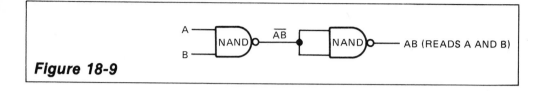

Figure 18-9

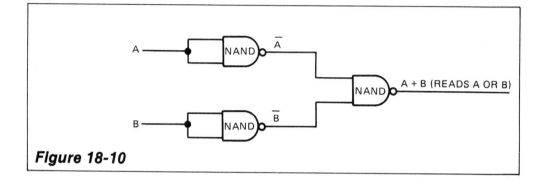

Figure 18-10

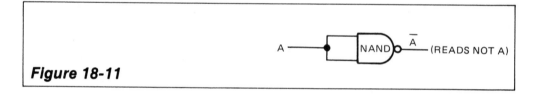

Figure 18-11

A.

B.

C.

INPUT A	INPUT B	OUTPUT $\overline{A + B}$
0	0	1
0	1	0
1	0	0
1	1	0

Figure 18-12

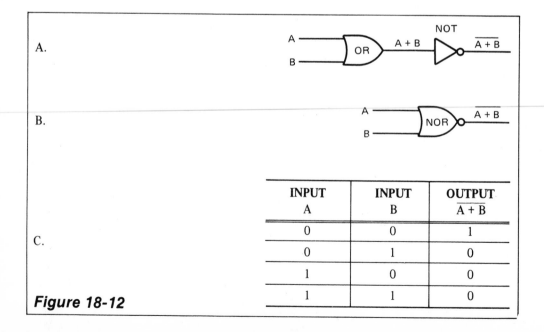

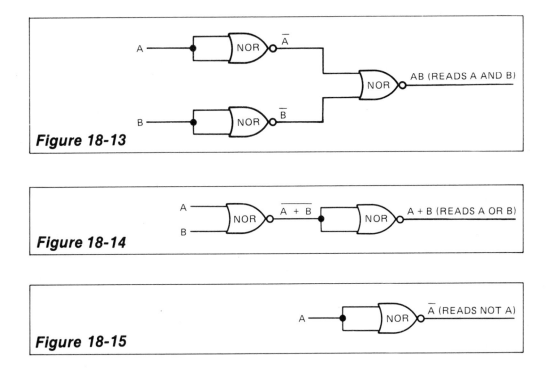

Figure 18-13

Figure 18-14

Figure 18-15

by using two NOR gates. The inputs to the second NOR gate are joined together so that only a single input, $\overline{A + B}$, is provided. The output from the second NOR gate is an OR function (A + B).

In Figure 18-15, the NOT or INVERTER gate is obtained with a single NOR gate by joining the inputs so that only a single input, A, is provided. The output is a NOT function ($\overline{A}$).

In our explanation of the OR logic gate, it is noted that when there is a HIGH or 1 at A or at B, there is an output A + B. It is also true that when there is a HIGH or 1 at A and a HIGH or 1 at B, there is an output A + B. Contrast this to the following explanation of the EXCLUSIVE-OR gate.

Another important logic gate is the EXCLUSIVE-OR gate. This gate produces an output when input A is HIGH or 1 or when input B is HIGH or 1, but not when both input A and input B are HIGH or 1.

The development of the EXCLUSIVE-OR logic gate is shown in Figure 18-16A. It is made up of two NOT or INVERTER gates, two AND gates, and one OR gate. Note that the output is written in a manner similar to that of the OR output, except that a circle is drawn around the + sign (+). This is usually read *not AB,* or *A not B.*

Figure 18-16B shows the symbol for the EXCLUSIVE-OR logic gate. It is

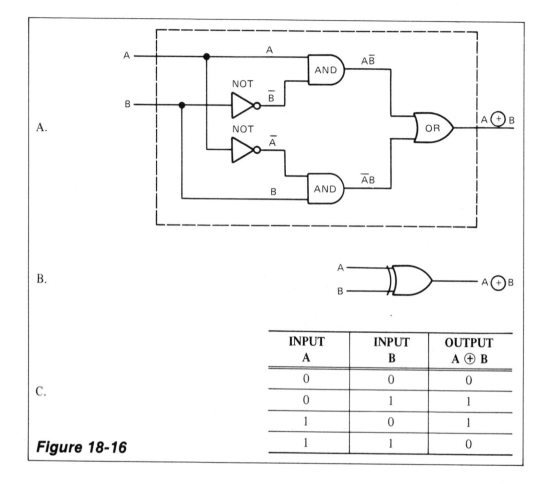

A.

B.

C.

Figure 18-16

INPUT A	INPUT B	OUTPUT A $\oplus$ B
0	0	0
0	1	1
1	0	1
1	1	0

similar to the OR symbol, except that an additional arc precedes the OR symbol.

The truth table for the EXCLUSIVE-OR logic gate functions is shown in Figure 18-16C. This table demonstrates that an output exists when there is an input at A or at B, but not at both A and B.

18.3 CIRCUIT APPLICATIONS

The control circuits using solid-state symbols in this section show the solid-state control circuit as well as a word description of the operation required and a ladder-type diagram using electromechanical components.

It is generally better to design the solid-state control circuit from a word description of the operation. Working directly from a ladder-type electromechanical control circuit may reveal redundancies in the relay circuit, but it is also important to understand the relay ladder-type control circuit. The knowledge and use of the ladder-type circuit becomes of even greater importance when dealing with programmable control.

Changes may be of concern with solid-state control as compared to the ladder-type electromechanical circuit in the area of input and output signals.

Problems are sometimes encountered when standard industrial components are used to provide input information to a logic system operating at a low direct-current voltage level. These components, such as push-button switches, limit switches, and pressure switches are generally designed to operate at 120 V alternating current. When used at a low dc voltage level, such factors as noise, dry switching, and contact bounce may enter into the control system to create problems.

Where these problems exist, 120 V ac can be used in the input or information section. An interface consisting of a signal converter module can be used to convert the input signal into signals compatible with the low voltage dc.

Figure 18-17 shows a typical relay equivalent circuit for a signal converter module.

In industrial plants, a voltage of 120 V ac is generally available on machines from an isolated secondary control transformer. Therefore, a power supply module is used to reduce and rectify the 120 V ac to a low-level dc required in the logic section.

Figure 18-18 shows a typical power supply module. A circuit showing this arrangement is illustrated in Figure 18-19.

In the output or work section of the solid-state control system, sufficient power must be available to operate components such as solenoids, contactors, and motor starters. Unless the power requirements are low, it is advisable to energize these components at 120 V ac.

The output interface (connection) to the power component may be a reed relay or an ac static switching unit. Many output components are inductive, so care must be taken that the interface device can handle a relatively high inrush current. In some cases, this current may reach 50 A.

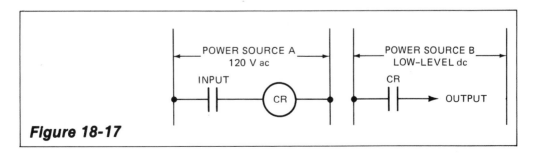

Figure 18-17

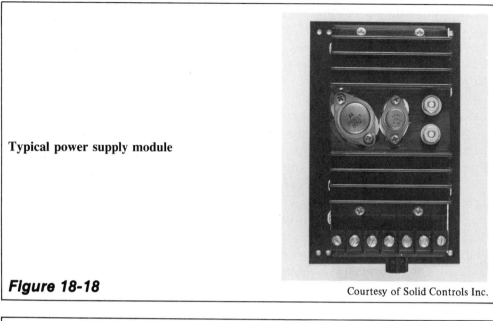

Typical power supply module

Figure 18-18

Courtesy of Solid Controls Inc.

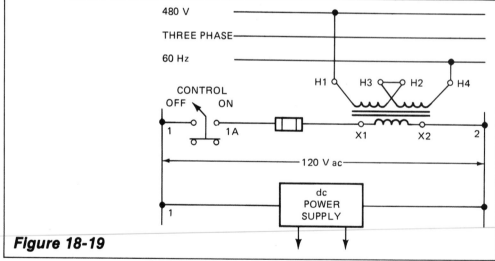

Figure 18-19

Figure 18-20 shows the symbol for an ac static switch. This module provides a solid-state means of switching ac power. It is made up of a bidirectional triode ac switch.

In the following circuits, a voltage of 120 V ac is available from the isolated secondary of a step-down control transformer. Low-voltage dc is available from a dc power supply, as shown.

Figure 18-20

Circuit #1

A cylinder and piston assembly is shown in Figure 18-21A. The piston is to move to the right until it engages and operates limit switch 1LS. At that point, the piston is to return to its start position. Provision must be made to return the piston from any point in its forward travel. A single-solenoid, spring-return operating valve is used to supply fluid power to the cylinder. Figure 18-21B shows the electromechanical circuit used to operate this system.

Figure 18-21C shows the solid-state circuit used to operate the system. The inputs in this circuit operate at a low-level dc. The outputs operate at 120 V ac. An interface module is shown in the output section of the circuit.

Up to this point in explaining the various logic module symbols, the names have been added to the symbols to help associate each symbol with its name. In the following circuits, the names are omitted; the symbol is identified only by its individual shape or form.

Note that when a low-level dc is given, the voltages generally range from 4 V to 12 V for most manufacturers of solid-state equipment. The input signals are obtained through an information component such as a push-button switch, limit switch, pressure switch, or temperature switch. The source is one side of the

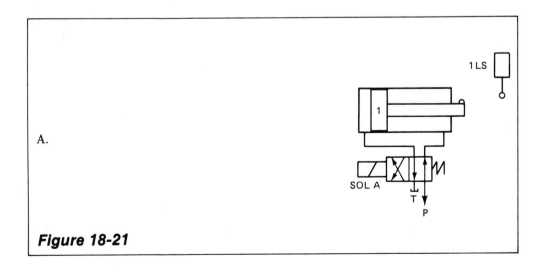

A.

Figure 18-21

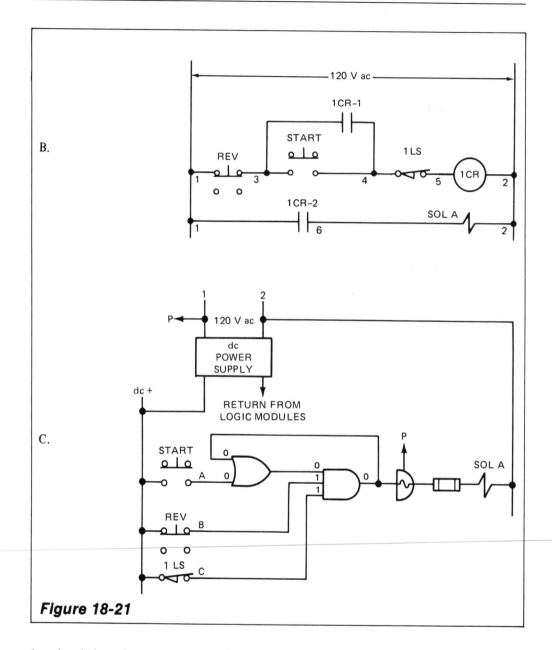

Figure 18-21

low-level dc voltage power supply. The return side of the dc power supply is not drawn in, but is indicated on the circuit.

The operation of this solid-state circuit is explained in steps to help visualize the changes that take place. The digit 1 is used to indicate a HIGH, and the digit 0 is used to indicate a LOW.

A two-input OR gate and a three-input AND gate are used in the logic circuit. An ac static switch is used in the output so that 120 V ac can be applied to the solenoid.

Step 1. In the explanation of the logic module symbols, it is shown that to obtain a HIGH or 1 output from a two-input OR gate, there must be a HIGH or 1 input at one of the inputs.

In Figure 18-21C, showing the deenergized or start condition, the START push-button switch has not been operated. There is thus a LOW (0) at one of the inputs. There is also a LOW (0) at the other input. (The reason for this will be clear later.) With a LOW (0) at both inputs of the OR gate, there is a LOW (0) output.

In a three-input AND gate, to obtain a HIGH (1) from the AND gate there must be a HIGH (1) at each of the three inputs.

While there is a HIGH (1) at B (REVERSE push button) as well as a HIGH (1) at C (limit switch 1LS), there is a LOW (0) input from the OR gate. This results in a LOW (0) output from the AND gate. Note that this LOW (0) is connected back as one of the OR gate inputs.

With a LOW (0) output from the AND gate, the ac static switch is not operated. Therefore, the solenoid is deenergized.

Step 2. In Figure 18-22A, the START push-button switch is operated. This provides a HIGH (1) input on the OR gate. There will now be a HIGH (1) on the OR gate output. This HIGH (1) signal on the input to the three-input AND gate now completes the conditions for a HIGH (1) output from the AND gate. This signal is connected into the static switch, providing 120 V ac to energize solenoid A. Note that the HIGH (1) output from the AND gate is connected back to one of the OR gate inputs. This is known as a *latching circuit;* its purpose will be evident in Step 3.

Step 3. In Figure 18-22B, the START push button has been released, resulting in a LOW (0) signal from A into the two-input OR gate. However, since the OR gate input was latched from the AND gate output (Step 2), the conditions are satisfied to continue a HIGH (1) output from the OR gate. Note that latching provides a function similar to that of the interlock circuit in the electromechanical circuit. That is, it allows the output to remain even though the initial energizing path has been opened. In this case, the path was through the START push-button switch. The remaining conditions are the same as in Step 2, with solenoid A energized.

Step 4. In Figure 18-22C, the piston has reached and operated the normally closed limit switch contact 1LS. This action opens the NC limit switch contact. The signal C now goes to LOW (0). Since this signal is one of the three inputs to the AND gate, the output of the AND gate goes to LOW (0). The static switch

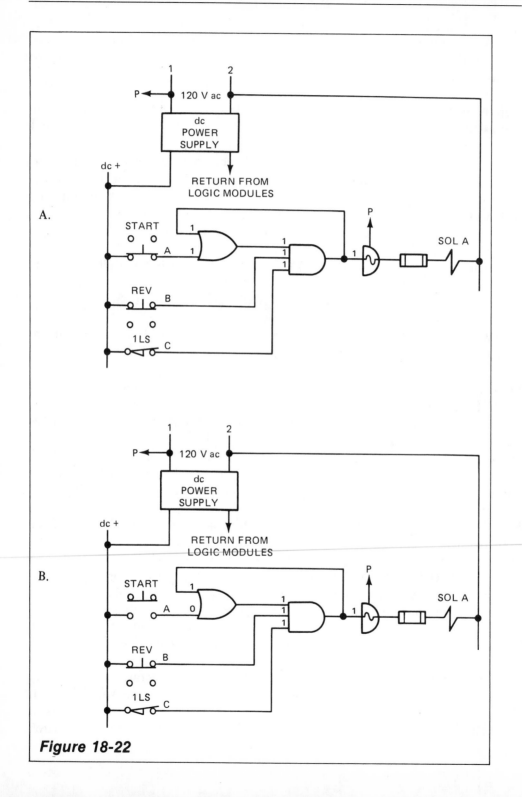

Figure 18-22

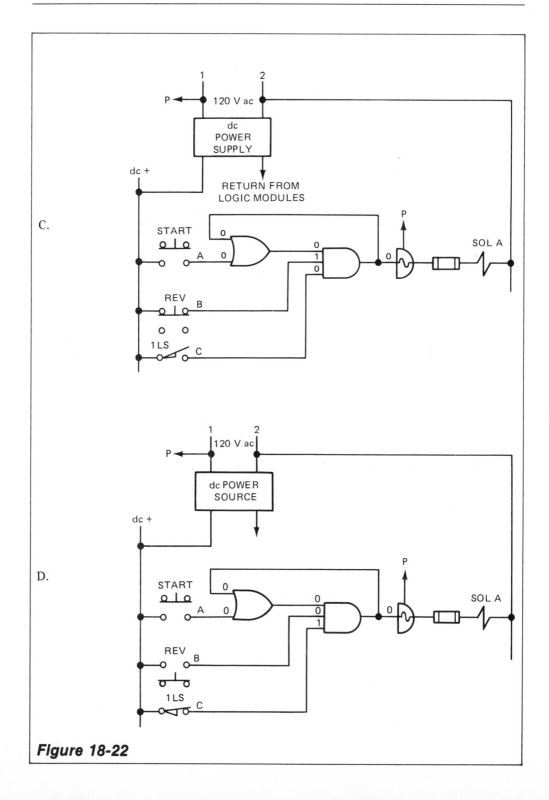

Figure 18-22

now opens, deenergizing solenoid A. Note that the OR gate is unlatched when the output of the AND gate goes to LOW (0).

Step 5. The circuit requirements state that at any point in the forward travel of the piston, before operating limit switch 1LS, it must be possible to reverse the direction of the piston travel. Figure 18-22D shows the circuit. This circuit operation is the same as that explained in Step 4, except that here the LOW (0) signal comes from the open contact on the REVERSE push-button switch.

In the Circuits #2–#6 that follow, the input signals are designated by letters. They start with the letter A and proceed alphabetically. While the names of the logic gates do not appear on the symbol, they are designated as:

> **AND** - A1, A2, etc.
> **OR** - 01, 02, etc.
> **NOT** - N1, N2, etc.

With this information, the descriptions of the circuit sequence of operation can be followed.

Circuit #2
A cylinder and piston assembly is shown in Figure 18-23A. The piston must be at position 1, operating limit switch 2LS, for start conditions. The piston is to move to the right until it engages and operates limit switch 1LS. At this point the piston is to return to position 1. Provision must be made to return the piston from any point in its forward travel. A single-solenoid, spring-return operating valve is used to supply fluid power to the cylinder. Figure 18-23B shows the electromechanical circuit.

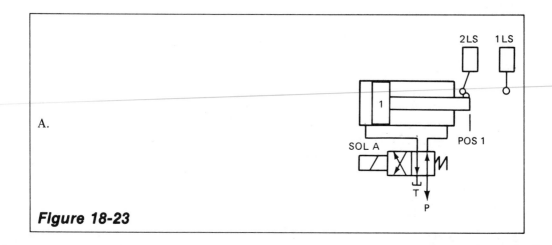

A.

Figure 18-23

The solid-state circuit for Circuit #2 is shown in Figure 18-23C. Line numbers 1 and 2 show 120 V ac; this is generally from the secondary of an isolated secondary transformer. The low voltage-level dc for the logic modules is shown

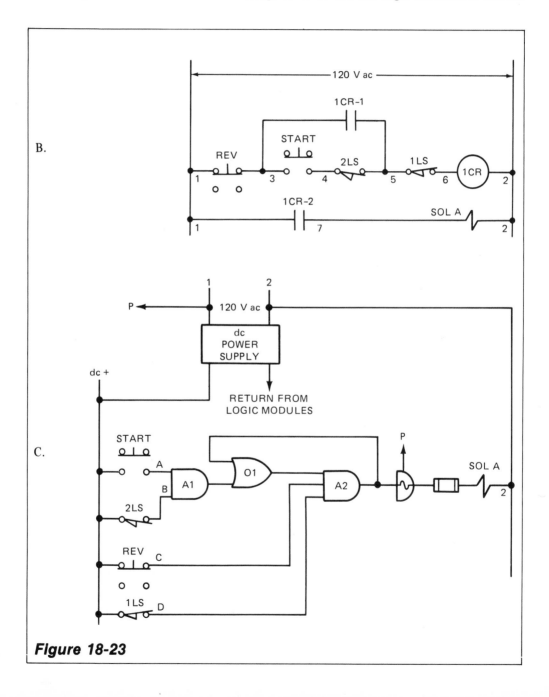

Figure 18-23

with one side connected directly to an input or information component. The other side of the dc supply indicates that it is the dc return from the logic modules.

This general arrangement is used on the following circuits so that it is not necessary to repeat for each circuit.

The sequence of operations proceeds as follows.

(See Figure 18-23D.)
1. Operate the START push-button switch.
2. HIGH (1) signal at A.
3. HIGH (1) signal at B through normally open, held-closed 2LS limit switch contact.
4. HIGH (1) output from A1.

(See Figure 18-23E.)
5. HIGH (1) output to O1.
6. HIGH (1) output from O1.

(See Figure 18-23F.)
7. HIGH (1) input at C from normally closed REVERSE push-button switch.

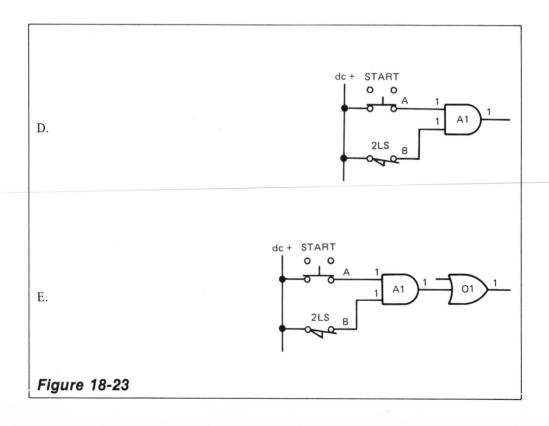

Figure 18-23

8. HIGH (1) input at D from normally closed limit switch 1LS.
9. HIGH (1) output from A2.

(See Figure 18-23G.)
10. The ac static switch operates, energizing solenoid A.
11. HIGH (1) output from A2 latches input to O1.

(See Figure 18-23H.)
12. The START push button is released. A signal goes to LOW (0).

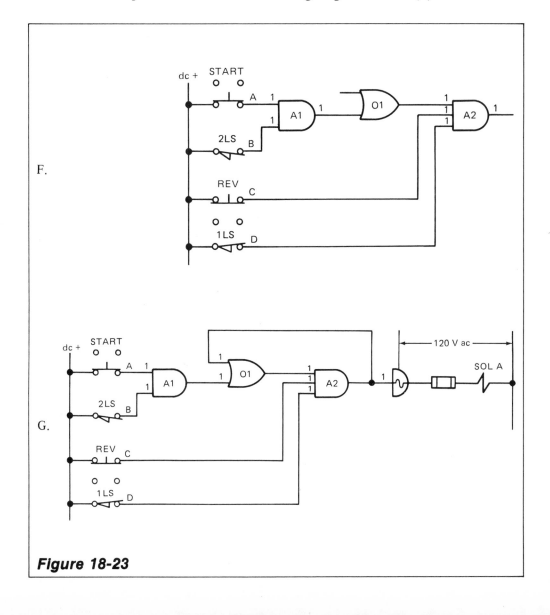

Figure 18-23

13. The piston starts forward, releasing limit switch 2LS. B signal goes to LOW (0).
14. A1 output signal goes to LOW (0).

(See Figure 18-23I.)
15. The piston engages and operates limit switch 1LS.
    - D goes to LOW (0).
    - A2 output goes to LOW (0).
    - O1 is unlatched.
    - Solenoid A is deenergized.
    - The piston starts to return to position 1.

(See Figure 18-23J.)
16. If at any time during the forward travel of the piston the REVERSE push-button switch is operated:
    - C goes to LOW (0).
    - A2 goes to LOW (0).
    - 01 is unlatched.
    - Solenoid A is deenergized.
    - The piston starts to return to position 1.

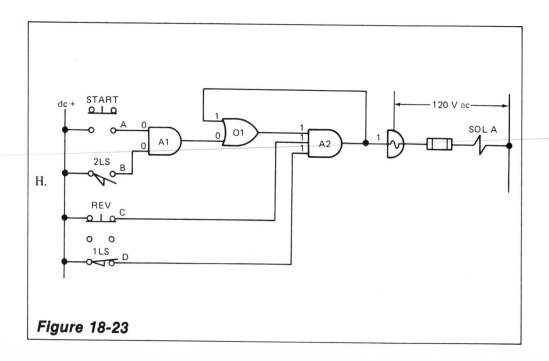

Figure 18-23

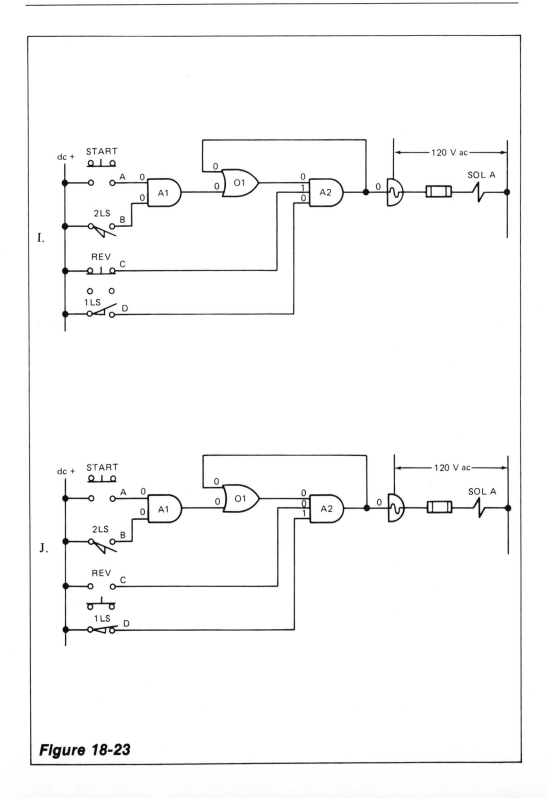

Figure 18-23

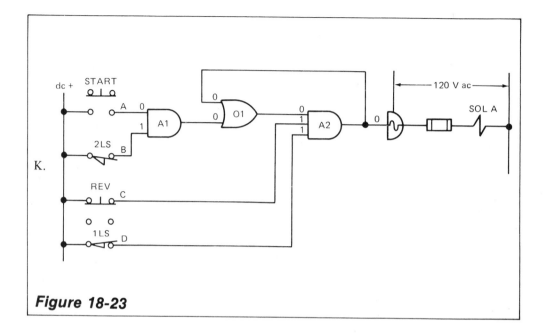

Figure 18-23

(See Figure 18-23K.)

17. The piston reaches position 1, operating limit switch 2LS; ready for the next cycle.

Circuit #3

A cylinder and piston assembly is shown in Figure 18-24A. The piston must be at position P1, operating limit switch 2LS, for a start condition. The piston is to move to the right until it engages and operates limit switch 1LS. At this point the piston is to stop and return to the start position P1. Provision must be made

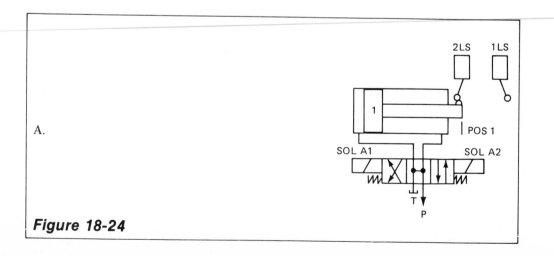

Figure 18-24

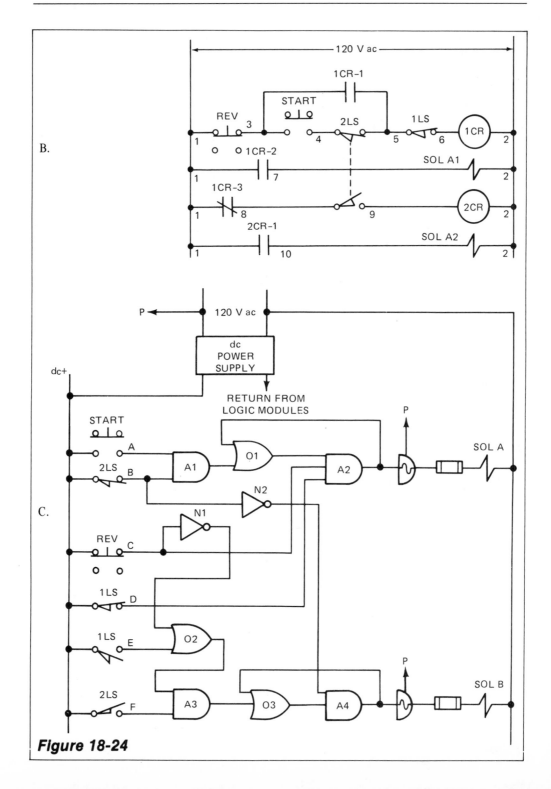

Figure 18-24

to return the piston from any point in its forward travel. A double-solenoid, center open, operating valve is used to supply fluid power to the cylinder. Figure 18-24B shows the electromechanical circuit for this system.

Note that the basic difference between Circuit #2 and Circuit #3 is that in Circuit #3, a double-solenoid operating valve is used in the fluid power circuit. With a double-solenoid operating valve, there is no pressure on either side of the piston head in the deenergized condition of both solenoids. The pressure is bypassed through the valve to the tank.

The solid-state circuit for Circuit #3 is shown in Figure 18-24C.

The sequence of operations proceeds as follows.

(See Figure 18-24D.)
1. Operate the START push-button switch.
2. HIGH (1) signal at A.
3. HIGH (1) signal at B through normally open, held-closed 2LS limit switch contact. Note that in this circuit both NO and NC contacts on 2LS are used.
4. HIGH (1) output from A1.

(See Figure 18-24E.)
5. HIGH (1) input to O1.
6. HIGH (1) output from O1.

(See Figure 18-24F.)
7. HIGH (1) input at C from normally closed REVERSE push-button switch.
8. HIGH (1) input at D from normally closed limit switch contact 1LS. Note that in this circuit both NO and NC contacts on 1LS are used.
9. HIGH (1) output from A2.

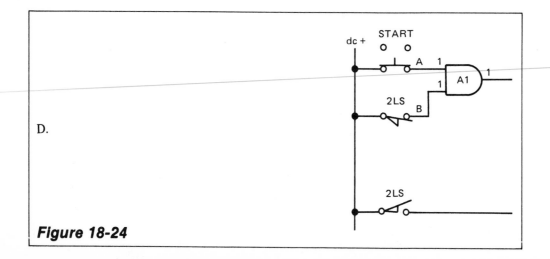

D.

Figure 18-24

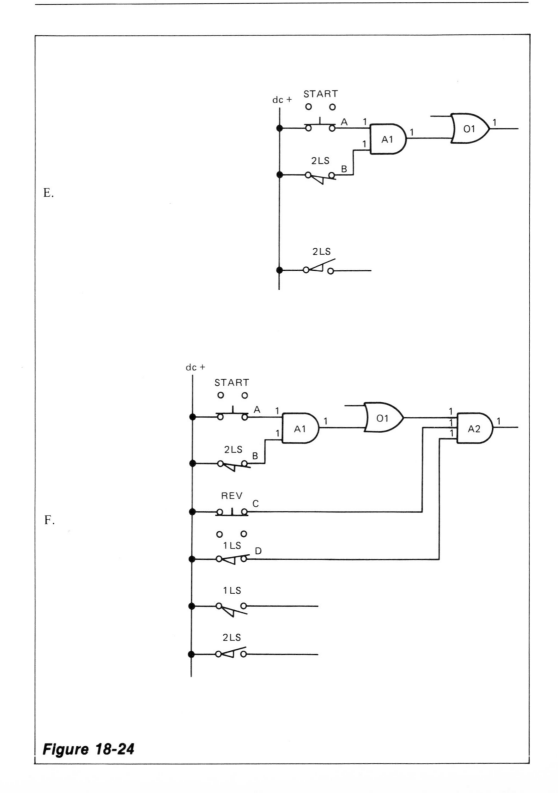

Figure 18-24

(See Figure 18-24G.)

10. The ac static switch operates, energizing solenoid A.
11. HIGH (1) from A2 latches input to O1.

(See Figure 18-24H.)

12. The START push button is released. A signal goes to LOW (0).
13. The piston starts forward, releasing limit switch contact 2LS. B signal goes to LOW (0).
14. A1 output signal goes to LOW (0).

(See Figure 18-24I.)

15. The piston engages and operates limit switch 1LS.
16. D signal goes to LOW (0).
17. A2 output goes to LOW (0).
18. O1 is unlatched.
19. Solenoid A1 is deenergized.

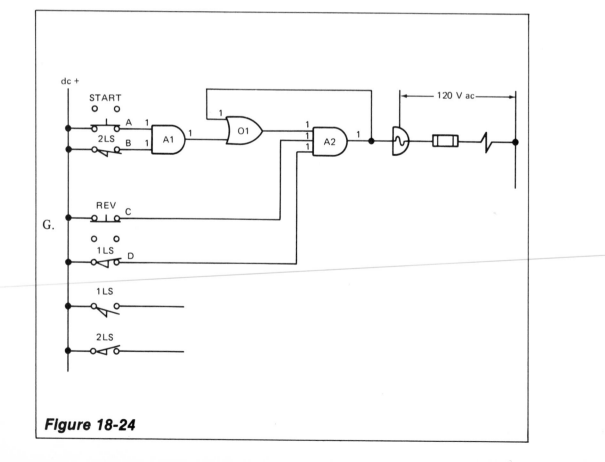

Figure 18-24

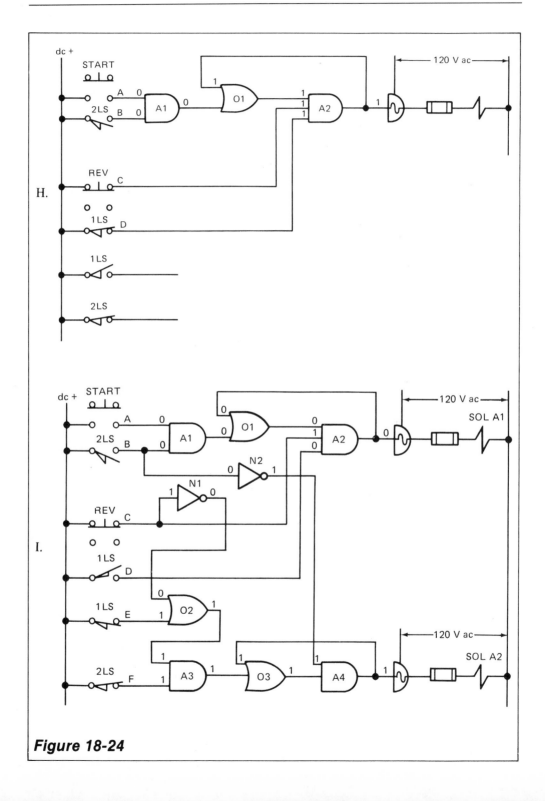

Figure 18-24

20. F goes to HIGH (1) when normally closed 2LS is released.
21. When limit switch 1LS is operated, E goes to HIGH (1).
22. This supplies an output of HIGH (1) from O2.
23. With this HIGH (1) from O2 and a HIGH (1) from 2LS, A3 output goes to HIGH (1).
24. This gives a HIGH (1) on the output of O3.
25. When limit switch 2LS is released, B signal goes to LOW (0). This input into N2 supplies a HIGH (1) output from N2.
26. The HIGH (1) from N2 and the HIGH (1) from O3 on the input to A4 give a HIGH (1) output for A4. O3 is latched by the HIGH (1) output from A4.
27. The static switch is now operated, energizing solenoid A2. The piston starts to return.

(See Figure 18-24J.)

28. The REVERSE push-button switch is operated on the forward stroke of the piston and before operating limit switch 1LS, resulting in the following.
29. C signal goes to LOW (0).

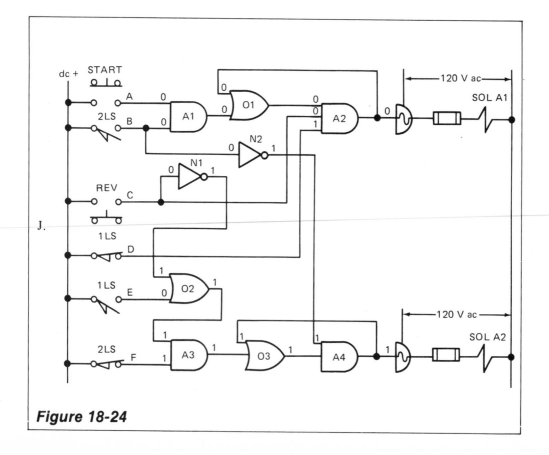

Figure 18-24

30. Output N1 goes to HIGH (1).
31. The output of O2 goes to HIGH (1).
32. With two HIGH (1) signals on A3, the A3 output goes to HIGH (1).
33. A HIGH (1) on O3 provides a HIGH (1) on A4. With the HIGH (1) from N2, the A4 output is HIGH (1). The static switch operates, energizing solenoid A2. The piston returns.
34. The piston returns to position 1, operating limit switch 2LS.
35. When limit switch 2LS is operated, the normally closed contact opens and the normally open contact closes. B goes to HIGH (1) and N2 output goes to LOW (0), thus causing A4 to go to LOW (0), dropping out solenoid A2.

(See Figure 18-24K.)

36. The circuit is now in the start condition for the next cycle. The signals present in this condition are as shown in Figure 18-24K.

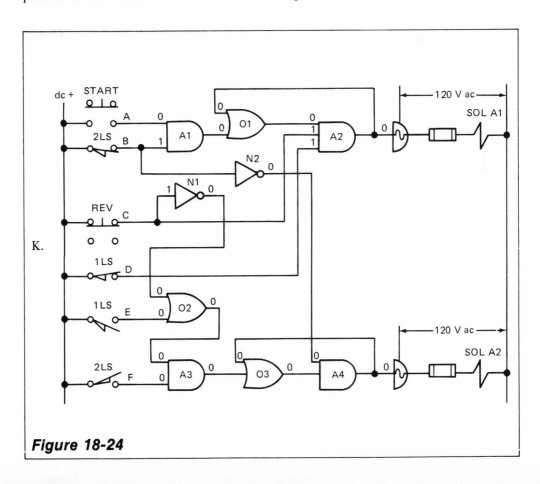

Figure 18-24

Circuit #4

Two piston and cylinder assemblies are shown in Figure 18-25A. The #1 piston must be in position P1, and piston #2 must be in position P4, for start conditions. As shown, #1 piston is to move to the right. At the same time, piston #2 is to move to the left. When #1 piston reaches a predetermined position P2, it is to stop and reverse to position P1. When #2 piston reaches a predetermined position P3, it is to stop and reverse to position P4. Provision must be made to reverse either piston from any point in their forward travel. Two double-solenoid, center-bypass operating valves are used to supply fluid power to the cylinders. Figure 18-25B shows the electromechanical circuit.

The solid-state circuit is shown in Figure 18-25C.

The sequence of operations proceeds as follows.

(See Figure 18-25D.)

1. Operate the START push-button switch.
2. HIGH (1) signal at A (operated START push-button switch).
3. HIGH (1) signal at B (normally open 2LS contact held operated).
4. HIGH (1) signal at C (normally open 4LS contact held operated). There is now a HIGH (1) output from A1.
5. HIGH (1) input to O1 provides a HIGH (1) output.
6. HIGH (1) signal at D (normally closed 1LS contact).
7. HIGH (1) signal at E (REVERSE push button — normally closed contact).
8. HIGH (1) output from A2.
9. O1 is latched.
10. The static switch operates, energizing solenoid A1.
11. HIGH (1) at O2 from A2 output.
12. HIGH (1) input to A3 from O2.
13. HIGH (1) at F (normally closed 3LS contact) with three inputs to A3 HIGH (1) — output HIGH (1).
14. O2 is latched.
15. The static switch operates, energizing solenoid B1. Piston #1 moves to the right; piston #2 moves to the left.

(See Figure 18-25E.)

16. The START push button is released. As the pistons move off limit switches 2LS and 4LS, the normally open contacts return to their open condition.
17. All inputs to A1 go to LOW (0).
18. O1 retains the HIGH (1) output as it is latched.

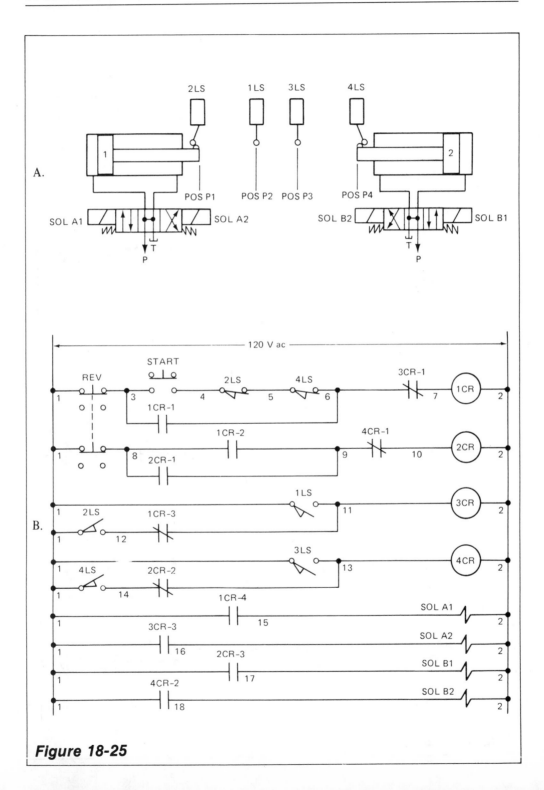

Figure 18-25

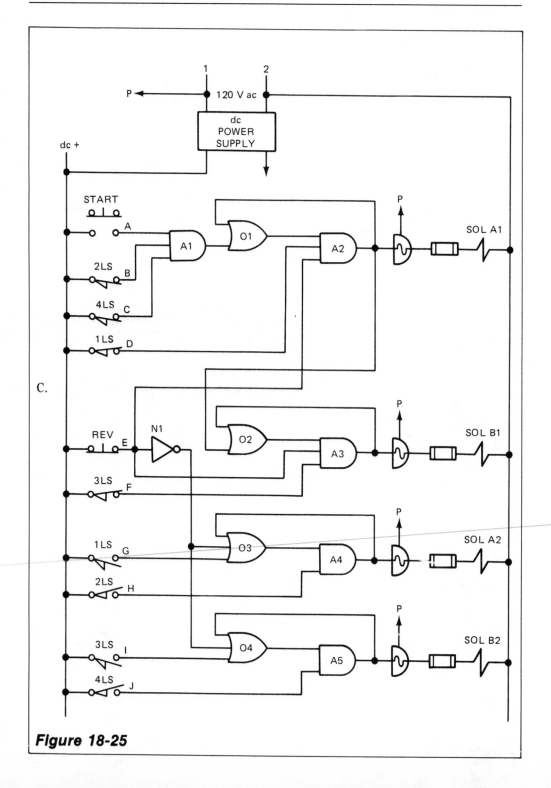

Figure 18-25

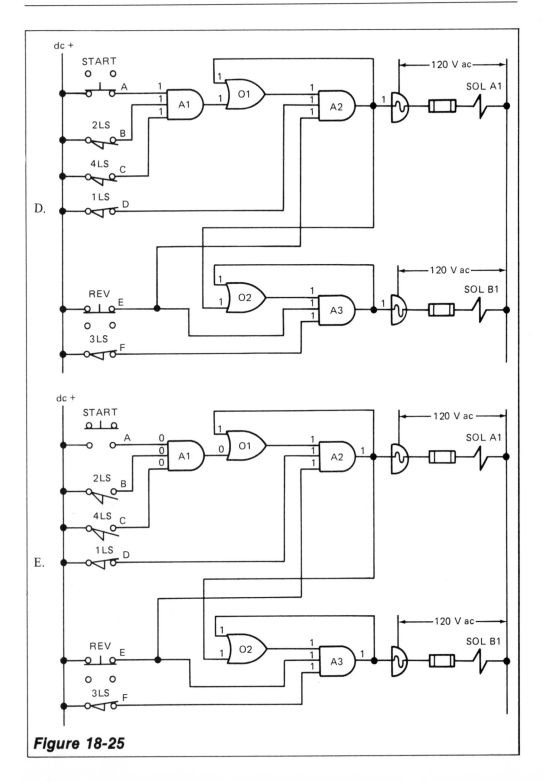

Figure 18-25

(See Figure 18-25F.)

19. Piston #1 engages and operates limit switch 1LS.
20. The normally closed contact on limit switch 1LS opens. Signal D goes to LOW (0).
21. The output from A2 goes to LOW (0). Solenoid A1 is deenergized.
22. O1 is unlatched.
23. O2 input from A2 output goes to LOW (0).
24. Piston #2 engages and operates limit switch 3LS.
25. The normally closed contact on limit switch 3LS opens; F signal goes to LOW (0).
26. The output from A3 goes to LOW (0). Solenoid B1 is deenergized.
27. O2 is unlatched.

(See Figure 18-25G.)

28. The normally open contact on 1LS is closed.
29. G signal goes to HIGH (1).
30. O3 output goes to HIGH (1).

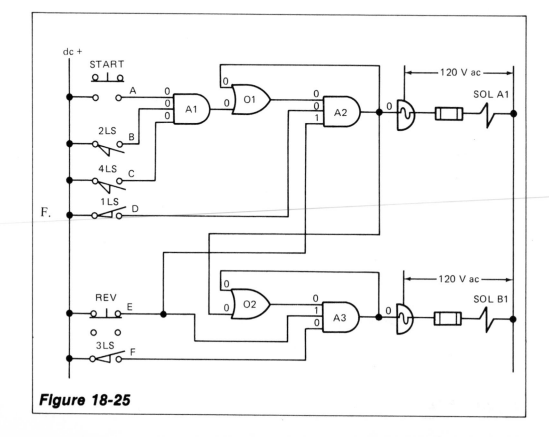

Figure 18-25

31. H signal is now HIGH (1); 2LS is released when #1 piston moves off the limit switch.
32. A4 output goes to HIGH (1).
33. O3 is latched.
34. The static switch operates. Solenoid A2 is energized.
35. The normally open contact on 3LS is closed.

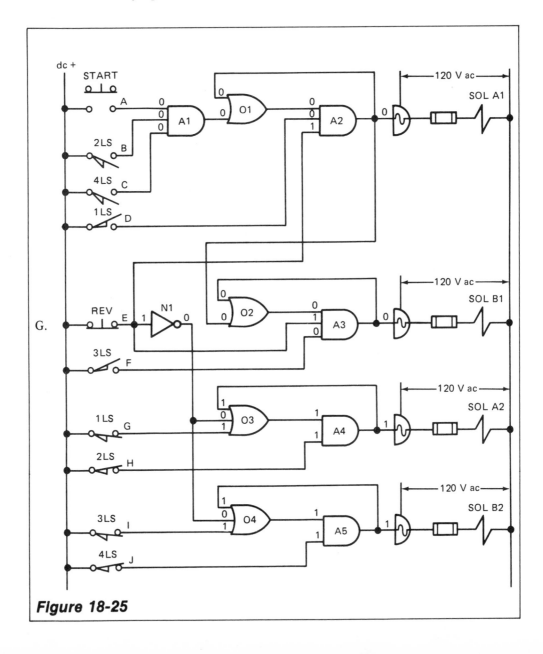

Figure 18-25

36. I signal goes to HIGH (1).
37. O4 output goes to HIGH (1).
38. J signal is now HIGH (1); 4LS is released when piston #2 moves off the limit switch.
39. A5 output goes to HIGH (1).
40. O4 is latched.
41. The static switch operates. Solenoid B2 is energized.

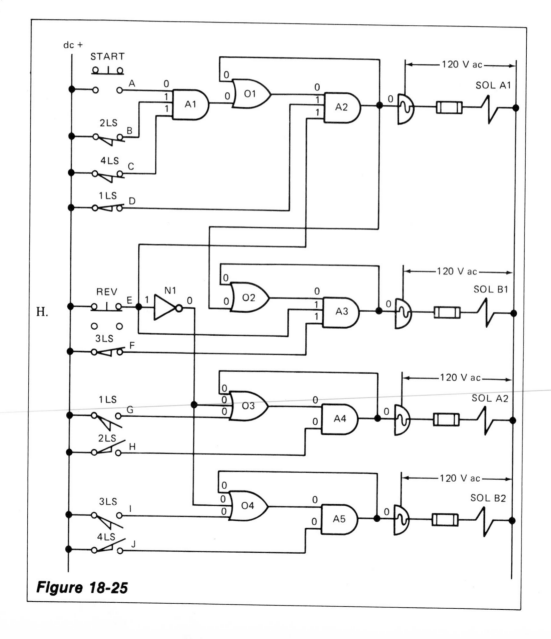

Figure 18-25

(See Figure 18-25H.)

42. Piston #1 starts to return to position 1.
43. Piston #1 operates limit switch 2LS.
44. Signal H goes to LOW (0).
45. A4 output goes to LOW (0). Solenoid A2 is deenergized.
46. O3 is unlatched.
47. Piston #2 starts to return to position 2.

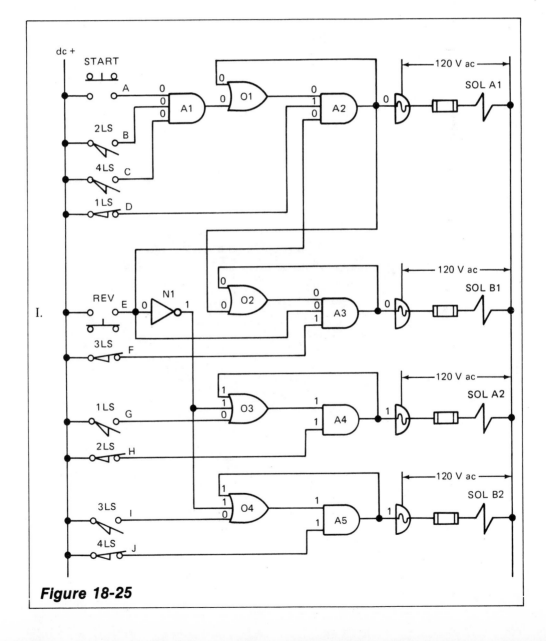

Figure 18-25

48. Piston #2 operates limit switch 4LS.
49. Signal J goes to LOW (0).
50. A5 output goes to LOW (0). Solenoid B2 is deenergized.
51. O4 is unlatched. The circuit is now in the start condition, ready for the next cycle.

(See Figure 18-251.)

52. At any time during the forward travel of the pistons, either or both pistons can be reversed by operating the REVERSE push button. Remember that the normally closed limit switch contacts on 2LS and 4LS are in a normal condition after the pistons start forward.
53. Operate the REVERSE push-button switch.
54. E signal goes to LOW (0).

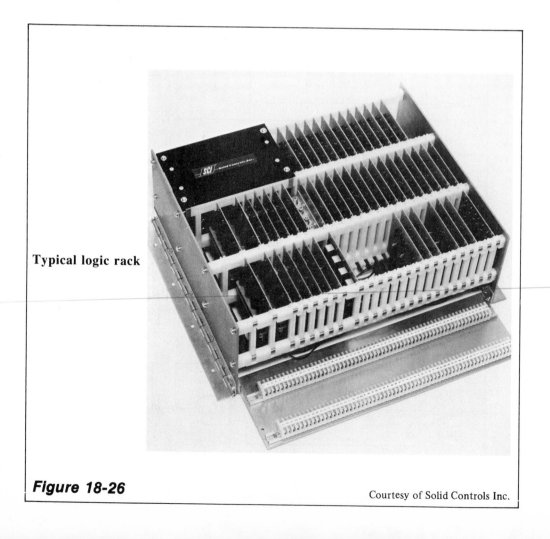

Typical logic rack

Figure 18-26

Courtesy of Solid Controls Inc.

55. A2 output goes to LOW (0). Solenoid A1 is deenergized.
56. O1 is unlatched.
57. A3 output goes to LOW (0). Solenoid B1 is deenergized.
58. O2 is unlatched.
59. The output from N1 goes to HIGH (1).
60. O3 output goes to HIGH (1).
61. A4 output goes to HIGH (1). Solenoid A2 is energized.
62. O4 output goes to HIGH (1).
63. A5 output goes to HIGH (1). Solenoid B2 is energized. The pistons return to their start positions. The solenoids are deenergized as described in Figure 18-25H.

In practice, applications of control such as shown in Circuits 1–4 are accomplished through the use of individual logic module cards mounted in a rack. The power unit is generally included as part of the rack. This type of assembly is sometimes referred to as a "mother board."

In the card shown in Figure 18-3, two AND functions are mounted on one board. Figure 18-26 shows a typical mother board with many functions included. The number of functions, of course, varies with the circuit requirements on any given machine.

ACHIEVEMENT REVIEW

1. Draw the symbol for each of the following:
   a. Three-input AND gate
   b. Two-input OR gate
   c. NOT or INVERTER gate

2. Draw the symbol for the EXCLUSIVE-OR gate, and explain how the EXCLUSIVE-OR gate differs from the OR gate.

3. Write the truth table for a three-input AND gate.

4. What does a digit 1 or a digit 0 mean when placed at the input or output of a gate?

5. If the electromechanical circuit in Figure 18-21B required two START pushbutton switches, draw the solid state circuit which would reflect this change. Assume you have an additional two input AND gate to use.

6. What is the meaning of the bar over the output signal letter in the NOT or INVERTER gate?

7. Explain the difference between the notations AB and A + B.

8. Why is the NOT gate used in Figure 18-25C?

9. What is the reason for using a latching circuit?

10. The NOT INVERTER gate is obtained with
    a. two NOTs.
    b. two ANDs.
    c. a single NAND.

11. The EXCLUSIVE OR gate produces an output when
    a. INPUT A is high.
    b. INPUT B is high.
    c. INPUT A and INPUT B are high.

12. Why can Boolean algebra be used to analyze and express logic statements?

CHAPTER 19

PROGRAMMABLE CONTROL

OBJECTIVES

After studying this chapter, the student will be able to:

- Show the relationship between the input section—control processing unit and the output section.
- List several advantages gained through the use of a programmable controller.
- Explain the need for a well-regulated power supply for the programmable controller.
- List four important factors considered in the input/output section.
- Understand the binary, octal and hexadecimal numbering systems.
- List two general classes of memory and explain the difference.
- Explain the scanning process.
- List several peripheral devices and the uses for each.
- Explain how a hard copy printout can reduce the cost of operation.
- Gain knowledge of a few typical industrial applications for PCs.
- List the information required for ac synchronous motor application.
- List some of the advantages in using a dc stepping motor.

19.1 INTRODUCTION

In Chapters 1 through 17, electromechanical control has been explained. Following this, the solid-state control was explained in Chapter 18. Solid-state

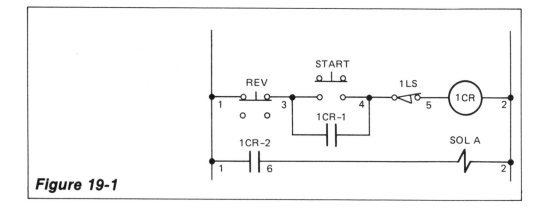

Figure 19-1

control brought out many advantages over electromechanical control. However, with solid-state control came the use of a new set of symbols and a new type of circuit diagram. Electricians and maintenance personnel who had become acquainted with the ladder-type diagram and symbols used with electromechanical control found problems in making the change.

A brief review of a few simple control circuits using either electromechanical or solid-state modules is shown in Figures 19-1 through 19-3. These circuits are called fixed circuits. That is, physical wiring and/or component changes must be made to change or alter the circuit.

In Figure 19-1, the operation of the circuit proceeds as follows:

1. Operate the start push-button switch.
2. Relay coil 1CR is energized.
   a. Relay contact 1CR-1 closes, interlocking around the start push-button switch.
   b. Relay contact 1CR-2 closes, energizing solenoid A.

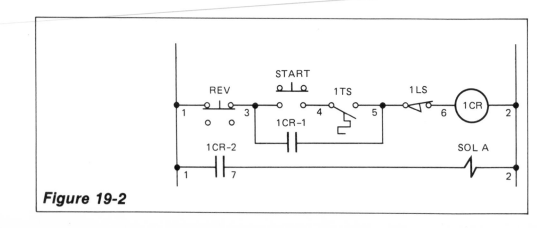

Figure 19-2

Assume that this circuit is now changed to add the requirement that the temperature of the operating fluid must be at a given level for start conditions. This means that the connecting wiring must be changed and a component (temperature switch) must be added. The temperature switch 1TS is added in series with the start push-button switch, Figure 19-2.

The same concern of physically reconnecting elements when a circuit is changed is true when using solid-state modules. For example, the electromechanical circuit shown in Figure 19-2 is now shown in Figure 19-3, using solid-state modules.

These circuits show that whether using electromechanical or solid-state modules physical changes must be made to change or alter the circuit. These changes may involve only wiring. However, in many cases, components will have to be added, changed, or omitted. Costs increase as changes are made as labor is now required to "fit" the changes in.

A machine using electromechanical or solid-state control is now shown in box form, Figure 19-4. This shows the relationship between the machine, input sensing devices, logic decision making and the output work units.

In the late 1960s and early 1970s, the programmable controller was introduced into industry. It came first to the mass production industry and later into

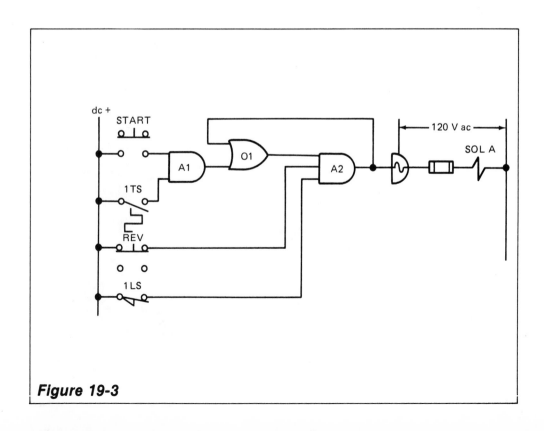

Figure 19-3

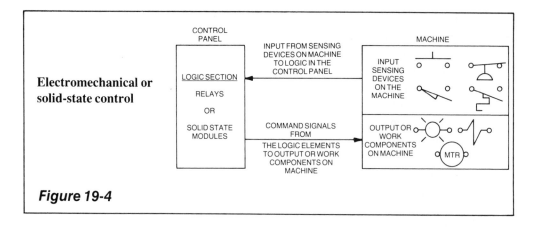

Electromechanical or solid-state control

Figure 19-4

the general machine tool industry.

NEMA (The National Electrical Manufacturers Association) defines a programmable controller as a "digital electronic apparatus with a programmable memory for storing instructions to implement specific functions such as logic, sequencing, timing, counting and arithmetic to control machines and processes." Two programmable controllers on the market today are shown in Figures 19-5A and 19-5B.

In both electromechanical and the solid-state chapters, it was shown that all control is divided into three sections. The division of control into three parts does not change with the use of programmable control, Figure 19-6.

The relationship as shown in Figure 19-6 is now shown in block form in Figure 19-7.

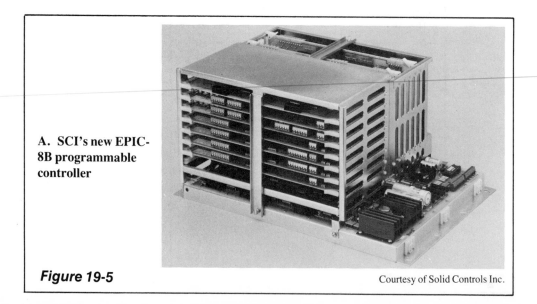

A. SCI's new EPIC-8B programmable controller

Figure 19-5

Courtesy of Solid Controls Inc.

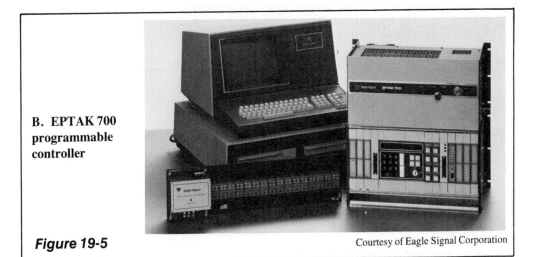

B. EPTAK 700 programmable controller

Figure 19-5

Courtesy of Eagle Signal Corporation

INFORMATION OR INPUTS	DECISION MAKING OR LOGIC	WORK OR OUTPUT
Pushbuttons and selector switches	Central Processor Unit (consisting of Processor, Memory, Power Unit)	Solenoids (fixed and proportional)
Limit switches or linear transducers		Relays
Pressure switches or hydraulic transducers		Contactors
Temperature switches or thermocouples		Lights
Motor starter contacts		Horns
Relay contacts		Heating elements
Photoelectric eyes		Fans
		Motor starter

Figure 19-6

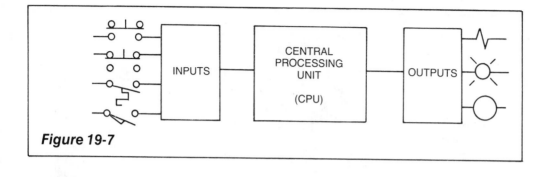

Figure 19-7

19.2 INPUTS/OUTPUTS (I/O)

The input section performs as an interface between the various information components on the machine and the controller. The signals from these components must be conditioned for use in the central processing unit (CPU). The components may consist of such items as push-button switches, pressure switches, temperature switches, analog sensors, thumbwheel switches and selector switches.

The output section receives signals from the CPU. These signals are used to control such devices as solenoid valves, lights, motor starters and position valves. In some input conditions the signal supplied to the input section is of an analog nature. These signals may originate from variables such as speed, pressure, temperature and position.

An analog signal is one that is continuous and depends directly on magnitude (voltage or current), to represent some condition.

As the processor accepts and uses only digital signals (1 or 0), the analog signal must be converted to a digital signal. This is done through the use of an analog-to-digital converter.

In the output section, the digital signals (1 or 0), are often required to be converted to analog signals for control of valve positioners, speed controllers or signal amplifiers. Here, the digital-to-analog converter is used.

There is one input associated with temperature sensing that may be considered in a special class. This is because it deals with a very low level voltage signal (millivolts). Typically, the type-J thermocouple is a sensor in this class. As with other analog signals, an analog-to-digital converter must be used.

TTL (Transistor-Transistor Logic) input modules are used with photoelectric sensors, sensing instruments and some 5-volt dc level control devices.

Four important factors are considered in the INPUT-OUTPUT (I/O) section of the programmable controller. They are:

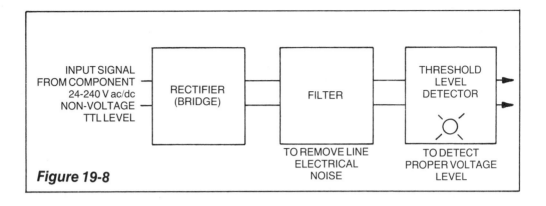

Figure 19-8

1. Conditioning
2. Isolation
3. Indication
4. Termination

Field-supplied voltages for both the inputs (information) and the outputs (work) components may cover a range of 24–240 volts ac/dc. The input signal must be converted to a dc logic level that the processor can use. This generally consists of a bridge rectifier and series resistors. Figure 19-8 shows in block form the changes that must be made in the power section for the conditioning of the incoming signals.

An indicating light (LED) may be included in the level detection circuit to indicate that a valid signal has been detected.

When a signal has reached this point, it must be electrically isolated from the processor. This helps to prevent large voltage spikes from damaging the processor. Manufacturers vary on the method used but it is generally a pulse transformer or optical coupler. Figure 19-9 shows its relation to the logic section and the threshold detector.

In addition to the isolation section, there is always a possibility of electrical noise coupling in external wiring. To help this condition, the wiring connected

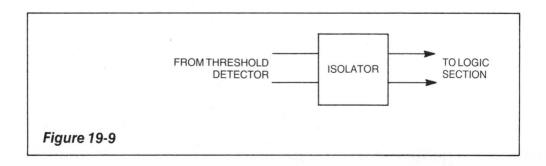

Figure 19-9

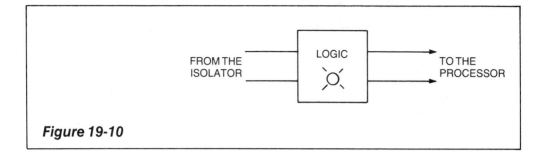

Figure 19-10

with the input and output devices should be run in separate conduits. An alternative to this method is to run a shielded, twisted-pair conductor to the inputs and outputs. The shield should be grounded at one end only.

The logic section, as shown in Figure 19-10, now holds the proper dc signal that can be used in the processor. For example, assume that a normally open push-button switch has been closed, putting a 120 V ac signal on the input. The signal has passed through the bridge rectifier, filter, threshold detector and logic section. The signal is now at the logic level dc voltage. The presence of logic 1 is indicated by an indicating light (LED).

Terminals are provided to wire (connect) to the various input components located on the machine, Figure 19-11.

The output interface is similar to the input interface except in reverse sequence. When the program that has been placed in the processor indicates an output is available to satisfy the program, a digital signal (logic 1), is present

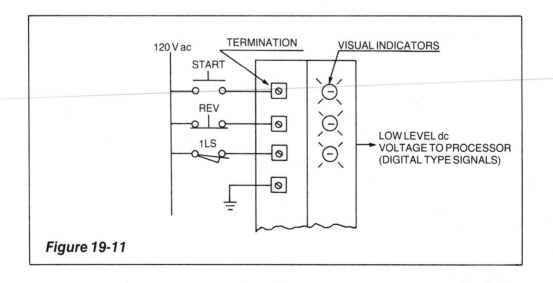

Figure 19-11

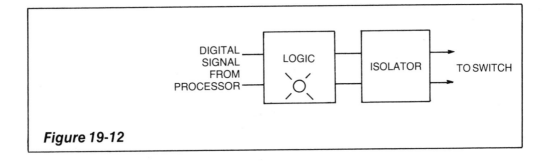

Figure 19-12

in the output logic section. This signal must be isolated from the output power line, Figure 19-12. Note indication is provided by an indicating light (LED).

This signal now gates a triac. The triac is equivalent to two SCRs (silicon-controlled rectifiers), connected in parallel (back to back). A gate controls the switching state once the breakdown voltage is reached on the gate. The triac conducts in either direction.

A filter is used in the output circuit to prevent electrical noise from affecting the circuit operation. A fuse is always used in the output circuit to prevent excessive current from damaging the ac switch, Figure 19-13. Note indication is provided by an indicating light (LED).

Termination is provided to connect to the various work components on the machine, Figure 19-14.

Most manufacturers of programmable controllers can supply a wide range of I/O modules to fit almost any application. I/O racks can be located at the controller or thousands of feet away. A typical example of a remote I/O adaptor is shown in Figure 19-15.

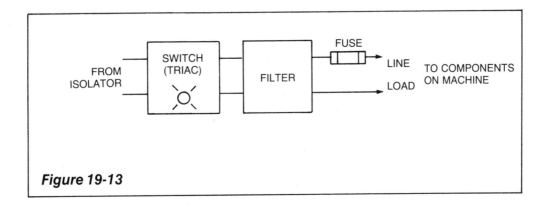

Figure 19-13

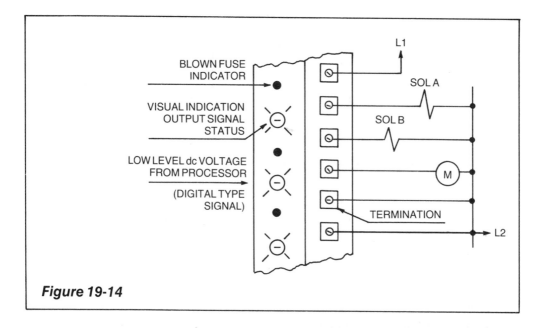

Figure 19-14

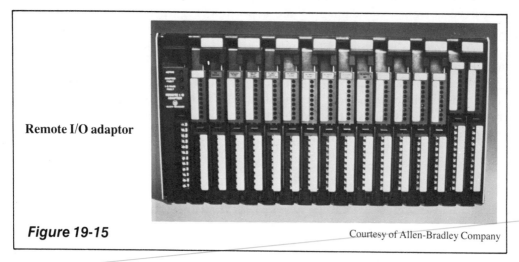

Remote I/O adaptor

Figure 19-15

19.3 THE CENTRAL PROCESSING UNIT

Figure 19-16 shows the *central processing unit* (CPU) and its division into three parts. The *input/output* sections just covered are shown in relief to indicate their relationship to the CPU.

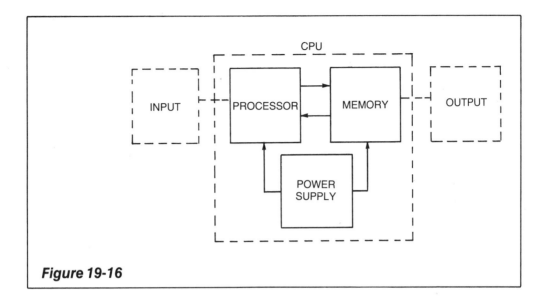

Figure 19-16

The central processing unit is made up of three parts: the *processor, memory* and *power supply*.

Manufacturers may vary as to how these three units are arranged physically. For example, they may be combined in one package or in separate units. For any one manufacturer this arrangement may vary in accordance with the size or model.

The *programmable controller* is a solid-state device and operates at a low level dc voltage. Therefore, it also will require the use of a power unit. The power supply unit supplies all the necessary voltages for the proper operation of the CPU.

The *power unit* for the *programmable controller* must supply a well-regulated power and protection for the other system components. In some industrial environments, an unstable line voltage condition may exist. In these cases the use of a constant voltage transformer may be indicated. This type of transformer will compensate for voltages on the primary side to maintain steady voltage on the secondary side, Figure 19-17.

In environments where voltage instability is not a problem but electromagnetic interference is, the controller could be connected to a separate isolation transformer. Where electrical noise or voltage spikes are present, a transient suppressor is often used.

The principal function of the processor is to command and govern the activities of the entire system. Again, manufacturers will vary as to process arrangements. For example, hardwired processors may be used in conjunction with microprocessors. The hardwired processor is used to execute the ladder diagram program logic while the microprocessor is used for the more difficult jobs such

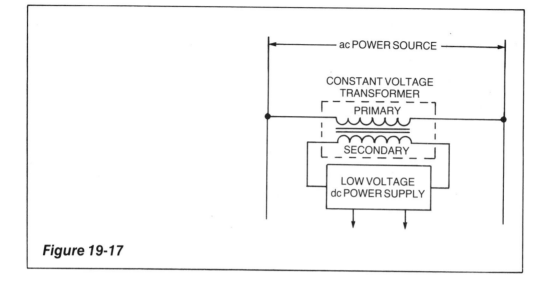

Figure 19-17

as solving math problems.

In the CPU with the processor is a memory section and circuits necessary to store and retrieve information from the memory. Communication circuits are required for the processor to interface with outside devices such as a programmer, printer and other peripheral devices.

Figure 19-18 shows a programmable controller with programmer and CRT.

During the operation of the PC, the processor first examines all the inputs to determine if they are all in the proper condition according to the program. The program has been entered into the memory section from the ladder diagram or network. With the input component in its proper condition the output is then

**Programmable controller
with programmer and CRT**

Figure 19-18   Courtesy of Westinghouse Electric Corporation

energized or deenergized according to this program. This procedure then continues to examine the inputs and updating the outputs as a machine being controlled moves through a cycle. This process of continuing to examine the inputs and updating the outputs in accordance with a stored program is called *scanning*. The time for a complete scan is very fast, for example, from 1 msec to 100 msec for a complete cycle. In many cases, the scan time can be important. It indicates how fast the controller can react to the connected inputs and solve the control logic.

Before we discuss the memory section of the CPU, it may be helpful to look at the numbering systems used. Electricians and maintenance personnel may have little direct contact with this part of the work. However it does help in understanding how the processor and memory operate. The PC operates on binary numbers in one form or another to represent various codes or quantities.

Returning for a moment to the electromechanical or solid-state control, remember that a switch had only two conditions, it was either open or closed. In like manner, the solid-state module had only two conditions. The inputs and outputs to a logic module were either 1 or 0.

The PC operates in the same general manner, accepting either a 1 or 0. To translate all the different conditions that exist in the machine inputs and outputs, each separate condition must be represented by a code or symbol that can be transmitted to the PC in binary code form (1 or 0).

One numbering system that all are familiar with is known as the decimal system. When a numbering system uses ten digits (0 through 9) it is called base 10. Take for example the decimal number 1111. The development of this number is shown in Figure 19-19.

THOUSANDS	HUNDREDS	TENS	UNITS
$Base^3 = 10^3$	$Base^2 = 10^2$	$Base^1 = 10^1$	$Base^0 = 10^0$
$10^3 = 10 \times 10 \times 10 = 1000$	$10^2 = 10 \times 10 = 100$	$10^1 = 10 \times 1$	$10^0 = 1$
$1000 \times 1 = 1000$	$100 \times 1 = 100$	$10 \times 1 = 10$	$1 \times 1 = 1$

	Thousands	1000
	Hundreds	100
Adding the	Tens	10
	Units	1
		1111

Figure 19-19

In programmable control it has been noted that it is necessary to deal with only two digits (1 or 0). This is in keeping with the signals that are either ON or OFF (switches closed or open). It is necessary, therefore, to use a system that incorporates only two digits. This can be found in the binary system. When a numbering system uses only two digits (1 or 0) it is called base 2.

In binary counting, when any column holding a 1 receives another count it goes back to 0 and develops a carry count to the next significant column to the left. Examples of counting in binary are shown in Figure 19-20.

$Decimal_{10}$	$Binary_2$
0	0000
1	0001
1	0001
+ 1	+ 0001
= 2	= 0010
2	0010
+ 1	+ 0001
= 3	= 0011
3	0011
+ 1	+ 0001
= 4	= 0100

Using the above development of BINARY equivalents of digits (0–4) all of the decimal digits of 0 through 9 are then expressed as follows:

Decimal Notation	Binary Notation
0	0000
1	0001
2	0010
3	0011
4	0100
5	0101
6	0110
7	0111
8	1000
9	1001

Figure 19-20

Note that in the decimal notation it required only one column. In binary notation it requires four columns. If the decimal notation required three columns, for example 173, the binary notation will require eight columns.

Using this larger decimal number 173_{10}, the binary number is 10101101_2. Note the decimal number is to the base 10 while the binary number is to the base 2.

The binary number can be converted to the decimal number as shown in Figure 19-21.

A Binary number	1	0	1	0	1	1	0	1
B Place values	2^7	2^6	2^5	2^4	2^3	2^2	2^1	2^0
Expressed in powers of 2	= 128	= 64	= 32	= 16	= 8	= 4	= 2	= 1

Multiply A x B	128	128
	0	0
	32	32
	0	0
	8	8
	4	4
	0	0
	1	1

$$101011101_2 = 173_{10}$$

Figure 19-21

		REMAINDER	
$\frac{173}{2}$	= 86	REMAINDER	1
$\frac{86}{2}$	= 43	"	0
$\frac{43}{2}$	= 21	"	1
$\frac{21}{2}$	= 10	"	1
$\frac{10}{2}$	= 5	"	0
$\frac{5}{2}$	= 2	"	1
$\frac{2}{2}$	= 1	"	0
$\frac{1}{2}$	= 0	"	1

$$1\ 0\ 1\ 0\ 1\ 1\ 0\ 1_2$$

Figure 19-22

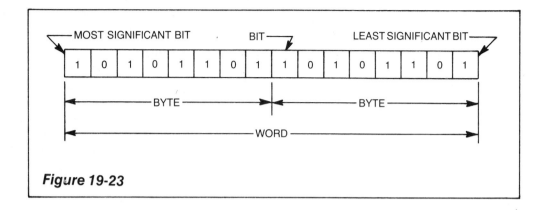

Figure 19-23

The decimal number 173 can now be converted back to the binary number by dividing by 2 (the base of the binary system). The remainder at each division (if any) then becomes the binary number in its proper position, Figure 19-22.

As has been stated, the binary system uses only two digits (1 and 0).

Each digit of a binary number is called a *bit*. This is taken from BInary digiT. The binary number 10101101 as shown in Figure 19-22 has 8 bits. A group of 8 bits is called a *byte*. A group of one or more bytes is called a *word*.

Figure 19-23 shows a binary number composed of 16 bits, with the least significant bit and the most significant bit.

In storing large numbers, it is more practical to convert to what is known as the *binary coded decimal* (BCD). In this system, each decimal number is represented by four binary digits. This system is identified by adding the letters BCD at the right of the units place, Figure 19-24.

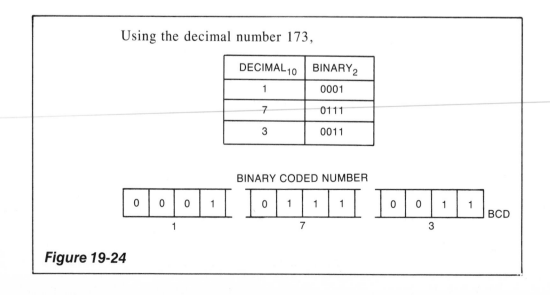

Figure 19-24

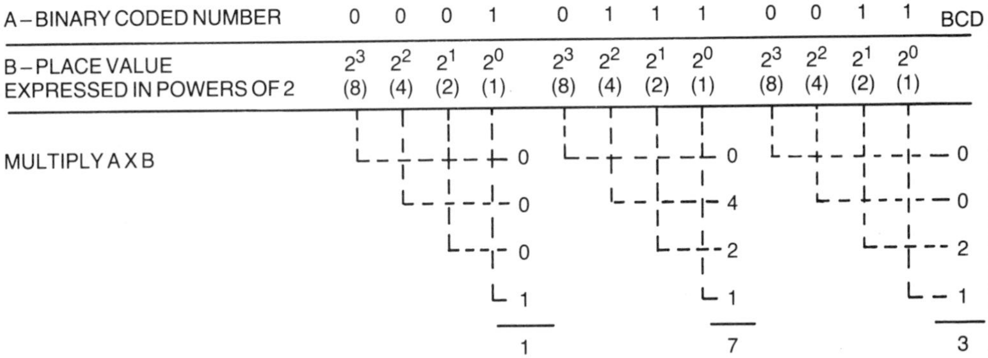

Figure 19-25

The BCD number can be converted to its decimal equivalent as shown in Figure 19-25.

Another binary system used is the *octal* system (base 8). This system is made up of eight digits, 0 through 7. The first digit to the left of the octal point has a power of 8^0. The next place to the left has a power of 8^1 or 8. The next place to the left has a power of 8^2 or 64. Any octal number will be designated by placing an 8 to the right of the units place, as shown in Figure 19-26.

The *binary coded octal* (BCO) system uses 3 binary digits to represent one octal (base 8) digit. For example, the number 173_8 is broken down into three

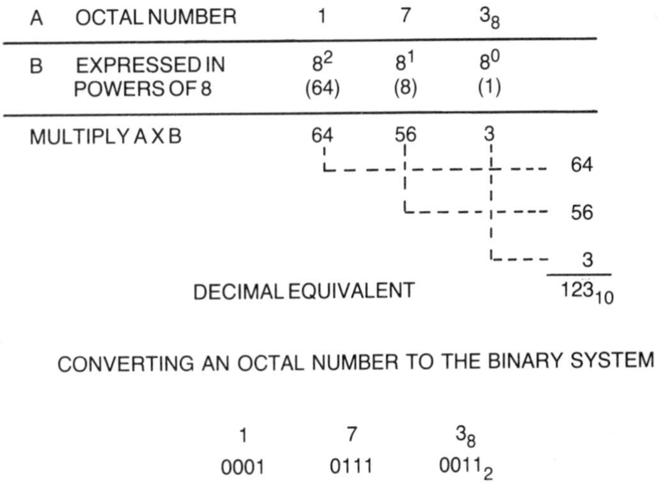

Figure 19-26

A	BCO NUMBER	0	0	1	1	1	1	0	1	1	BCO
B	PLACE VALUE EXPRESSED IN POWERS OF 2	2^2 (4)	2^1 (2)	2^0 (1)	2^2 (4)	2^1 (2)	2^0 (1)	2^2 (4)	2^1 (2)	2^0 (1)	

MULTIPLY A X B

$$0 + 0 + 1 \qquad 4 + 2 + 1 \qquad 0 + 2 + 1$$

0	4	0
0	2	2
1	1	1
1	7	3_8

Figure 19-27

groups (3 bits), (3 bits), and (3 bits) as shown in Figure 19-27.

The hexidecimal system (sometimes referred to as HEX) is another binary system and is now very popular. This system generally appears in manufacturer-supplied and after-market programming systems for programmable controllers.

The HEX system has a base of 16. It consists of 16 digits, numbers 0 through 9 and letters A through F. These letters are substituted for the numbers 10 through 15.

In the table shown in Figure 19-28, the decimal number, binary and hexadecimal equivalents are listed.

DECIMAL	BINARY	HEXADECIMAL
0	0000	0
1	0001	1
2	0010	2
3	0011	3
4	0100	4
5	0101	5
6	0110	6
7	0111	7
8	1000	8
9	1001	9
10	1010	A
11	1011	B
12	1100	C
13	1101	D
14	1110	E
15	1111	F

Figure 19-28

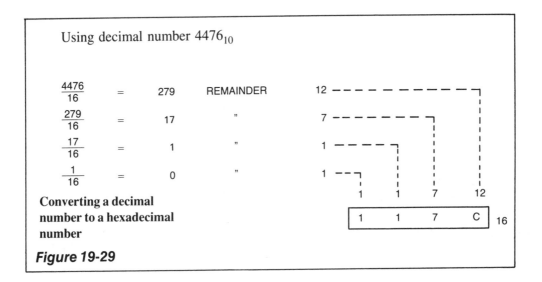

Using decimal number 4476_{10}

Converting a decimal number to a hexadecimal number

Figure 19-29

See Figure 19-29 on converting a decimal number to a hexadecimal number.

See Figure 19-30 on converting a hexadecimal number to a decimal number.

See Figure 19-31 on converting a hexadecimal number to a binary number.

In one additional numbering system letters, numbers and symbols are used. This is the American Standard Code for Information Interchange (ASCII). The entire alphabet of 26 letters (A through Z), numbers 0 through 9, and mathematical and punctuation symbols such as #, @, $, %, & and others are used.

This code can be made up of 6, 7, or 8 bits. The standard ASCII character sets use a 7-bit code. This provides all possible combinations of characters used

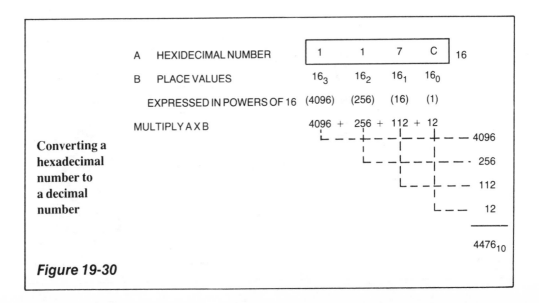

Converting a hexadecimal number to a decimal number

Figure 19-30

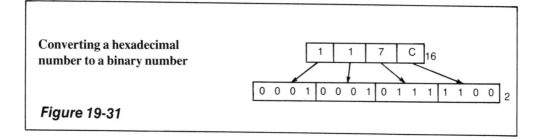

Converting a hexadecimal number to a binary number

Figure 19-31

when communicating with interfaces and peripheral equipment.

The manufacturers of programmable controllers will vary as to the numbering system they use to store information. The information is stored in the form of binary digits (bits) into the memory system of the CPU.

Regardless of the system used, it is important to remember that the information is always stored as 1s or 0s.

Now the *memory* section of the CPU can be discussed. It serves two important functions:

1. Remembers information that the processor may need to make decisions. This part of the memory is sometimes referred to as *storage memory* or *data table*. It is in this area where the status (ON-OFF)—(1-0) of all the discrete inputs and outputs are stored. Numeric values of timers and counters may also be stored in this memory.
2. Remembers the instructions given by the user, telling the programmable

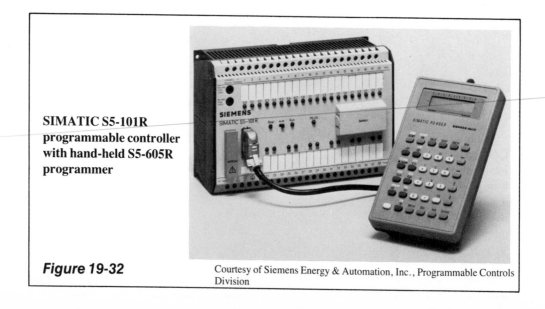

SIMATIC S5-101R programmable controller with hand-held S5-605R programmer

Figure 19-32

Courtesy of Siemens Energy & Automation, Inc., Programmable Controls Division

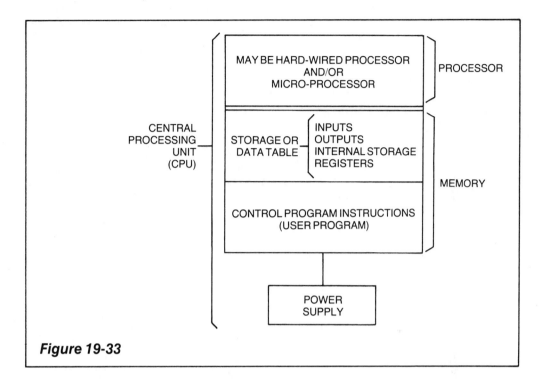

Figure 19-33

controller what to do. This may be referred to as the *user program memory* and contains ladder diagram instructions. This section of the memory is generally many times larger than the storage or data table. Instructions are placed in the memory by using a programming device, magnetic tape or systems computer. Figure 19-32 shows a programmable controller with a hand-held programmer.

Part of Figure 19-16 is repeated here in Figure 19-33 to help in remembering the place that the data table and the program storage have in the memory.

There are two general classes of memory:

1. Volatile
2. Non-volatile

The following describes the various types of memory within these two classes. It is important to learn how their characteristics affect the manner in which programmed instructions are retained or altered.

Volatile memory will lose its programmed contents if operating power is removed or lost. It is therefore necessary to have battery backup power at all times. One type of volatile memory is the RAM. This means read/write, solid-state random access memory. RAM information stored in the memory can be retrieved or read. Write indicates that the user can program or write information into the memory. Random access refers to the ability of any location in the memory to be accessed or used.

There are several types of RAM memory. Two of these are:

1. MOS (metal oxide semiconductor)
2. CMOS (complimentary metal oxide semiconductor)

The non-volatile memory retains its information when power is lost. It therefore does not require a battery backup. A common type of non-volatile memory is the ROM (read only memory).

As with the volatile memory there are several types of ROM memory.

1. PROM
2. EPROM
3. EAROM
4. EEPROM

PROM means *programmable read only memory*. It is a special type of ROM and is generally programmed by the manufacturer. It has the disadvantage of requiring special programming equipment and once programmed cannot be erased or altered.

EPROM means *erasable programmable read only memory*. A program can

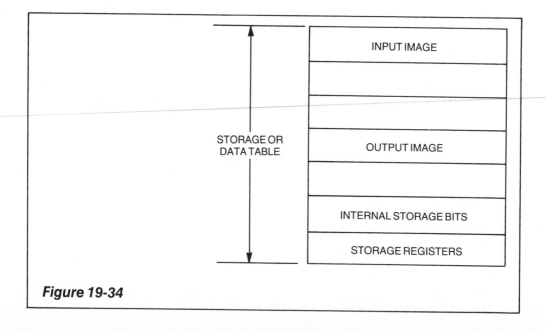

Figure 19-34

be completely erased by the use of an ultraviolet light source. After the program chip is completely erased, program changes can be made.

EAROM means *electrically alterable read only memory*. An erasing voltage applied to the proper pin of an EAROM chip will completely erase the program.

EEPROM means *electrically erasable programmable read only memory*. While it is a non-volatile memory it offers the same programming flexibility as the RAM memory. The program can be easily changed with the use of the standard CRT or manual programming unit.

Programmable controllers may have different memory sizes. The actual size will depend on the user's application. The sizes are generally expressed in K (1000) values. They may be 2K, 4K, 16K, etc. These numbers represent the number of words available. The range may be from 256 words for small PCs to 64K for large PCs.

Referring back to Figure 19-33, the data table is part of the memory. In this data table is the input image table and the output image table, Figure 19-34.

The *input image table* reflects the status of the input terminals. The *output image table* reflects the status of the output terminals. In these tables are stored the status of discrete input and output devices, preset and accumulated values of timers and counters, internal I/O relay equivalents, and numerical values for arithmetic functions, etc.

The data table can now be shown as in Figure 19-35 as it is subdivided into separate "cells." Each cell contains the information on a single bit. This signal is represented by either a 1 or a 0. The 1 indicates that the input or output device is ON or closed. The 0 indicates that the signal is OFF or open.

In Figure 19-23 it was shown that a group of 8 bits was called a "byte." When it is necessary for the processor to handle more than one bit, it is more efficient to handle a group of bits or a byte. A byte is the smallest group of bits

1	0	0	1	1	0	0	0
1	1	1	0	1	1	1	0
1	1	0	1	1	1	0	1
0	0	1	1	0	1	0	0
1	1	0	0	1	0	0	1
1	0	1	1	0	0	1	0

THIS BIT IS ON

THIS BIT IS OFF

Figure 19-35

the processor can handle at one time. A third division of bits, as shown in Figure 19-23, is the "word." The word is a fixed group of bits and is the unit that a processor uses when data is to be operated on or an instruction is to be performed.

In Figures 19-19 through 19-31 several of the numbering systems used were explained. Now it is necessary to organize the memory so that each bit representing a unit of information can be located. This is accomplished by assigning word and bit addresses. This procedure varies with each manufacturer of the programmable controller. An arrangement of one such organization is shown in Figure 19-36.

Note that the normally open push-button switch has been given an address of I0114. This means that input information from the push-button switch will be located in bit 14 of word address I01. Likewise the normally open limit switch (held operated) is given an address of I0112. This means that input information for the limit switch will be located in bit 12 of word address I01.

Since there is no signal available from the normally open push-button switch, bit 14 which holds the information from this switch will register 0. Looking at input I0112 (a normally open limit switch held operated), there is a signal (voltage) available. Therefore, bit 12 which holds the information from this switch will register a 1.

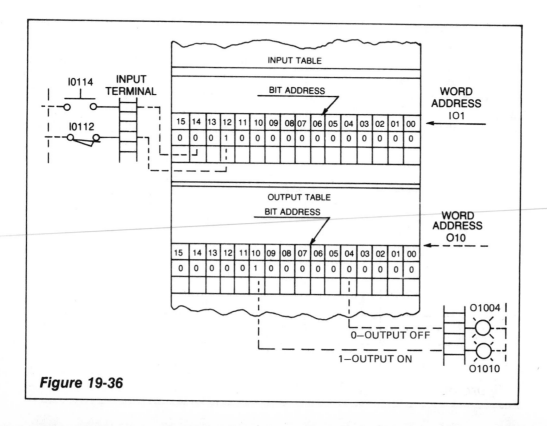

Figure 19-36

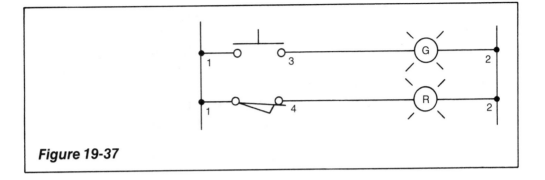

Figure 19-37

If an electromechanical ladder-type circuit diagram is drawn using the components shown in Figure 19-36, it would appear as in Figure 19-37.

If a program is written for the circuit as drawn, it would say: If the push-button switch is operated, then the green light will be energized. If the limit switch remains operated, then the red light remains energized.

However, in drawing the circuit for a programmable controller, the programmers do not have the symbols as used in electromechanical circuit diagrams. (Push-button switches, limit switches, pressure switches, etc.). While programming devices will vary with the manufacturer as to the characters they use, one uses normally open relay contact symbols for all component switches. The circuit from Figure 19-37 will now appear as shown in Figure 19-38.

The program storage or user section of the memory has been given an instruction that says that if the limit switch is closed, then the red light should be energized. With a 1 in bit 12 of word I01 from the input, there will now be a 1 in bit 04 of word 010 and the red light will be energized.

Another instruction has also been given the program storage. The green light is to be energized when the push-button switch is operated. Therefore, when the push-button is operated, bit 14 of word I01 will receive a 1. This then changes bit 04 of word 010 to a 1 and the green light will be energized.

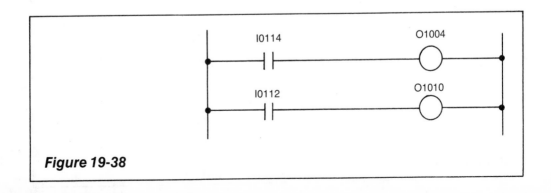

Figure 19-38

A program statement then consists of a condition and an action. For example, "If the limit switch is closed then energize the red light."

In general then, it can be said that each action is represented by a specific instruction. The instruction tells the processor to do something with the information stored in the data or user table.

Another electromechanical ladder-type circuit diagram is presented in Figure 19-39. The main function of any ladder-type diagram is to present the conditions that are available to control outputs based on inputs.

The conditions existing in the circuit shown in Figure 19-39 are as follows:

1. The normally closed stop push-button switch is not operated.
2. The normally open start push-button switch is not operated.
3. The normally closed limit switch is not operated.
4. Relay coil 1CR is not energized.
5. Relay contact 1CR-1 is open.

An instruction then, could be to operate the start push-button switch. The sequence of operation is then:

1. Relay coil 1CR is energized.
   a. Relay contact 1CR-1 closes, interlocking around the start push-button switch.
2. Relay coil 1CR remains energized.

From this it can be stated that for an output to be energized or activated, there must be a complete path for electrical energy from 1 to 2 in the top rung of the ladder. Or, it can be expressed as logic continuity.

If the circuit shown in Figure 19-39 were programmed for a programmable

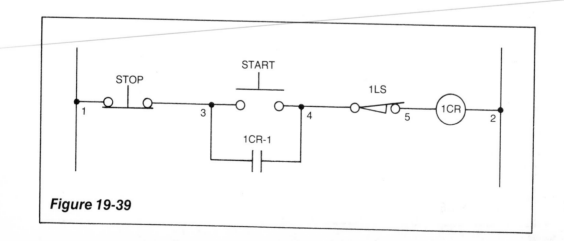

Figure 19-39

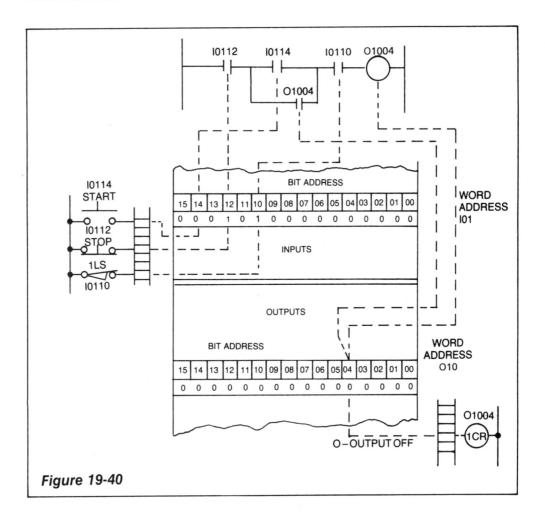

Figure 19-40

controller, it would appear as shown in Figure 19-40. The sketch combines the inputs, outputs and data tables in the memory.

Each contact and coil is given an address that references it to the data table. This can be either a connected input or output, or an internal stored bit output.

In the electromechanical circuit a contact on relay 1CR was used to interlock around the start push-button switch. In programmable control the holding contact 01004 will have the same address as the output. That is, they are on the same bit (04 of word 010) in the I/O register. The logic equivalent of the holding contact is set to 1 when the output is set to 1.

To start the process in the operation of this circuit as programmed, the processor scans the circuit for a self-check. This is to determine if all the connected inputs and outputs are in the condition as shown in the circuit.

The scanner checks the status of all discrete input devices for OFF or ON

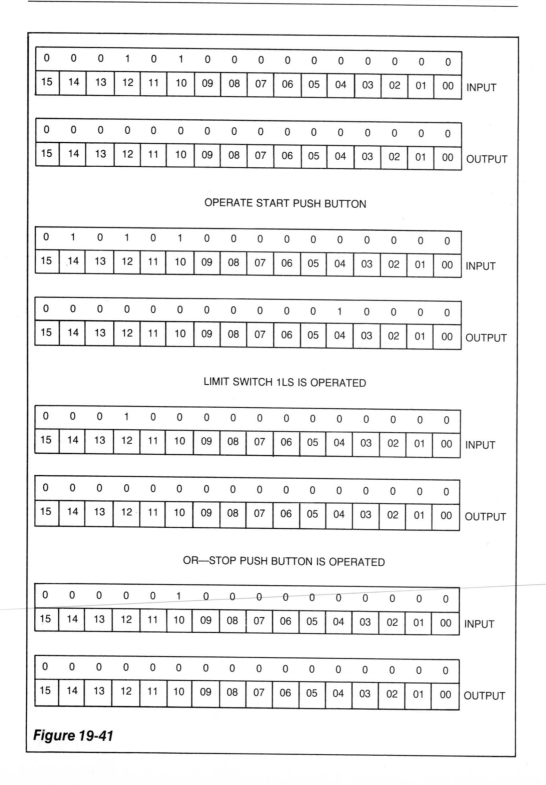

Figure 19-41

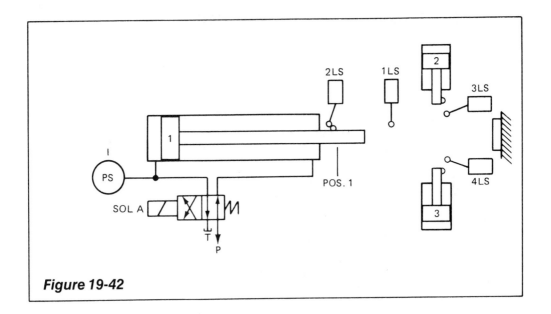

Figure 19-42

(0 or 1). It then scans the user program and solves the logic. The next step is to set the discrete output devices ON if conditions indicate this is the resulting action required by the program.

The process continues—examine inputs—solve logic—determine output status.

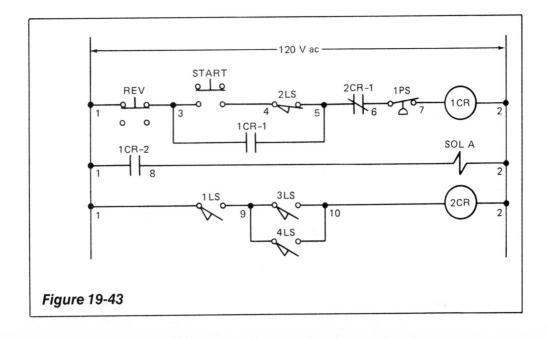

Figure 19-43

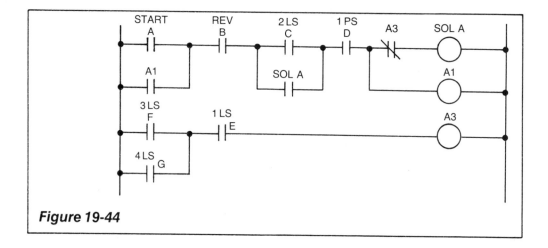

Figure 19-44

The bit status of each input and output device is shown for start conditions in Figure 19-40. As each event occurs through a complete cycle of the machine, the bit status changes to reflect the status of the input and/or the output device.

In Figure 19-41 these changes in bit status (1 or 0) are displayed for each event.

We can now look at an industrial application involving a plastic injection molding machine with cores in the mold.

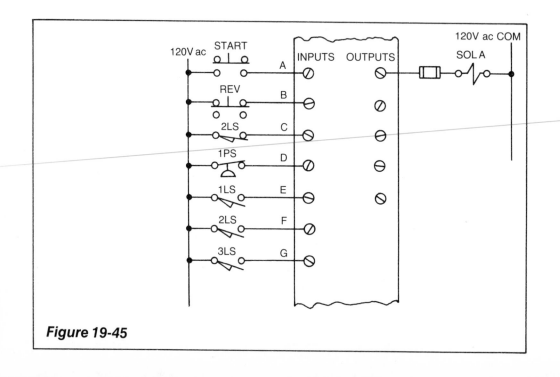

Figure 19-45

A piston and cylinder assembly is shown in Figure 19-42. Fluid power is supplied to the cylinder through a single-solenoid, spring-return operating valve. Piston and cylinder assemblies #2 and #3 (associated with the cores) can be operated with fluid power, pneumatic power, or a mechanical device. Their motion must be detected by limit switches.

Piston #1 must be in position 1 for start conditions. The piston is to move to the right as shown. At the position that #1 piston engages and operates limit switch 1LS, if either #2 or #3 piston has moved to a position operating limit switches 3LS or 4LS, then #1 piston stops and returns to position 1. If limit switches 3LS or 4LS are not operated when #1 piston operates 1LS, the piston continues its travel and builds pressure to a preset level on a work piece. The piston then returns to position 1. Provision must be made to return the piston from any point in its forward travel.

Figure 19-43 shows the electromechanical control circuit.

Figure 19-44 shows the same circuit arranged for a programmable controller.

Figure 19-45 shows the connections of the "real world" components as they would be connected into the PC's I/O devices.

19.4 PERIPHERAL DEVICES

One of the more important factors in the use of programmable controllers is the wealth of devices that can be added. The information that is stored in the memory can be reached and increase the overall capability of the control systems.

Several of the peripheral devices are:

1. Programmers
   a. Desk top (including CRT—cathode ray tube)
   b. Hand-held or miniprogrammer
2. Hard copy printout of programs
3. Program loaders
4. Computer

PROGRAMMERS

The desk top programmer generally consists of a CRT (cathode ray tube), similar to a television screen, a keyboard (similar to a typewriter keyboard) and electronic circuits and memory to use in loading a program into the processor memory, Figure 19-46. The CRT will give a visual display of the program.

There are two types of programming devices:

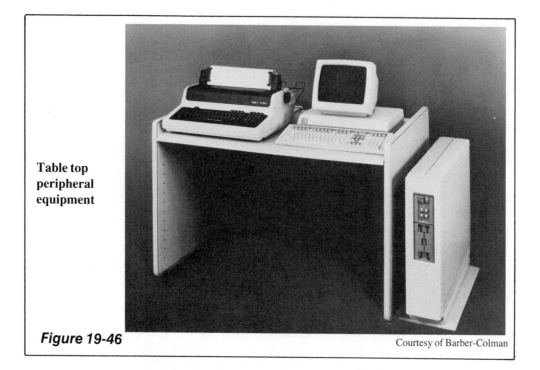

**Table top
peripheral
equipment**

Figure 19-46   Courtesy of Barber-Colman

1. Those that are not microprocessor based. This unit must be connected to the PC to enter or edit the control program. It generally can be used with programmable controllers of different manufacturers.
2. Those that are microprocessor based. With this unit the program can be stored without being connected to the PC. They generally use a disk or tape to store the program. They have the disadvantage of not being compatable with different manufacturers of PCs.

The hand held or miniprogrammer has limited display capabilities, but they are lightweight, portable, rugged in construction and less expensive.

One area where they are used is in start-up, changing and monitoring. Again, as with the desk-top type, one available unit is not microprocessor based and the other is microprocessor based. It follows then that the microprocessor based unit has greater capabilities. They have diagnostic functions such as memory, display and communications.

HARD COPY PRINTOUT

Hours of time that leads to costly operation can be saved with a printer. This prints out the program using familiar symbols and becomes a useful tool in

Loader/monitor

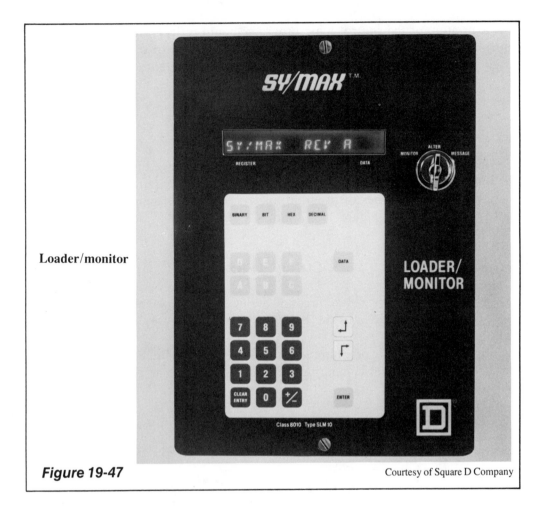

Figure 19-47   Courtesy of Square D Company

troubleshooting. The printouts are then available at some future time for reference. It also helps in gaining a better understanding of the particular program.

19.4.1 Documentation   Recorded information concerning the operation of the machine with the hardware and software components used in the control system is important.

The information should start with a complete description of how the machine should operate. Following this a sketch of the machine should be drawn showing the major hardware and software components such as the CPU, subsystems and peripherals.

An I/O connection diagram and the address assignments should be made. This also includes the Internal I/O address assignments.

A printout of the control logic program should reflect what is stored in the controller's memory. When changes are made in the control scheme, the records should be updated so that the stored program is kept current and accurate.

PROGRAM LOADERS

The program loaders are generally used only to load or reload a program into the processor. With the program finalized and debugged in the programmer it is transmitted to the program loader and stored. It can be retrieved at any time. There are two basic types of program loaders: electronic memory modules and the cassette recorders. Figure 19-47 shows a typical loader/monitor unit.

COMPUTERS

In some cases the manufacturers of the PCs have personal computers available. They have special software that allows them to be used as a programming device.

In addition to their programming capability, the computer provides ability for such tasks as general computing when it is not in use as a programmer.

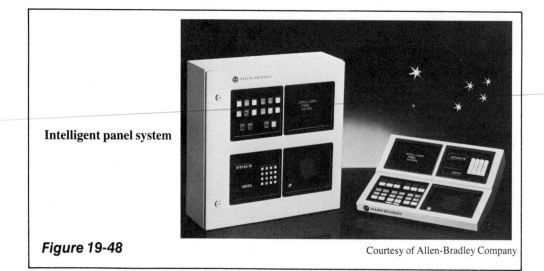

Intelligent panel system

Figure 19-48

Courtesy of Allen-Bradley Company

CRT DATA PORTS

Recently, machine programmable controllers include a CRT and data entry device. This equipment is used to enter machine operating parameters such as time and count setpoints, positions, temperatures, pressures/speeds, set-up conditions and occasionally process control parameters. The CRT is also useful for machine diagnostics and process monitoring. Many of today's systems include electronic storage of the above setpoints. This can allow quick machine set-up for process changes.

INTELLIGENT PANEL SYSTEM

Another more recent introduction into the general field of peripheral equipment is the *intelligent panel system*. This is a new microprocessor-based, modular approach to custom control systems. The intelligent panel system translates operator's commands into standard communications which are transmitted to the programmable controller. The heart of the intelligent panel system's flexibility is a powerful microcomputer which controls all module functions and communications between the host controller or programmable controller. See Figure 19-48 for a view of this system.

19.5 INDUSTRIAL APPLICATIONS

19.5.1 Information on cost savings when using a Programmable Controller/ Motor Control Center Combination is given courtesy of Allen-Bradley.

Multiple-motor installations are often used and the operator's control is mounted remotely in one or more consoles. The use of a *programmable controller/ motor control center* (PC/MCC) can result in considerable savings in installation costs. The savings increase as the distance between the motor starter and the operator's control increases.

There are also other savings in costs. The PC can perform functions that would require auxiliary contacts, timers and counters. Another valuable feature is on-board diagnostics. This can reduce the debugging and troubleshooting time drastically.

Figure 19-49A is a circuit showing the cost problem of interconnecting conductors between a motor starter and a remotely located operator's station. Figure 19-49B shows the same circuit using a PC.

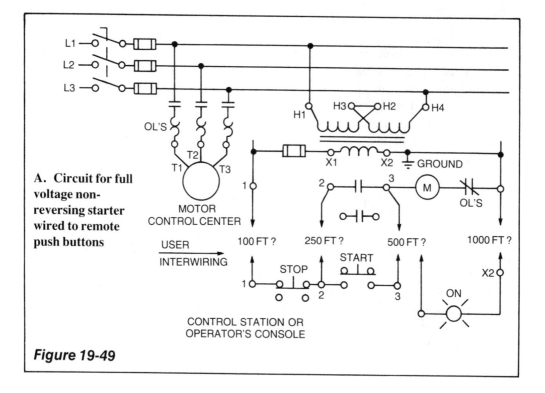

A. Circuit for full voltage non-reversing starter wired to remote push buttons

Figure 19-49

Figure 19-50 shows a motor control center with I/O racks mounted in a vertical section.

A communications cable leads from the PC processor to an I/O adaptor module located in the I/O rack. The outputs are connected in series with the starter coils. The inputs in the I/O rack monitor the starter's auxiliary contact, which verifies to the processor that the starter has been energized. Additional inputs can monitor overload relays, verify that the control circuit has power, or any number of other functions. Outputs can also be commanded to perform other functions.

Input signals are sent to the processor, where they are compared with the program that has been designed. If actual input conditions match those that have been programmed, output signals are sent to the output modules to energize the motor starter. If not, no processor action is taken. The process is completely automatic and any logic changes can be made at any time by programming new commands in the processor memory through the industrial terminal.

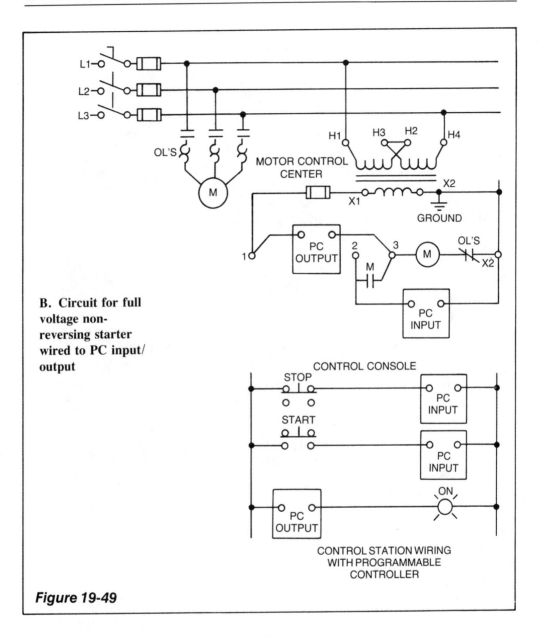

B. Circuit for full voltage non-reversing starter wired to PC input/output

Figure 19-49

19.5.2 Information on retrofit of controls for Extrusion Blow Molding Machine by using a Programmable Controller is given courtesy of Eagle Signal Division of Gulf Western Manufacturing Company.

A blow molding machine was originally controlled with electromechanical timers, relays and discrete temperature controls. A new system was installed

**Sy/MAX
Model 300
PC with
standard
I/O**

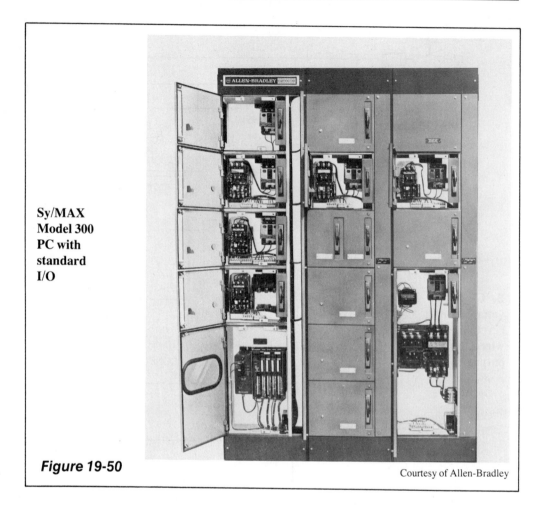

Figure 19-50

Courtesy of Allen-Bradley

using general purpose programmable controllers, Figure 19-51.

The machine required controls that would handle several process variables. Heater temperatures were controlled by 14 analog loops configured for that purpose.

A second PC was used to handle cycling control as well as water and oil temperature control. The PCs are also capable of handling future expansion or controlling additional features at a later time.

Safety interlocks were also very important in this system to prevent any chance of hazard to the operators or damage to the machine during automatic operation. Both requirements were incorporated into the PC software.

Maintenance mechanics use the visible I/O modules that indicate the circuits to troubleshoot. The operators will appreciate the safety features and the ease of operation.

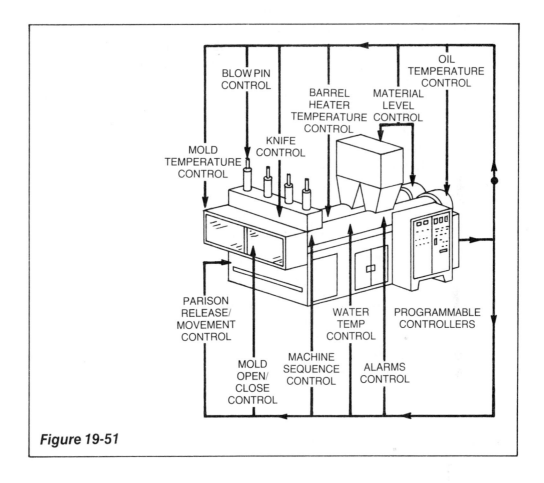

Figure 19-51

19.5.3 Information on controlling Palletizers in a Cookware Plant is given courtesy of Allen-Bradley.

It was decided to replace the manual control system with an automatic system. One important item was to reduce labor costs in building pallets and to keep an orderly flow of material through the plant. This move also helped to ensure maximum output.

Once the cookware has been cerammed, it is unloaded and placed on a continuously moving overhead monorail system which conveys the ware to the Final Inspection area. In this area, there are seven accumulating conveyors which empty onto a common trunk line conveyor. Each of the seven accumulating conveyor lines has two work stations, where Q.C. Inspectors hand-examine the product for defects, separate the acceptable cookware from the rejected ones, and then load the acceptable cookware into large corrugated containers called hampers which are about the size of orange crates. The inspector then places the hampers of inspected ware on a metering belt, which feeds onto an incline conveyor which conveys the hamper 10 feet up to the overhead conveyor system. The entire conveying system is overhead, allowing use of the floor area below.

Each of the seven accumulating conveyors can accommodate a queue of hampers 50 feet long. The hampers are lined up as they are fed from the inspection station and are prevented from entering the trunk line by an air-operated stop, located at the front of each accumulating conveyor. When a pallet load of 12 hampers has been accumulated, a photoeye, wired to a PC input, detects the presence of the last hamper, darkens, and another stop is raised to prevent other hampers from entering the primary zone. If necessary, a secondary zone behind it can handle another six hampers in an overflow area.

The darkening of the photoeye signals the Mini-PLC-2 Processor to check the trunk line for any hampers from the other six conveyors now accumulating. If none are found, one queue of 12 hampers is discharged onto the trunk line where it travels to one of the two palletizers for processing. When all hampers released have been counted by a photoeye at the palletizer, the diverter arm swings to allow the next load to enter the opposite palletizer. At this point, the trunk line is clear, and another accumulating conveyor can be emptied. The Mini-PLC-2 controller automatically controls the release of the accumulating lines and the position of the diverter arm.

Each of the two palletizers is designed to allow three levels, four hampers per level per pallet to be processed sequentially. An operator stationed on a platform between the two palletizers uses a foot switch to control the flow of hampers. The operator manually orients four hampers in one operation on an indexing platen, and then retracts it, lowering the boxes on the pallet below. This process is repeated a total of three times for each pallet and then a chain-operated conveyor, monitored by a tier counter, detects a full pallet and lowers it to floor level where it passes a final inspection station. There, the accept/audit status, as determined by a statistical sampling plan, is transmitted to Inspection Technicians by red and green lamps, controlled by the second Mini-PLC-2 controller. Once the final quality check has been made, fork lifts transport the pallets to the dock for shipment to the packaging and distribution facility.

In this application, a 64-I/O rack was used for each Mini-PLC-2 controller with all AC input and output modules wired to photoeyes, limit switches and other standard control devices. This system has run for 8 months, 24 hours a day, 7 days a week with no major maintenance problems.

19.5.4   Information on Metal Extrusion Control is given courtesy of Eagle Signal, division of Gulf-Western Manufacturing Company.

The application of the EPTAK 700 controller to metal extrusion control can provide significant economies of operation, productivity and quality improvements. A microprocessor-based programmable controller designed specifically for process control such as metal extrusion, the EPTAK 700 Controller offers advanced capabilities normally found only in more sophisticated and expensive computer systems. As an industrial controller it is both highly efficient and reliable. Yet, it does not share a computer system's complexity or its special environmental requirements.

With modular hardware and software tailored to the application, the EPTAK controller is low in cost compared to process computer systems or systems marrying computers with less sophisticated programmable controllers. The EPTAK system is designed to operate in plant-floor environment as well as in control room installations. It requires a minimum of factory space.

Advanced capabilities found in the EPTAK 700 Controller include the availability of 255 PID (Proportional Integral Derivative) loops for process control; a high-speed, floating-point arithmetic capability; a color-graphic software package and a diskette-based CRT programming terminal.

Customers can choose from three experience-oriented programming languages. Data display and computations in engineering units common to metal extrusion process simplifies the operator interface. Offering more analog capability than a programmable controller, the EPTAK system can also control more than 2,000 external digital devices.

As a result, using the EPTAK controller, the plant manager and engineering personnel responsible for metal extrusions can realize benefits of:

- A more economical approach than manually controlled processes or more complex general purpose computers requiring a variety of interface devices and programs.

- An effective combination of the easy programming and the ruggedness of a programmable controller with the advanced analog control and math capability of a process computer.

CONTROLLING METAL EXTRUSION PRODUCTION

The extrusion process, using the direct (conventional) or indirect (Hooker process) method, can be reliably controlled with the EPTAK system. The EPTAK system provides machine control, material handling control and information management used in the metal extrusion production process. These are some of the activities normally controlled:

- Machine control of equipment such as billet saws, rolling mills, casting equipment, picklers, furnaces and conveyors.

Additional or auxiliary operations provided by the system include:

• Inventory of raw materials and finished product.

• Improved equipment maintenance through tracing of faulty components.

• Metering of energy usage.

• Logging and reporting functions.

• Operator interface.

• Communications (with other controllers or plant computers.)

• Safety and security monitoring.

An entire metal extrusion process can be controlled with an EPTAK system. For example, an EPTAK system controls a complete brass extrusion cycle including control of a billet saw, conveyors, gas furnace, induction furnaces, press cycle, and automatic pickle line. It also prints production reports.

19.5.5 Information on control of a Rubber Forming Press is given courtesy of Solid Controls Inc.

Complete enclosure assembly

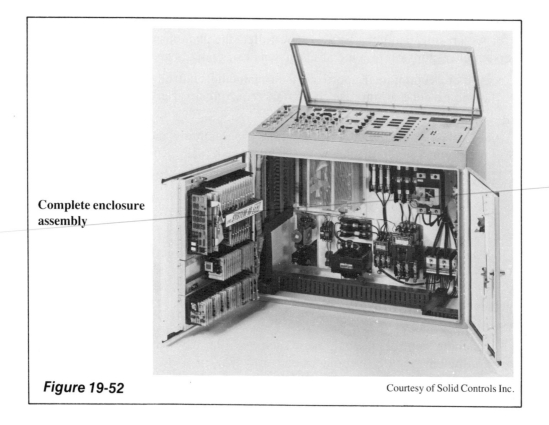

Figure 19-52   Courtesy of Solid Controls Inc.

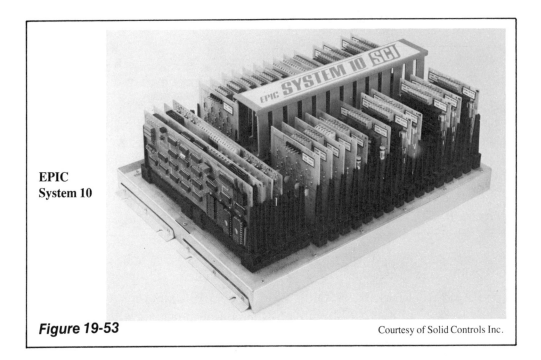

**EPIC
System 10**

Figure 19-53

Courtesy of Solid Controls Inc.

The system used was designed to control a rubber forming press with extruder head, overhead injection unit, clamp and shuttle.

The application requirements included:

- Fast I/O sequence decision making control
- Timing and counting
- Function and mode selection control
- Multizone P.I.D. closed-loop temperature control
- Interface control for separate analog process controller
- LED sequence indicators
- Machine/system diagnostics

The cabinet photo, Figure 19-52, shows the complete enclosure assembly. Housed inside are all electronic and electrical power components needed for machine control. Two SCI PC architectures are used for control. The EPIC System 10, Figure 19-53, provides all the functions listed above except for temperature control which is provided by the EPIC-8A architecture (screened-in against back panel). Figure 19-54 shows the operator's control panel.

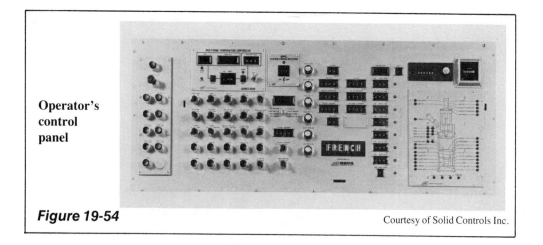

Operator's control panel

Figure 19-54

Courtesy of Solid Controls Inc.

Additional specifications for the system included:

- BASE MICROPROCESSORS: EPIC System 10—Motorola MC14500 1 bit processor, EPIC-8A—Motorola MC6809 8 bit processor

- PROGRAMMING LANGUAGES: EPIC System 10—EPIC (SCI's 1 bit assembly language). EPIC-8A—Motorola MPL compiler

- PROGRAMMING DEVICES: EPIC System 10—SCI's Mobile Programming Terminal, EPIC Data Base Programming System or IBM Personal Computer (with SCI's compatible software), EPIC-8A, any program development system capable of supporting MPL.

19.5.6 Information on retrofitting a Plastic Injection Molding Machine is given courtesy of Solid Controls Inc.

The application of retrofitting a plastic injection molding machine required the following:

- Fast I/O sequence control decision making control
- Timing and counting
- Function and mode selection control
- Hydraulic pump selection control
- Interface to separate analog closed-loop process controller
- Machine/system diagnostics

The two cabinet photos, Figures 19-55A and 19-55B, show the operator control panel and the inside of the enclosure. Figure 19-56 shows the SCI's EPIC System 10 PC which is the heart of the system.

A. Operator's control panel

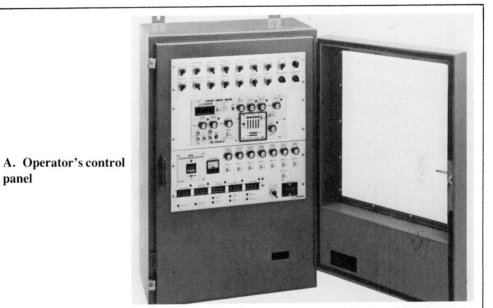

Courtesy of Solid Controls Inc.

B. View of inside of enclosure

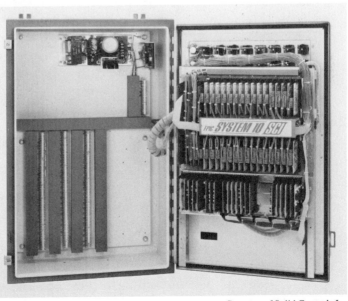

Figure 19-55

Courtesy of Solid Controls Inc.

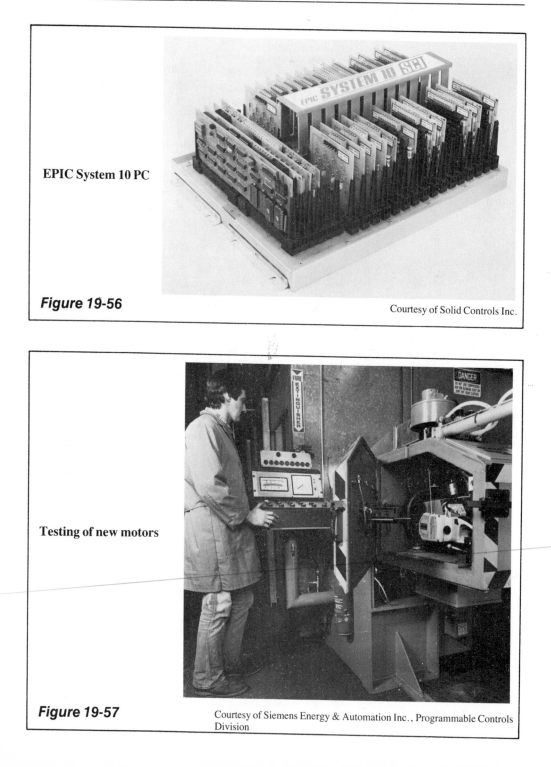

EPIC System 10 PC

Figure 19-56

Courtesy of Solid Controls Inc.

Testing of new motors

Figure 19-57

Courtesy of Siemens Energy & Automation Inc., Programmable Controls Division

PC and CRT programmer used to control and monitor a depalletizing process

Figure 19-58

Courtesy of Siemens Energy & Automation Inc., Programmable Controls Division

Additional specifications include:

- Base Microprocessor: Motorola MC14500 1-bit processor
- Programming Language: EPIC (SCI's 1 bit assembly language)
- Programming Devices: SCI's EPIC Mobile Programming Terminal, SCI's EPIC Data Base Programming System, IBM Personal Computer (with SCI's compatible software)

19.5.7 A Siemens S5-115U programmable controller controls the testing of saw motors at an in-line test stand of a chain saw manufacturer. Use of the PC increased the number of units tested by eight times over manual testing—from 15 per hour to 120, Figure 19-57.

19.5.8 At an aluminum can manufacturing plant, a Siemens S5-115U programmable controller and PF 675 CRT programmer are used to control and monitor a depalletizing process, Figure 19-58.

19.6 MOTION CONTROL

In many areas of industry today, motion control is an important consideration and/or requirement.

The motion required for a specific application must be accurate in very small increments of change. It may be linear bidirectional, in one or more axis of travel. It also may involve rotary motion.

The applications for motion control are many. They range from machining, positioning of tools, testing, inspection, welding, assembly, etc. In each application a drive motor and control combination are designed to satisfy the requirements.

One supplier of motion control equipment offers an ac synchronous motor and a dc stepping motor made under the trade mark of *SLO-SYN®. SLO-SYN® ac synchronous motors are permanent magnet ac motors having extremely rapid starting, stopping and reversing characteristics. They have a basic shaft speed of 72 or 200 RPM, depending on the motor selected. See Figure 19-59 for an exploded view.

Three lead types need only a single-pole, three-position switch for complete forward, reverse and OFF control. When operating a SLO-SYN® motor from a single-phase source, a phase-shifting network consisting of a resistor and capacitor is used. See Figure 19-60.

Standard SLO-SYN® motors are available in torque ratings from 22–1500 ounce-inches. The output torque varies with changes in the input voltage.

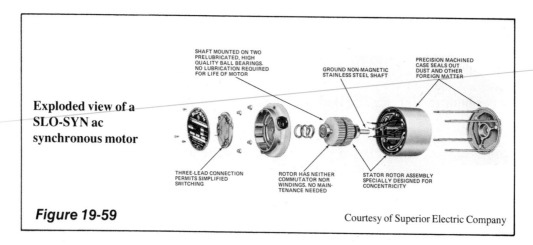

Exploded view of a SLO-SYN ac synchronous motor

SHAFT MOUNTED ON TWO PRELUBRICATED, HIGH QUALITY BALL BEARINGS. NO LUBRICATION REQUIRED FOR LIFE OF MOTOR

GROUND NON-MAGNETIC STAINLESS STEEL SHAFT

PRECISION MACHINED CASE SEALS OUT DUST AND OTHER FOREIGN MATTER

THREE-LEAD CONNECTION PERMITS SIMPLIFIED SWITCHING

ROTOR HAS NEITHER COMMUTATOR NOR WINDINGS. NO MAINTENANCE NEEDED

STATOR ROTOR ASSEMBLY SPECIALLY DESIGNED FOR CONCENTRICITY

Figure 19-59   Courtesy of Superior Electric Company

*Information on SLO-SYN® motors is supplied through the courtesy of Superior Electric Co.

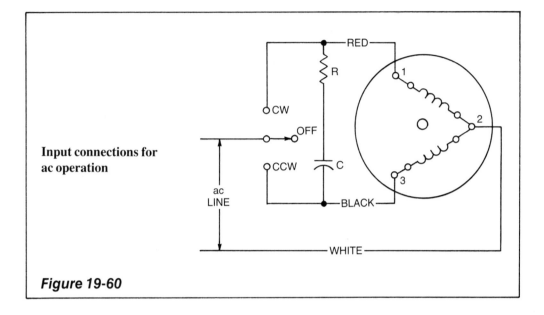

**Input connections for
ac operation**

Figure 19-60

When a SLO-SYN® motor is energized, ac current flows only through the windings. Current does not flow through the rotor or through brushes, since the motor is of a brushless construction. Therefore, it is not necessary to consider high inrush currents when designing a control for a SLO-SYN® motor since starting and operating current are, for all practical purposes, identical.

The permanent magnet construction of a SLO-SYN® motor provides a small residual torque which holds the motor shaft in position when the motor is deenergized. When holding torque is required, a dc voltage can be applied to one winding when the ac voltage is removed. If necessary, dc voltage can be applied to both windings with resulting increase in holding torque. See Figure 19-61.

In determining the ac synchronous motor required for a specific application, the following information must be known:

1. **Motor shaft speed.** SLO-SYN® motors are available in shaft speeds of 72 and 200 RPM at 60 Hertz. Gearing can be used to obtain other speeds. It is also necessary to consider the effects of gears (if used) on torque and inertial load characteristics of the load.

2. **Load characteristics.** Examples that follow show how to determine the torque and moment of inertia characteristics of the load.
   a. Torque in ounce-inches = Fr (See Figure 19-62.)

$$\text{Where } F = \text{force in ounces}$$
$$r = \text{radius in inches}$$

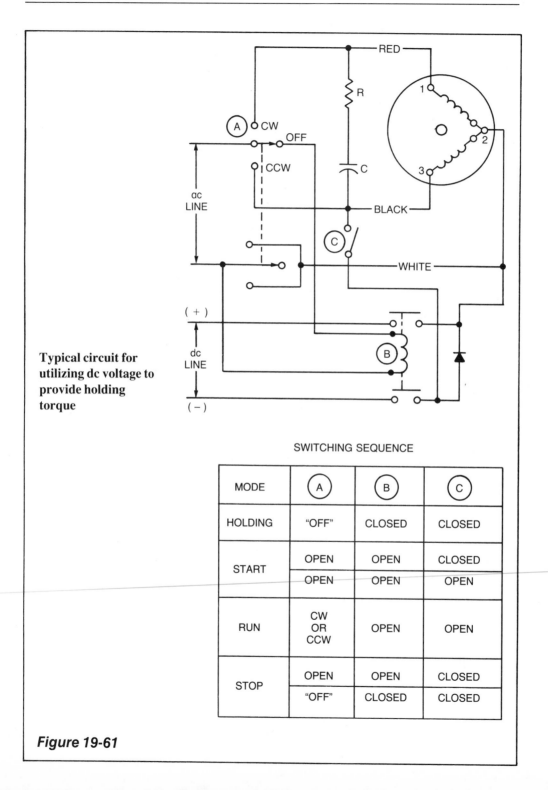

Typical circuit for utilizing dc voltage to provide holding torque

SWITCHING SEQUENCE

MODE	(A)	(B)	(C)
HOLDING	"OFF"	CLOSED	CLOSED
START	OPEN	OPEN	CLOSED
	OPEN	OPEN	OPEN
RUN	CW OR CCW	OPEN	OPEN
STOP	OPEN	OPEN	CLOSED
	"OFF"	CLOSED	CLOSED

Figure 19-61

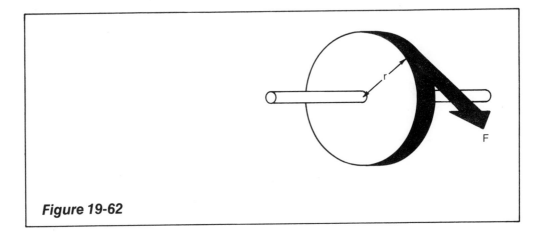

Figure 19-62

Using a 4" diameter pulley it is found that a 2 pound pull on the scale is required to rotate the pulley.

$$F = 2 \text{ pounds} = 32 \text{ ounces}$$
$$r = \frac{4}{2}$$
$$= 2"$$

The torque is then 32 x 2 = 64 ounce-inches

b. Moment of Inertia in (lb-in.^{2}) (See Figure 19-63.)

$$I(\text{lb-in.}^2) = \frac{Wr^2}{2} \text{ for a disc}$$

OR

$$I(\text{lb-in.}^2) = \frac{W}{2} (r_1^2 + r_2^2) \text{ for a cylinder}$$

Where W = weight in pounds
r = radius in inches

Using a load of a 8" diameter gear weighing 8 ounces:

$$W = \frac{8}{16} = 0.5 \text{ lb}$$

$$r = \frac{8}{2} = 4 \text{ in.}$$

$$\text{Moment of inertia} = \frac{0.5 \times (4)^2}{2} = 4 \text{ lb-in.}^2$$

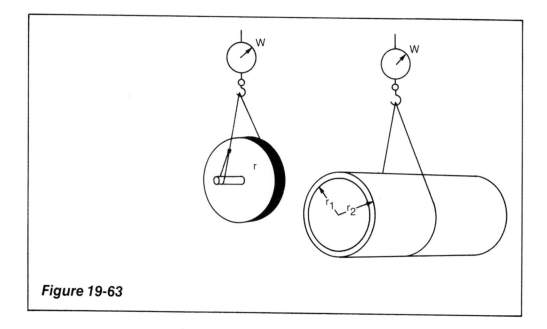

Figure 19-63

With voltage and frequency known, the user should refer to the manufacturer's tables of torque and moment of inertia to select the motor that best suits the requirements.

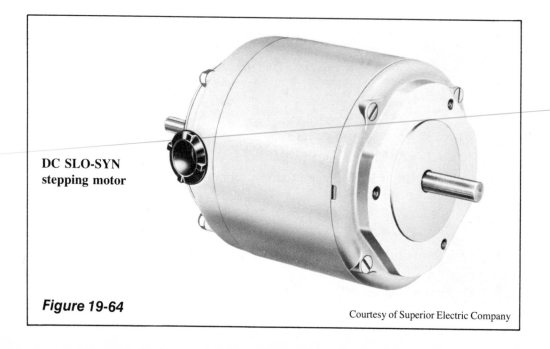

DC SLO-SYN stepping motor

Figure 19-64

Courtesy of Superior Electric Company

Another motor offered by the same company is the SLO-SYN® dc stepping motor. It is a permanent magnet motor that converts electronic signals into mechanical motion. See Figure 19-64.

SLO-SYN® stepping motors operate on phase-switched dc power. Each time the direction of current in the motor windings is changed, the motor output shaft rotates a specific angular distance. The motor shaft can be driven in either direction and can be operated at very high stepping rates.

The motor shaft advances 200 steps per revolution (1.8° per step), when a four-step input sequence (full-step mode) is used, and 400 steps per revolution (0.9° per step) when an eight-step input sequence (half-step mode) is used. Power transistors connected to flip-flops or other logic devices are normally used for switching as shown in the wiring diagram, Figure 19-65. The four-step input sequence is also shown here.

Since current is maintained on the motor windings when the motor is not being stepped, a high holding torque results.

A SLO-SYN® dc stepping motor offers many advantages as an actuator in a digital-controlled positioning system. It is easily interfaced with a microcomputer or microprocessor to provide opening, closing, rotating, reversing, cycling and highly accurate positioning in a variety of applications.

Mechanical components such as gears, clutches, brakes and belts are not needed since stepping is accomplished electronically.

A SLO-SYN® dc stepping motor applies holding, or detent torque at standstill to help prevent an unwanted motion.

The motors are available in a range of frame sizes and with standard step angles of 0.72°, 1.8°, and 5°. Step accuracies are 3% or 5%, non-cumulative. They can be driven at rates to 20,000 steps per second with a minimum of power input.

In determining the motor required for a specific dc stepping motor application, the following information must be determined:

- Output speed in steps per second

- Torque in ounce-inches

- Load inertia in lb-in.²

- Required step angle

- Time to accelerate in milli-seconds

- Time to decelerate in milli-seconds

- Type of drive system to be used

- Size and weight considerations

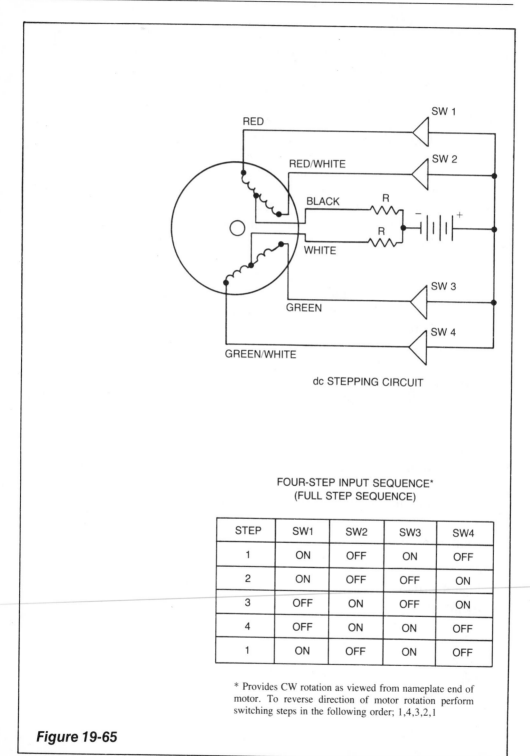

dc STEPPING CIRCUIT

FOUR-STEP INPUT SEQUENCE*
(FULL STEP SEQUENCE)

STEP	SW1	SW2	SW3	SW4
1	ON	OFF	ON	OFF
2	ON	OFF	OFF	ON
3	OFF	ON	OFF	ON
4	OFF	ON	ON	OFF
1	ON	OFF	ON	OFF

* Provides CW rotation as viewed from nameplate end of
motor. To reverse direction of motor rotation perform
switching steps in the following order; 1,4,3,2,1

Figure 19-65

1. Torque: (Oz − in.) = Fr
   Where F (in ounces) required to drive the load
      r (in inches) radius

2. Moment of inertia

$$I(\text{lb-in.}^2) = \frac{Wr^2}{2} \text{ for a disc}$$

$$= \frac{W}{2} (r_1^2 + r_2^2) \text{ for a cylinder}$$

   Where W = weight in pounds
      r = radius in inches

3. Equivalent inertia
   A motor must be able to:
   a. Overcome any frictional load in the system.
   b. Start and stop all inertia loads including that of its own rotor.
   The basic rotary relationship is:

$$T = \frac{I\,\alpha}{24}$$

   Where T = Torque in ounce-inches
      I = Moment of inertia
      α = Angular acceleration in radians per second^2
      24 = Conversion factor from ounce-inch-sec^2 x 24 = pound-in.^{2}

(Measuring angles in radians: the angle at the center of a circle which embraces an arc equal in length to the length of the radius. The value of the radian in degrees equals 180/57.296 degrees. Then, 3.1416 radians denotes an angle of 180 degrees. Note it is often convenient to measure angles in radians when dealing with angular velocity. If ω = angular velocity per second of the revolving body, in radians; V = velocity of a point on the periphery of the body, in feet per second; and r = radius in feet)

$$\omega = \frac{V}{r}$$

for example, if the velocity of a point on the periphery is 10 feet per second and the radius is 1 ft:

$$\omega = \frac{10}{1} = 10 \text{ radians}$$

Angular acceleration (α) is a function of the change in velocity (ω) and the time required for the change.

$$\alpha = \frac{\omega_2 - \omega_1}{t}$$

Or if starting from 0:

$$\alpha = \frac{\omega}{t}$$

Where ω = angular velocity in rad/sec − t = time in seconds.
   Since:

$$\omega = \frac{\text{steps per second}}{\text{steps per revolution}} \times 2\pi$$

Angular velocity and angular acceleration can also be expressed in steps per second (ω') and steps per second^2 (α'), respectively.

SAMPLE CALCULATIONS

1. To calculate the torque required to rotationally accelerate an inertia load. See Figure 19-66.

$$T = 2 \times I_o \left(\frac{\omega'}{t} \times \frac{\pi \, \theta}{180} \times \frac{1}{24} \right)$$

Where T = Torque required in ounce-inches
   I_o = Inertial load in lb-in.^{2}
   π = 3.1416
   θ = Step angle in degrees
   ω' = Step rate in steps per second

Example #1: Assume the following conditions:
   • Inertia = 9.2 lb-in.^{2}
   • Step angle = 1.8°
   • Acceleration = 0–1000 steps per second in 0.5 seconds
Then:

$$T = 2 \times 9.2 \times \frac{1000}{0.5} \times \frac{\pi \times 1.8}{180} \times \frac{1}{24}$$

$$T = 48.2 \text{ oz-in. to accelerate inertia}$$

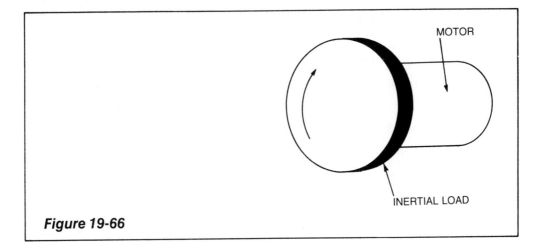

Figure 19-66

2. To calculate the torque required to accelerate a mass moving horizontally and driven by a rack and pinion or similar device. See Figure 19-67.

   The total torque which the motor must provide includes the torque required to:

   a. Accelerate the weight, including that of the rack
   b. Accelerate the gear
   c. Accelerate the motor rotor
   d. Overcome frictional forces

To calculate the rotational equivalent of the weight:

$$I_{eq} = Wr^2$$

Where W = weight in pounds
       r = radius in inches

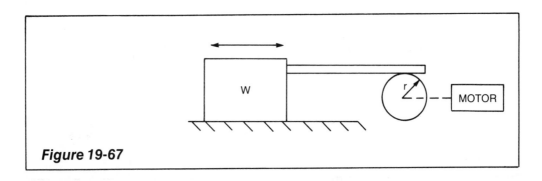

Figure 19-67

Example #2: Assume the following:

- Weight = 5 lb
- Gear pitch diameter = 3"
- Gear radius = 1.5"
- Velocity = 15 ft per sec
- Time to reach velocity = 0.5 sec
- Pinion inertia = 4.5 lb-in.^{2}
- Motor rotor inertia = 2.5 lb-in.^{2}

$$I_{eq} = Wr^2 = 5(1.5)^2 = 11.25 \text{ lb.-in.}^2$$
$$\text{Pinion} = \quad 4.5 \quad " \quad "$$
$$\text{Rotor} = \quad \underline{2.5} \quad " \quad "$$
$$\text{Total} = 18.25 \quad " \quad "$$

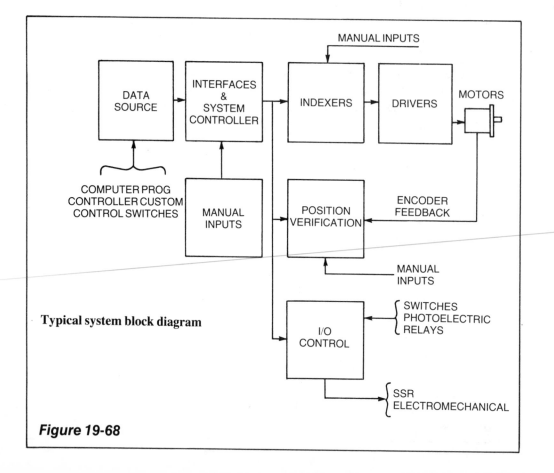

Typical system block diagram

Figure 19-68

Velocity is 15 feet per second, with a 3 inch pitch diameter gear. Therefore, speed:

a. $= \dfrac{15 \times 12}{3 \times 3.1416} = 19.1$ revolutions per second

b. The motor step angle is 1.8° (200 steps per revolution)

c. Velocity in steps per second $\omega' = 19.1 \times 200 = 3820$ steps per second

d. To calculate torque to accelerate the system:

$$T = 2 \times I_o \times \dfrac{\omega}{t} \times \dfrac{\pi\,\theta}{180} \times \dfrac{1}{24}$$

$$= 2 \times 18.25 \times \dfrac{3820}{0.5} \times \dfrac{3.1416 \times 1.8}{180} \times \dfrac{1}{24} = 364 \text{ oz-in.}$$

To calculate torque needed to slide the weight, assume a frictional force of 6 ounces. Then: $T_{friction} = 6 \times 1.5 = 9$ oz-in. Therefore total torque required $= 364 + 9 = 373$ oz-in.

There are many ways in which these motors can be controlled. In general, the design of the control will depend on the particular application.

A typical system block diagram is shown in Figure 19-68. This illustrates the versatile building block approach provided by Superior Electric's MODULYNX motor controls. This allows the user to select the combination of options and features needed for a specific application.

Drivers are available at various power ratings up to 3.5 HP. These drivers, coupled with control capabilities of MODULYNX indexers, microstepping indexers and microstepping translators provide flexible and inexpensive motor control capability. With MODULYNX motor controls, external control operations associated with overall system operation can be easily integrated into the motion control requirement.

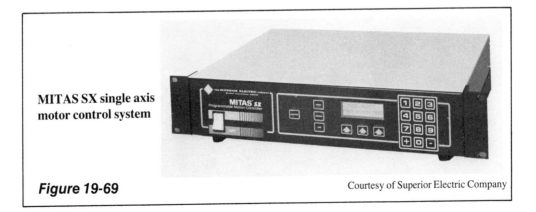

MITAS SX single axis motor control system

Figure 19-69   Courtesy of Superior Electric Company

From the many systems offered by Superior Electric, two are shown here with a brief discussion of their capabilities. The MITAS SX, single-axis motor control system is shown in Figure 19-69. It has a non-volatile memory which provides storage of multiple programmed operations. The user controls such things as speed, move distance and acceleration/deceleration rates. Some of the features include:

- Program protect
- Subroutine capability
- Edit capability simplifies program changes or corrections
- Programmable I/O
- Absolute or incremental programming
- Electrically isolated I/O

A second system is the MITAS two-axis controller (MODULYNX Intelligent Two-Axis System), Figure 19-70. This system offers the users a simple-to-install packaged system providing sophisticated operating and control features. The system consists of a two-axis controller which can be interfaced with various drives and motors. This depends on the speed and torque requirements of the specific application.

The controller utilizes a 16-bit microprocessor and has a non-volatile (E^2 PROM) memory. This type of memory does not lose stored data when power is removed. Controller and drives operate from ac power lines, eliminating the need for additional power supplies. The drive unit can be located up to 1000 feet from the controller. Some of the many controller features are:

- Packaged unit simplifies installation and eliminates assembly requirements
- Can be operated from the front panel or remote switches

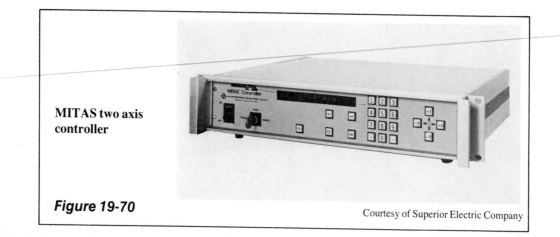

MITAS two axis controller

Figure 19-70

Courtesy of Superior Electric Company

- Front panel key switch provides system security
- Teach mode. This mode represents a major innovation in stepping motor control. In the teach mode, the controller can be "taught" an absolute or incremental program move by positioning the motors. This move is then stored in the controller memory. This information can be retrieved from the front panel or by Superior Electric's R232 device for execution or external storage of the system memory.
- Eight relay controls for operating external equipment
- Unit interlock for multi-axis control
- Absolute or incremental programming

INDUSTRIAL MOTION CONTROL APPLICATIONS AND SOLUTIONS

A few of the many varied problems that arose in industry are described and the solution that was used is given.

WRAPPING MACHINES

Problem

A motor with a clutch/brake mechanism was used to precision-wrap labels on various consumer products. Labels were fed from a continuous roll and the brand name was centered on the product. The process was cumbersome and the clutch/brake assembly required heavy maintenance.

Solution

The SLO-SYN® stepping motor eliminated the clutch/brake system, reducing initial costs. The reliability and performance of the SLO-SYN® stepper motor reduced overall maintenance costs. Production hours increased since no maintenance is required for the stepping motor system.

HOLE PUNCHING

Problem

A company manufactured flat drive belts indexed every .0375" for the tractor-type mechanism of computer printers. An air cylinder advance mechanism and two belt-driven pulleys produced the necessary drive ratio. The manufacturer wanted to improve his production method.

Solution

A SLO-SYN® stepper motor and preset indexer easily accomplished the indexing of these drive belts with greater accuracy and reliability than the previous method. Scrap was reduced 7%, less compressed air was needed, and the belt production was increased 21%.

ROTARY TABLE INDEXER

Problem

Various products are clamped on a rotary table and indexed for machining on three specific angles. Frequent changeovers demanded fast, accurate set up to maintain production rates.

Solution

The operation was automated with a Superior Electric indexer, SLO-SYN® motors, three thumbwheel set-up switches and sequencing circuitry. Labor content was lowered while production was increased by 18%.

TIRE MANUFACTURING

Problem

In the manufacturing of radial tires, two rubber strips must be precisely overlaid. The previous brake and brake system generated substantial waste.

Solution

A SLO-SYN® stepping motor with a 10,000 steps/second translater programs the exact number of steps so that the strips are accurately positioned. Production rates have been increased and scrap reduced.

ANTENNA POSITIONING

Problem

A 40-hour antenna test sequence required the full attention of an operator to position an antenna along 3 axes of rotation with a servo system. Also, operators found that motor leads would wind up as the fixture turned.

Solution

MODULYNX programmable control, driving three SLO-SYN® stepping motors, automated the 40-hour test sequence thus saving operator time and minimizing test error. Cost of the MODULYNX system was 75% less than the servo-driven test system. Problems with the motor leads winding up were corrected with the use of high-quality slip rings.

ACHIEVEMENT REVIEW

1. Using a block diagram, show the relationship between the input section-central processing unit-output section and the power unit.

2. State four advantages to be gained through the use of a programmable controller.

3. Explain why the power unit for the programmable controller must supply a well-regulated power supply and protection for the other system components.

4. What are four important factors to be considered in the input/output section of a programmable controller?

5. Convert the binary number—10101101_2—to a decimal number.

6. Explain the difference between a decimal number, binary number, octal number and the hexadecimal number.

7. Explain the difference between the volatile memory and the non-volatile memory.

8. What is the difference between the input image table and the output image table?

9. What is meant by the term "Scanning"?

10. List a few of the peripheral devices that are available to use with the programmable controller.

11. CRT means:
    a. critical return time.
    b. control resistance technology.
    c. cathode ray tube.

12. CRT data ports are useful for:
    a. mathematical calculations.
    b. machine diagnostics and process monitoring.
    c. replacing relays, timers and counters.

13. Using a 6-inch diameter pulley it is found that a 3-pound pull on the scale is required to rotate the pulley. Calculate the torque in ounce-inches.

14. You have a cylinder that weighs 50 pounds. The radius of the outside is 6 inches. The radius of the inside is 5 inches. Calculate the moment of inertia in lb-in.^{2}.

GLOSSARY

AC Input Interface (Module)—An input circuit that conditions various ac signals from connected devices to logic levels that are required by the processor

AC Output Interface (Module)—An output circuit that switches the user-supplied control voltage required to control connected ac devices

Accessible (As applied to equipment)—Admitting close approach because not guarded by locked doors, elevation or other effective means

Across-the-Line Starter—A motor starter that connects the motor to full voltage supply

Actuator—A cam, arm or similar mechanical or magnetic device used to trip limit switches

Address—A reference number assigned to a unique memory location. Each memory location has an address and each address has a memory location

Alternating Current (ac)—Electric current that periodically changes direction and magnitude

Alternation—One-half cycle in alternating current, either positive or negative half

Ambient Conditions—The condition of the atmosphere which is adjacent to the electrical apparatus; the specific reference may apply to temperature, contamination, humidity, etc.

Ampacity—The current-carrying capacity, expressed in amperes

Ampere (A)—The unit of current

Analog Device—Apparatus that measures continuous information (voltage—current). The measured analog signal has an infinite number of possible values. The only limitation on resolution is the accuracy of the measuring device

Analog Input Interface—An input circuit that employs an analog-to-digital converter to convert an analog value, measured by an analog measuring device to a digital value that can be used by the processor

Analog Output Interface—An output circuit that employs a digital-to-analog converter to convert a digital value, sent from the processor to an analog value that will control a connected analog device

Analog Signal—One having the characteristic of being continuous and changing smoothly over a given range, rather than switching suddenly between certain levels as with discrete signals

AND—An operation that yields a logic "1" output if all the inputs are "1" and a logic (0) if any of the inputs are (0)

Apparatus—The set of control devices used to accomplish the intended control functions

Arithmetic Capability—The ability to perform such math functions as addition, subtraction, multiplication, division, square roots, etc. A given controller may have some or all of these functions

Auxiliary Contacts—Contacts in addition to the main circuit contacts in a switching device. They function with the movement of the main circuit contacts

Auxiliary Device—Any electrical device other than motors and motor starters necessary to fully operate the machine or equipment

Available Short-Circuit Current—The maximum short-circuit current that can flow in an unprotected circuit

Binary Coded Decimal—A binary number system in which each decimal digit from 0 to 9 is represented by four binary digits (bits). The four positions have a weighted value of 1, 2, 4, and 8 respectively, starting with the least significant bit. A thumbwheel switch is a BCD device and when connected to a programmable controller, each decade requires four wires. For example: Decimal 9 = 1001 BCD

Binary Number System—A number system that uses two numerals, (Binary digits) "0" and "1". Each digit position for a binary number has a place value of 1, 2, 4, 8, 16, 32, 64, 128 and so on, beginning with the least significant (right hand) digit. It is called base 2

Binary Word—A related grouping of ones and zeros having coded meaning assigned by position or as a group; has some numerical value. 10101101 is an eight bit binary word, in which each bit could have a coded significance or as a group represent the number 173 in decimal

Bit—One binary digit. The smallest unit of binary information. Bit is an abbreviation for BInary digiT. A bit can have a value of 1 or 0

Block Diagram—A diagram which shows the relationship of separate sub-units (blocks) in the control system

Bonding—The permanent joining of metallic parts to form an electrical conductive path which will assure electrical continuity and the capacity to conduct safely any current likely to be imposed (NFPA 79-1984, National Electrical Code)

Branch Circuit—That portion of a wiring system extending beyond the final overcurrent device protecting the circuit. (A device not approved for branch circuit protection, such as a thermal cutout or motor overload protective device is not considered as the overcurrent device protecting the circuit)

Breakdown Voltage—The voltage at which a disruptive discharge takes place, either through or over the surface of insulation

Breaker—An abbreviated name for the circuit breaker

Bus—Power distribution conductors

Byte—A group of adjacent bits usually operated upon as a unit, such as when moving data to and from the memory. One byte consists of eight bits

Capacitor—Two conductors separated by an insulator

Cassette Recorder—A peripheral device for transferring information between the PC memory and magnetic tape. In the record mode it is used to make a permanent record of a program existing in the processor memory. In the playback mode it is used to enter a previously recorded program into the processor memory

Cassette Tape—A magnetic recording tape permanently enclosed in protective housing

Cathode Ray Tube (CRT)—A vacuum tube with a viewing screen as an integral part of its envelope

Central Processing Unit (CPU)—That part of the programmable controller that governs systems activities, including interpretation and execution of programmed instructions. In general, the CPU consists of a logic unit, timing and control circuitry, program counter, address stack and an instruction register

Character—One symbol of a set of elementary symbols such as a letter of the alphabet or a decimal number. Characters can be expressed in many binary codes

Chassis—A sheet metal box, frame or simple plate on which electronic components and their associated circuitry can be mounted

Chip—A very small piece of semiconductor material on which electronic components are formed. Chips are generally made of silicon and are generally less than one-quarter inch square and one thousandth inch thick

Circuit Breaker—A device which opens and closes a circuit by non-automatic means or which opens the circuit automatically on a predetermined overload of current, without damage to itself when properly applied within its rating

Circuit Interrupter—A non-automatic, manually operated device designed to open, under abnormal conditions, a current-carrying circuit without damage to itself

CMOS—An abbreviation for complimentary metal oxide semiconductor. A family of very low power, high-speed integrated circuits

Combination Starter—A magnetic motor starter which has a manually operated disconnecting device built into the same enclosure with the motor starter. Protection is always added. A control transformer may be added to provide 120 volts for control. START-STOP pushbuttons and a pilot light may be added in the enclosure door

Command—In data communication, an instruction represented in the control field or a frame and transmitted by the primary device. It causes the addressed secondary to execute a specific data link control function

Compartment—A space within the base, frame or column of the equipment

Compatability—The ability of various specified units to replace one another with little or no reduction in capability

Complement—A logical operation that inverts a signal or bit. The complement of 1 is 0. The complement of 0 is 1

Component—*See* Device

Conductor—A substance that easily carries an electrical current

Conduit, Flexible Metal—A flexible raceway of circular cross-section specially constructed for the purpose of pulling in or withdrawing wires or cables after the conduit and its fittings are in place

Conduit, Flexible Non-Metallic—A flexible raceway of circular cross-section specially constructed (of non-metallic material) for the purpose of pulling in or withdrawing wires or cables after the conduit and fittings are in place

Conduit, Rigid Metal—A raceway specially constructed for the purpose of pulling in or withdrawing wires or cables after the conduit and fittings are in place. It is made of metal pipes of standard weight and thickness, permitting the cutting of standard threads

Contact—One of the conducting parts of a relay, switch or connector that are engaged or disengaged to open or close an electrical circuit. Also when considering software, it is the junction point that provides a complete path when closed

Contact Symbology—A set of symbols used to express the control program using conventional relay symbols

Contactor—A device for repeatedly establishing and interrupting an electrical power circuit

Continuity—A complete conductive path for an electrical current from one point to another in an electrical circuit

Continuous Rating—A rating which defines the substantially constant load which can be carried for an indefinite period of time

Control Circuit—The circuit of a control apparatus or system which carries the electrical signals directing the performance of the controller but does not carry the main power circuit

Control Circuit Transformer—A voltage transformer used to supply a voltage suitable for the operation of control devices

Control Circuit Voltage—The voltage provided for the operation of shunt coil magnetic devices

Control Compartment—A space within a base, frame or column of the machine used for mounting the control panel

Control Logic—The program. Control plan for a given system

Control Panel—*See* Panel

Control Station—*See* Pushbutton Station

Controller, Electric—A device (or group of devices) which serves to govern, in some predetermined manner, the electric power delivered to the apparatus to which the device is connected

CRT—The abbreviation for cathode ray tube which is an electronic display tube similar to the familiar TV picture tube

CRT Programmer—A programming device containing a cathode ray tube (CRT). This programming device is primarily used to create and monitor the control program. It can also be used to display data

Current-Limiting Fuse—A fuse which will limit both the magnitude and duration of current flow under short-circuit conditions

Delta Connection—Connection of a three-phase system so that the individual phase elements are connected across pairs of the three phase power leads. A-B, B-C, C-A

Derate—To reduce the current, voltage or power rating of a device to improve its reliability or to permit operation at high ambient temperatures

Device (Component)—An individual device used to execute a control function

Dielectric—The insulating material between metallic elements of any electrical or electronic component

Digital—The representation of numerical quantities by means of discrete numbers. It is possible to express in binary form all information stored, transferred or processed by dual state conditions; for example, ON-OFF, CLOSED-OPEN

Disconnect Switch (Motor Circuit Switch)—A switch intended for use in a motor branch circuit. It is rated in horsepower and is capable of interrupting the maximum operating overload current of a motor of the same rating at the same rated voltage

Disconnecting Means—A device which allows the current-carrying conductors of a circuit to be disconnected from their source of supply

Documentation—An orderly collection of recorded hardware and software information covering the control system. These records provide valuable reference data for installation, de-bugging and maintenance of the programmable controller

Downtime—The time when a system is not available for production due to required maintenance either scheduled or unscheduled

Drop-Out—The current, voltage or power value that will cause energized relay contacts to return to their normal deenergized condition

Dual-Element Fuse—Often confused with time delay. Dual element is a manufacturer's term describing fuse element construction

EAROM—Electrically alterable read only memory (Chips can have the stored program erased electrically)

EEPROM—Electrically erasable programmable read only memory (Provides permanent storage of the program and can be easily changed with the use of a manual programming unit or standard CRT)

Electrical Equipment—The electromagnetic, electronic and static apparatus as well as the more common electrical devices

Electrical Optical Isolator—A device which couples input to output using a light source and detector in the same package. It is used to provide electrical isolation between input circuitry and output circuitry

Electrical-Mechanical—A term applied to any device in which electrical energy is used to magnetically cause mechanical movement

Electrical System—An organized arrangement of all electrical and electromechanical components and devices in a way that will properly control the machine or industrial equipment

Electronic Control—Electronic, static, precision and associated electrical control equipment

Elementary (Schematic) Diagram—A diagram in which symbols and a plan of connections are used to illustrate in simple form the scheme of control

Enclosure—A case, box or structure surrounding the electrical equipment which protects it from contamination; the degree of tightness is usually specified. (Such as NEMA 12)

Energize—To apply electrical power

EPROM—Erasable programmable read only memory (PROM that can be erased with ultra-violet light and then reprogrammed)

Exposed (As applied to electrically live parts)—Capable of being inadvertently touched or approached nearer than a safe distance by a person. It is applied to parts not suitably guarded, isolated or insulated (NFPA 70-1984 National Electric Code)

External Control Devices—Any control device mounted external to the control panel

Fail-Safe Operation—An electrical system designed so that the failure of any component in the system will prevent unsafe operation of the controlled equipment

Fault—An accidental condition in which a current path becomes available which bypasses the connected load

Feeder—The circuit conductors between the service equipment or the generator switchboard of an isolated plant and the branch circuit overcurrent device

Fuse—An overcurrent protective device containing a calibrated current-carrying member which melts and opens a circuit under specified overcurrent conditions

Fuse Element—A calibrated conductor which melts when subjected to excessive current. The element is enclosed by the fuse body and may be surrounded by an arc-quenching medium such as silica sand. The element is sometimes referred to as a link

Gate—A circuit having two or more input terminals and one output terminal, where an output is present when, and only when, the prescribed inputs are present

Grounded—Connected to earth or to a conducting body which serves in place of the earth

Grounded Circuit—A circuit in which one conductor or point (usually the neutral or neutral point of a transformer or generator windings) is intentionally grounded (earthed), either solidly or through a grounding device

Grounded Conductor—A conductor which carries no current under normal conditions. It serves to connect exposed metal surfaces to an earth ground to prevent hazards in case of breakdown between current-carrying parts and exposed surfaces. If insulated, the conductor is colored green, with or without a yellow stripe

Guarded—Covered, shielded, fenced, enclosed or otherwise protected by suitable covers or casings, barriers, rails, screens, mats or platforms to prevent contact or approach of persons or objects to a point of danger

Hard Copy—A printed document of what is stored in the memory

Hardware—Includes all the physical components of the programmable controller. This also includes the peripherals

Hardwired Logic—Logic control functions that are determined by the way devices are interconnected, as contrasted to programmable control in which logic control functions are programmable and easily changed

Hermetic Seal—A mechanical or physical closure that is impervious to moisture or gas, including air

Hertz (Hz)—Cycles per second; the unit measuring the frequency of alternating current

Hexadecimal Numbering System—A number system that uses the numerals 0 through 9 and the letters A through F. The system uses the base 16

Hysteresis (Magnetic)—A term applied to a certain magnetic property of iron. It causes a power loss when iron is magnetized and demagnetized due to the fact that the change in magnetic flux in the iron lags behind the change in mmf that causes it

Image Table—An area in the PC memory dedicated to the I/O data. Ones and zeros represent ON and OFF conditions respectively. During every I/O scan each input controls a bit in the input image table. Each output is controlled by a bit in the output image table

Inching—*See* Jogging

Input—Information sent to the processor from connected devices. They come in through the input interface

Input Device—Any connected equipment that will supply information to the central processing unit such as switches, pushbuttons, sensors or peripheral devices. Each type of input device has a unique interface to the processor

Inrush Current—In a solenoid or coil, the steady state current taken from the line with the armature blocked in the rated maximum open position

Instruction—A command or order that will cause a PC to perform one certain prescribed operation

Interconnecting Wire—A term referring to connections between sub-assemblies, panels, chassis and remotely mounted devices; does not necessarily apply to internal connections of these units

Interconnection Diagram—A diagram which shows all terminal blocks in the system, with each terminal identified

Interface—A circuit that permits communication between the central processing unit and a field input or output device

Interlock—A device actuated by the operation of another device with which it is directly associated. The interlock governs succeeding operations of the same or allied devices and may be either electrical or mechanical

Interrupting Capacity—The highest current at rated voltage that a device can interrupt

I/O—Abbreviation for input/output

I/O Address—A unique number assigned to each input and output. The address number is used when programming, monitoring or modifying a specific input or output

I/O Module—A plug-in type assembly that contains more than one input or output circuit. A module usually contains two or more identical circuits. For example 2, 4, 8, 16 circuits

I/O Update—The continuous process of revising each and every bit in the I/O tables, based on the latest results from reading the inputs and processing the outputs according to the control program

Isolated I/O—Input and output circuits that are electrically isolated from any and all other circuits of a module. Isolated I/O are designed to allow for connecting field devices that are powered from different sources to one module

Isolation Transformer—A transformer used to isolate one circuit from another

Jogging (Inching)—A quickly repeated closure of the circuit to start a motor from rest to accomplish small movements of the driven machine

Joint—A connection between two or more conductors

Kilohertz (kHz)—1000 Hertz

Kilowatts (kW)—1000 Watts

Ladder Diagram—An industry standard for representing relay-logic control systems

Ladder Element—Any one of the elements that can be used in a ladder diagram. The elements include relays, switches, timers, counters, etc.

Ladder Program—A type of control program that uses relay-equivalent contact symbols as instructions

Language—A set of symbols and rules for representing and communicating information among people or between people and machines. The method used to instruct a programmable device to perform various operations

Latch—A ladder program output instruction that retains its state even though the conditions which caused it to latch ON may go OFF. A latched output must be unlatched. A latched output will retain its last state (ON or OFF) if power is removed

Latching Relay—A relay that can be mechanically latched in a given position manually, or when operated by one element and released manually or by the operation of a second element

LED—Abbreviation for light-emitting diode. A semi-conductor diode, the junction of which emits light when passing a current in the forward direction

Legend Plate—A plate which identifies the function of operating controls, indicating lights, etc.

Limit Switch—A switch operated by some part or motion of a power driven machine or equipment to alter the associated electric circuit

Line Printer—A hard copy device that prints one line of information at a time

Location—In reference to memory, a storage position or register identified by a unique address

Locked Rotor Current—The steady-state current taken from the line with the rotor locked and with rated voltage (and rated frequency in the case of alternating current motors) applied to the motor

Logic—A process of solving complex problems through the repeated use of simple functions that can be either true or false (ON or OFF)

Logic Control Panel Layout—The physical position or arrangement of the devices on a chassis or panel

Logic Diagram—A diagram which shows the relationship of standard logic elements in a control system; it is not necessary to show internal detail of the logic elements

Logic Level—The voltage magnitude associated with signal pulses that represent ones and zeros in digital systems

Machine Language—A program written in binary form

Magnetic Device—A device operated by electromagnetic means

Magnetostrictive Material—The phenomenon of magnetostriction relates to the stresses and changes in dimensions produced in a material by magnetization and the inverse effect of changes in the magnetic properties produced by mechanical stresses. Nickel, alloys of nickel and iron, invar, nichrome and various other alloys of iron exhibit pronounced magnetostrictive effects

Main Memory—The block of data storage location connected directly to the CPU

Master Terminal Box—The main enclosure on the equipment containing terminal blocks for the purpose of terminating conductors from the control enclosure. Normally associated with equipment requiring a separately mounted control enclosure

Memory—That part of the programmable controller where data and instructions are stored either temporarily or semi-permanently. The control program is stored in the memory

Microprocessor—A digital electronics-logic package capable of performing the program control, data processing functions and execution of the central processing unit

Micro-Second—One millionth of a second

Milli-Second—One thousandth of a second

Mnemonic—Aiding or designed to aid the memory

Motor Circuit Switch—*See* Disconnect Switch

Motor Junction (Conduit) Box—An enclosure on a motor for the purpose of terminating a conduit run and joining motor to power conductors

NAND—A logical operation that yields a logic "1" output if any input is "0" and a logic "0" if all inputs are "1"

Noise—Random, unwanted electrical signals normally caused by radio waves or electrical or magnetic fields generated by one conductor and picked up by another

Nominal Voltage—The utilization voltage (see the appropriate NEMA standard for device voltage ratings)

Nonvolatile Memory—A type of memory whose contents are not lost or disturbed if operating power is lost

NOR—A logical operation that yields a logic "1" output if all inputs are "0" and a logic "0" output if any input is "1"

Normally Closed, Normally Open—When applied to a magnetically operating switch device such as a contactor or relay or to the contacts thereof, these terms signify the position taken when the operating magnet is deenergized. The terms apply only to nonlatching types of devices

NOT—A logical operation that yields a logic "1" at the output if a logic "0" is entered at the input and a logic "0" at the output if a logic "1" is entered at the input

Octal Numbering System—A number system that uses eight numeral digits (0 through 7). Base 8

Ohm (Ω)—Unit of electrical resistance

Operating Floor—A floor or platform used by the operator under normal operating conditions

Operating Overload—The overcurrent to which electrical apparatus is subjected under normal operating conditions; such overloads are currents that may persist for a very short time only, usually a matter of seconds

Operator's Control Station—*See* Pushbutton Station

Optical Coupler—A device that couples signals from one circuit to another by means of electromagnetic radiation, usually visable or infrared

Optical Isolation—Electrical separation of two circuits with the use of an optical coupler

OR—A logical operation that yields a logic "1" output if one or any number of inputs is "1" and a logic "0" if all inputs are "0"

Oscilloscope—An instrument used to visually show voltage or current waveforms or other electrical phenomena, either repetitive or transient

Outline Drawing—A drawing which shows approximate overall shape with no detail

Output (PC)—Information sent from the processor to a connected device through an interface

Output Device (PC)—Any connected equipment that will receive information or instructions from the CPU. This may consist of solenoids, motors, lights, etc.

Overcurrent—Current in an electrical circuit which causes excessive or dangerous temperature in the conductor or conductor insulation

Overcurrent Protective Device—A device which operates on excessive current and which causes and maintains the interruption of power in the circuit

Overlapping Contacts—Combinations of two sets of contacts actuated by a common means; each set closes in one of two positions and is arranged so that its contacts open after the contacts of the other set have been closed (NEMA IC-1)

Overload—Operation of equipment in excess of normal full-load rating or a conductor in excess of rated ampacity, which, when it persists for a sufficient length of time would cause damage or overheating

Overload Relay—A device which provides overload protection for the electrical equipment

Panel—A sub-plate upon which the control devices are mounted

Panel Layout—The physical position or arrangement of the components on a panel or chassis

Parallel Circuit—A circuit in which two or more of the connected components or contact symbols in a ladder diagram are connected to the same set of terminals, so that current may flow through all the branches. This is in contrast to a series circuit where the parts are connected end to end so that current flow has only one path

Pendant (Station)—A push-button station suspended from overhead, connected by means of flexible cord or conduit but supported by a separate cable

Plugging—A control function which provides braking by reversing the motor line voltage polarity or phase sequence so that the motor develops a counter torque which exerts a retarding force (NEMA IC-1)

Plug-In Device—A plug arranged so that it may be inserted in its receptacle only in a predetermined position

Potting—A method of securing a component or group of components by encapsulation

Power Factor—A term used to describe the ratio between apparent power (volt-amperes) and actual or true power (watts). The value is always unity or less than unity

Power Supply—The unit that supplies the necessary voltage and current to the system circuitry

Precision Device—A device which operates within prescribed limits and consistently repeats operations within those limits

Pressure Connector—A conductor terminal applied with pressure to make the connection mechanically and electrically secure

Processor—*See* CPU

Program—A planned set of instructions stored in a memory and executed in an orderly fashion by the central processing unit

Programming Device—A device for inserting the control program into the memory. The programming device is also used to make changes to the stored program

PROM—Programmable read only memory. It can be programmed once and cannot be altered after that

Push-button Station—A unit assembly of one or more externally operable push-button switches. It sometimes includes other pilot devices such as indicating lights and selector switches in a suitable enclosure

Raceway—Any channel designed expressly for and used solely for the purpose of holding wires, cables or bus bars

RAM—Random access memory. Referred to as read/write as it can be written into as well as read from

Read/Write Memory—A type of memory that can be read from or written to. Can be altered quickly and easily by writing over the part to be changed or inserting a new part to be added

Readily Accessible—Capable of being reached quickly for operation, renewal or inspection without a worker having to climb over or to remove obstacles or use a portable ladder, etc.

Rejection Fuse—A current-limiting fuse with high interrupting rating and with unique dimension or mounting provisions

Relay—A device which is operative by a variation in the conditions of one electric circuit to affect the operation of other devices in the same or another electric circuit

ROM—Read only memory. A type of memory that permanently stores information

Rung—A ladder-program term that refers to the programmed instructions that drive one output

Scan Time—The time required to read all the inputs, execute the control program and update local and remote I/O

Schematic Circuit—*See* Elementary Diagram

SCR—Silicon controlled rectifier. A semiconductor device that functions as an electrically controlled switch for dc loads

Semiconductor—A device which can function either as a conductor or non-conductor, depending on the polarity of the applied voltage, such as a rectifier or transistor which has a variable conductance depending on the control signal applied

Semiconductor Fuse—An extremely fast-acting fuse intended for the protection of power semiconductors. Sometimes referred to as a rectifier fuse

Sequence of Operations—A detailed written description of the order in which electrical devices and other parts of the equipment should function

Series Circuit—A circuit in which all the components or contact symbols are connected end-to-end. All must be closed to permit current flow

Shielded Cable—A single- or multiple-conductor cable surrounded by a separate conductor (shield) in order to minimize the effects of other electrical circuits

Short-Time Rating—A rating that defines the load that can be carried for a short definitely specified time with the machine, apparatus or device being at approximately room temperature at the time the load is applied

Software—Any written documents associated with the system hardware. This can be the stored program or instructions

Solenoid—An electromagnet which has an energized coil, approximately cylindrical in form and an armature whose motion is reciprocating within and along the axis of the coil

Solid State—Circuitry designed using only integrated circuits, transistors, diodes, etc.

Starter—An electric controller which accelerates a motor from rest to normal speed. (A device for starting a motor in either direction of rotation includes the additional function of reversing and should be designated a controller)

Static Device—As associated with electronic and other control or information handling circuits, a device with switching functions that has no moving parts

Status—The condition or state of a device; for example is it ON or OFF

Stepping Relay (Switch)—A multiposition relay in which wiper contacts mate with successive sets of fixed contacts in a series of steps, moving from one step to the next in successive operations of the relay

Subassembly—An assembly of electrical or electronic components, mounted on a panel or chassis, which forms a functional unit by itself

Subplate—A rigid metal panel on which control devices can be mounted and wired

Swingout Panel—A panel which is hinge mounted in such a way that the back of the panel may be made accessible from the front of the machine

Symbol—A widely accepted sign, mark or drawing which represents an electrical device or component thereof

Temperature Controller—A control device responsive to temperature

Terminal—A point of connection in an electrical circuit

Terminal Block—An insulating base or slab equipped with one or more terminal connectors for the purpose of making electrical connections

Termination—The load connected to the output end of a transmission line and provision for ending a transmission line and connecting to a bus bar or other terminating device

Three Phase—Three different alternating currents or voltages, 120 degrees out of phase with each other

Tie Point—A distribution point in circuit wiring other than a terminal connection where junction of leads are made

Time Delay Fuse—A fuse which will carry an overcurrent of a specified magnitude for a minimum specified time without opening. The current/time requirements are defined in the UL 198 fuse standards

Torque—The turning effect about an axis; it is measured in foot-pounds or inch-ounces and is equal to the product of the length of the arm and the force available at the end of the arm

Transient (Transient Phenomena)—Rapidly changing action occurring in a circuit during the interval between closing of a switch and settling to steady state condition or any other temporary actions occurring after some change in a circuit or its constants

Triac—A semiconductor device that functions as an electrically controlled switch for ac loads

Truth Table—A table listing that shows the state of a given output as a function of all possible input combinations

TTL—Abbreviation for transistor-transistor logic. A semiconductor logic family in which the basic logic element is a multiple-emitter transistor. This family of devices is characterized by high speed and medium power dissipation.

Undervoltage Protection—The effect of a device operative upon the reduction or failure of voltage which causes and maintains the interruption of power to the main circuit

Undervoltage Release—The effect of a device operative upon the reduction or failure of voltage which causes the interruption of power to the main circuit; voltage will return to the device when nominal voltage is re-established

User Memory—The memory where the application control program is stored

Ventilated—Provided with a means to permit circulation of air sufficient to remove excess heat, fumes or vapors (NFPA 70-1984, National Electric Code)

Viscosity—The property of a body in which, when flow occurs inside, forces a rise in such a direction as to oppose the flow

Volatile Memory—A memory whose contents are irretrievable when operating power is lost

Volt (V)—The unit of electrical pressure or potential

Watt (W)—The unit of electrical power

Wheatstone Bridge—A circuit employing four arms, in which the resistance of one unknown arm may be determined as a function of the remaining three arms which have known values

Wireway—A sheet metal trough with a hinged cover for housing and protecting electrical conductors and cable, in which conductors are laid in place after the wireway has been installed as a complete system

Wobble Stick—A rod extended from a pendant station which operates the STOP contacts; it functions when pushed in any direction

Word—The unit number of binary digits (bits) operated upon at any time by the central processing unit when it is performing an instruction or operating on data

Write—The process of putting information into a storage location

Wye Connection—A connection in a three-phase system in which one side of each of the three phases is connected to a common point or ground; the other side of each of the three phases is connected to the three-phase power line

APPENDIX A
SUMMARY OF
ELECTRICAL SYMBOLS

COMMON ELECTRICAL SYMBOLS AS USED IN THIS TEXTBOOK

CONDUCTORS CONNECTED	
CONDUCTORS NOT CONNECTED	
BATTERY	
THERMOCOUPLE	
RELAY COIL	CR
NO RELAY CONTACT	

NC RELAY CONTACT	
TIME-DELAY RELAY COIL	TR
INSTANTANEOUS CONTACT, NO	
INSTANTANEOUS CONTACT, NC	
NO TIMING CONTACT DELAY AFTER ENERGIZING	
NC TIMING CONTACT DELAY AFTER ENERGIZING	
NO TIMING CONTACT DELAY AFTER DEENERGIZING	
NC TIMING CONTACT DELAY AFTER DEENERGIZING	
TWO-CIRCUIT PUSH-BUTTON SWITCH	
VOLTMETER	V
AMMETER	A
SOLENOID COIL	

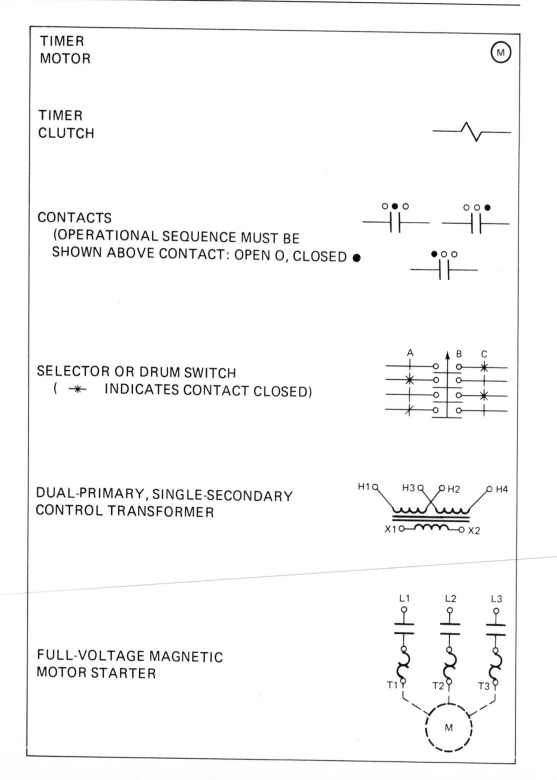

TIMER
MOTOR

TIMER
CLUTCH

CONTACTS
　(OPERATIONAL SEQUENCE MUST BE
　SHOWN ABOVE CONTACT: OPEN O, CLOSED ●

SELECTOR OR DRUM SWITCH
　(✳ INDICATES CONTACT CLOSED)

DUAL-PRIMARY, SINGLE-SECONDARY
CONTROL TRANSFORMER

FULL-VOLTAGE MAGNETIC
MOTOR STARTER

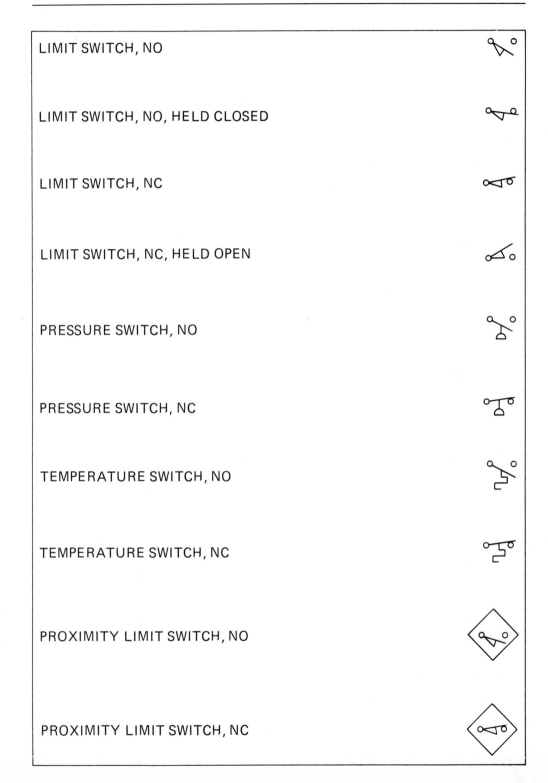

LIMIT SWITCH, NO

LIMIT SWITCH, NO, HELD CLOSED

LIMIT SWITCH, NC

LIMIT SWITCH, NC, HELD OPEN

PRESSURE SWITCH, NO

PRESSURE SWITCH, NC

TEMPERATURE SWITCH, NO

TEMPERATURE SWITCH, NC

PROXIMITY LIMIT SWITCH, NO

PROXIMITY LIMIT SWITCH, NC

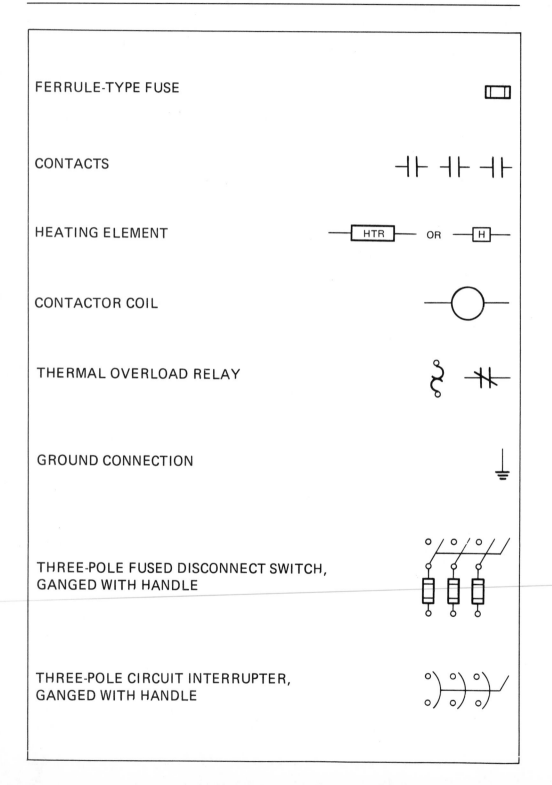

FERRULE-TYPE FUSE

CONTACTS

HEATING ELEMENT

CONTACTOR COIL

THERMAL OVERLOAD RELAY

GROUND CONNECTION

THREE-POLE FUSED DISCONNECT SWITCH, GANGED WITH HANDLE

THREE-POLE CIRCUIT INTERRUPTER, GANGED WITH HANDLE

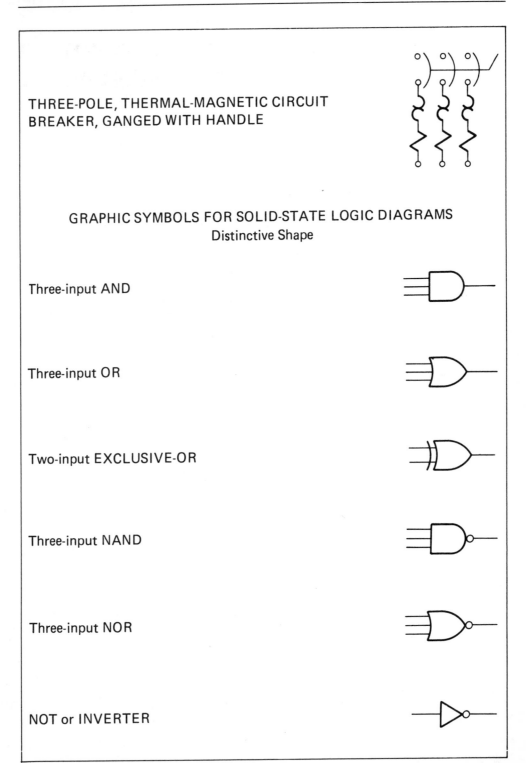

THREE-POLE, THERMAL-MAGNETIC CIRCUIT
BREAKER, GANGED WITH HANDLE

GRAPHIC SYMBOLS FOR SOLID-STATE LOGIC DIAGRAMS
Distinctive Shape

Three-input AND

Three-input OR

Two-input EXCLUSIVE-OR

Three-input NAND

Three-input NOR

NOT or INVERTER

APPENDIX B
UNITS OF MEASUREMENT

METRIC SYSTEM OF MEASUREMENT

Principal unit for length: Meter
Principal unit for capacity: Liter
Principal unit for weight: Gram

Prefixes Used for Subdivisions and Multiples

pico-(p) = 10^{-12} deka-(da) = 10^1
nano-(n) = 10^{-9} hecto-(h) = 10^2
micro-(μ) = 10^{-6} kilo-(k) = 10^3
milli-(m) = 10^{-3} mega-(M) = 10^6
centi-(c) = 10^{-2} giga-(G) = 10^9
deci-(d) = 10^{-1} tera-(T) = 10^{12}

Measures of Length

10 millimeters (mm) = 1 centimeter (cm)
10 centimeters (cm) = 1 decimeter (dm)
10 decimeters (dm) = 1 meter (m)
1000 meters (m) = 1 kilometer (km)

Square Measure

100 square millimeters (mm^2) = 1 square centimeter $(cm)^2$
100 square centimeters (cm^2) = 1 square decimeter (dm^2)
100 square decimeters (dm^2) = 1 square meter (m^2)

Cubic Measure

1000 cubic millimeters (mm^3) = 1 cubic centimeter (cm^3)
1000 cubic centimeters (cm^3) = 1 cubic decimeter (dm^3)
1000 cubic decimeters (dm^3) = 1 cubic meter (m^3)

Dry and Liquid Measure

10 milliliters (mL) = 1 centiliter (cL)
10 centiliters (cL) = 1 deciliter (dL)
10 deciliters (dL) = 1 liter (L)
100 liters (L)   = 1 hectoliter (hL)

1 liter = 1 dL^3 = the volume of 1 kilogram of pure water at a temperature of 4°C.

Measure of Weight

10 milligrams (mg) = 1 centigram (cg)
10 centigrams (cg) = 1 decigram (dg)
10 decigrams (dg)   = 1 gram (g)
10 grams (g)   = 1 dekagram (dag)
10 dekagrams (dag) = 1 hectogram (hg)
10 hectograms (hg) = 1 kilogram (kg)
1000 kilograms (kg) = 1 (metric) ton (t)

METRIC AND U.S. CUSTOMARY CONVERSIONS

Linear Measure

1 kilometer (km)   = 0.6214 mile
1 meter (m)   = 3.2808 feet
1 centimeter (cm) = 0.3937 inch
1 millimeter (mm) = 0.03937 inch

1 mile (mi) = 1.609 kilometer (km)
1 yard (yd) = 0.9144 meter (m)
1 foot (ft) = 0.3048 meter (m)
1 inch (in) = 2.54 centimeter (cm)

Square Measure

1 square kilometer (km^2) = 0.3861 square mile = 247.1 acres
1 hectare (ha^2) = 2.471 acres = 107.640 square feet
1 acre (a^2) = 0.0247 acre = 1076.4 square feet
1 square meter (m^2) = 10.764 square feet = 1.196 square yard
1 square centimeter (cm^2) = 0.155 square inch
1 square millimeter (mm^2) = 0.00155 square inch

1 square mile (mi^2) = 2.5899 square kilometers
1 acre (acre) = 0.4047 hectare = 40.47 ares
1 square yard (yd^2) = 0.836 square meter
1 square foot (ft^2) = 0.0929 square meter = 929 square centimeters
1 square inch (in^2) = 6.452 square centimeters = 645.2 square millimeters

Cubic Measure

1 cubic meter (m^3) = 35.314 cubic feet = 1.308 cubic yard
1 cubic meter (m^3) = 264.2 U.S. gallons
1 cubic centimeter (cm^3) = 0.061 cubic inch
1 liter (L) (cubic decimeter) = 0.0353 cubic foot = 61.023 cubic inches
1 liter (L) = 0.2642 U.S. gallon = 1.0567 U.S. quart

1 cubic yard (yd^3) = 0.7645 cubic meter
1 cubic foot (ft^3) = 0.02832 cubic meter = 28.317 liters
1 cubic inch (in^3) = 16.38716 cubic centimeters
1 U.S. gallon (gal) = 3.785 liters
1 U.S. quart (qt) = 0.946 liter

Weight

1 metric ton (t) = 0.9842 ton (of 2240 pounds) = 2204.6 pounds
1 kilogram (kg) = 2.2046 pounds = 35.274 ounces avoirdupois
1 gram (g) = 0.03215 ounce troy = 0.3527 ounce avoirdupois
1 gram (g) = 15.432 grains

1 ton (t) (of 2240 pounds) = 1.016 metric ton = 1016 kilograms
1 pound (lb) = 0.4536 kilogram = 453.6 grams
1 ounce (oz) avoirdupois = 28.35 grams
1 ounce (oz) troy = 31.103 grams
1 grain (gr) = 0.0648 gram

1 kilogram per square millimeter (kg/mm^2) = 1422.32 pounds per square inch
1 kilogram per square centimeter (kg/cm^2) = 14.223 pounds per square inch

1 kilogram-meter (kg·m) = 7.233 foot-pounds (ft-lb)
1 pound per square inch (psi) = 0.0703 kilogram per square centimeter
1 kilocalorie (kcal) = 3.968 British thermal units (Btu)

HEAT

Thermometer Scales

On the Fahrenheit (F) thermometer, the freezing point of water is marked at 32 degrees (°) on the scale, and the boiling point of water at atmospheric pressure is marked at 212°. The distance between these points is divided into 180°.

On the Celsius (C) thermometer, the freezing point of water is marked at 0° on the scale, and the boiling point of water is marked at 100°.

The following formulas are used for converting temperatures:

$$\text{Degrees Fahrenheit} = \frac{9 \times \text{degrees Celsius}}{5} + 32$$

$$\text{Degrees Celsius} = \frac{5 \times (\text{degrees Fahrenheit} - 32)}{9}$$

UNITS DERIVED FROM THE BASE ELECTRICAL UNIT (AMPERE)

	Measurement	Derived Unit of Measurement	Formula
Base Electrical Unit: Ampere (A)	Electrical potential	volt (V)	$V = W/A$
	Electrical resistance	ohm (Ω)	$\Omega = V/A$
	Electrical capacitance	farad (F)	$F = A \cdot (s/V)$
	Quantity of electricity	coulomb (C)	$C = A \cdot s$
	Electrical inductance	henry (H)	$H = Wb/A$
	Magnetic flux	weber (Wb)	$Wb = V \cdot s$

(s = seconds)

APPENDIX C
RULES OF THUMB FOR
ELECTRICAL MOTORS

Horsepower Versus Amperes

Here are some simple ways to take the mystery out of determining several important factors concerning electric motors. For example, using the following relationships, we come up with something useful. For three-phase motors we find that approximately:

1 hp - 575 V requires 1 A of current
1 hp - 460 V requires 1.25 A of current
1 hp - 230 V requires 2.50 A of current
1 hp - 2300 V requires 0.25 A of current

The key is 1 hp - 1 A on 575 V. All others then become a direct ratio of this figure, *i.e.*, 575/460 times 1.0 equals 1.25 A for 1 hp required on 460 V. Now, for example, with these facts at hand, you can say that a 150-hp motor on 230 V will require approximately 375 A. This is obtained as follows: 1 hp on 230 V requires 2.5 A per horsepower, so you simply multiply 2.5 times 150 hp.

Horsepower Revolutions Per Minute — Torque

Torque is simply a twisting force that causes rotation around a fixed point. For example, torque is something we all experience every time we pass through a revolving door. *Horsepower* is what is required when we pass through the door, because we are exercising torque at a certain revolution per minute (r/min). The faster we go through the revolving door, the more horsepower is required. In this

452

case, the torque remains the same. In relating this concept to motors, we use this rule of thumb:

1 hp at 1800 r/min delivers 3 ft-lb of torque
1 hp at 900 r/min delivers 6 ft-lb of torque

Using this simple rule, we can see that a 10-hp motor at 1800 r/min delivers 30 ft-lb of torque, and a 20-hp motor at 1800 r/min delivers 60 ft-lb of torque.

In another example, a 1-hp motor at 1200 r/min delivers 4.5 ft-lb of torque. Here is how to figure this. Multiply the torque at 1800 r/min by the ratio 1800/1200. Torque, then, is the inverse ratio of the speed. In other words, SPEED DOWN — TORQUE UP for the same horsepower.

A quick estimate for the torque of a 125-hp, 600-r/min motor can be figured by the following procedure. 125 hp times 3 ft-lb equal 375 ft-lb of torque for a 125-hp motor at 1800 r/min. Now, to convert to 600 r/min, multiply 375 times 1800/600 r/min or 1125 ft-lb.

This gives a rule that will enable you to quickly determine the torque a motor is capable of delivering, down to 10 r/min. Below this speed, other factors enter in that must be taken into consideration.

Shaft Size — Horsepower — Revolutions Per Minute

A number to remember here is 1150. This is easy to remember, and it stands for this: A 1-inch diameter shaft can transmit 1 hp at 50 r/min. As the shaft speed goes up, so does the horsepower, and by the same ratio. Therefore, if you double the speed, you double the horsepower capacity of the shaft. However, when you double the shaft diameter, the capacity of the shaft to transmit horsepower is increased 8 times. Thus, whatever the shaft size is in inches, cube it and multiply the resultant figure by the proper speed ratio, and the horsepower-transmitting ability of the shaft is determined. However, it is advisable to be conservative, so modify the results by 75% and use this resulting figure. To express this in a formula, use the following:

$$\text{Shaft horsepower} = \frac{(\text{Shaft diameter in inches}) \times \text{r/min}}{50}$$

APPENDIX D
ELECTRICAL FORMULAS

ELECTRICAL FORMULAS

1. Synchronous speed, frequency, and number of poles of ac motors and generators

$$r/min = \frac{120 \times f}{P} \qquad f = \frac{P \times r/min}{120} \qquad P = \frac{120 \times f}{r/min}$$

where: r/min = revolutions per minute
f = frequency, in cycles per second
P = number of poles

2. Power in ac and dc circuits

To Find	Alternating Current			Direct Current
	Three Phase	Two Phase *(Four Wire)	Single Phase	
Amperes when hp is known	$I = \dfrac{746 \times hp}{1.73 \times E \times Eff \times PF}$	$I = \dfrac{746 \times hp}{2 \times E \times Eff \times PF}$	$I = \dfrac{746 \times hp}{E \times Eff \times PF}$	$I = \dfrac{746 \times hp}{E \times Eff}$
Amperes when kW is known	$I = \dfrac{1000 \times kW}{1.73 \times E \times PF}$	$I = \dfrac{1000 \times kW}{2 \times E \times PF}$	$I = \dfrac{1000 \times kW}{E \times PF}$	$I = \dfrac{1000 \times kW}{E}$
Amperes when kVA is known	$I = \dfrac{1000 \times kVA}{1.73 \times E}$	$I = \dfrac{1000 \times kVA}{2 \times E}$	$I = \dfrac{1000 \times kVA}{E}$	—
Kilowatts input	$kW = \dfrac{1.73 \times E \times I \times PF}{1000}$	$kW = \dfrac{2 \times E \times I \times PF}{1000}$	$kW = \dfrac{E \times I \times PF}{1000}$	$kW = \dfrac{E \times I}{1000}$
Kilovolt-Amperes	$kVA = \dfrac{1.73 \times E \times I}{1000}$	$kVA = \dfrac{2 \times E \times I}{1000}$	$kVA = \dfrac{E \times I}{1000}$	—
Horsepower Output	$hp = \dfrac{1.73 \times E \times I \times Eff \times PF}{746}$	$hp = \dfrac{2 \times E \times I \times Eff \times PF}{746}$	$hp = \dfrac{E \times I \times Eff \times PF}{746}$	$hp = \dfrac{E \times I \times Eff}{746}$

I = Amperes
E = Volts
Eff = Efficiency, in Decimals
PF = Power Factor, in Decimals

kW = Kilowatts
kVA = Kilovolt-amperes
hp = Horsepower Output

*For two-phase, three-wire balanced circuits, the amperes in common conductor = 1.41 times that in either of the other two.

APPENDIX E
TABLES

It is recommended that these tables be used only for general information and reference. In specific industrial applications, the current and applicable codes and standards should be consulted. The tables that follow are reproduced from the NFPA 79 ELECTRICAL STANDARD FOR INDUSTRIAL MACHINERY 1985 through the courtesy of the NATIONAL FIRE PROTECTION ASSOCIATION INC.

Diagram 6-1   Protection of Machine Electrical Circuits
TYPICAL DIAGRAMS—CONSULT TEXT

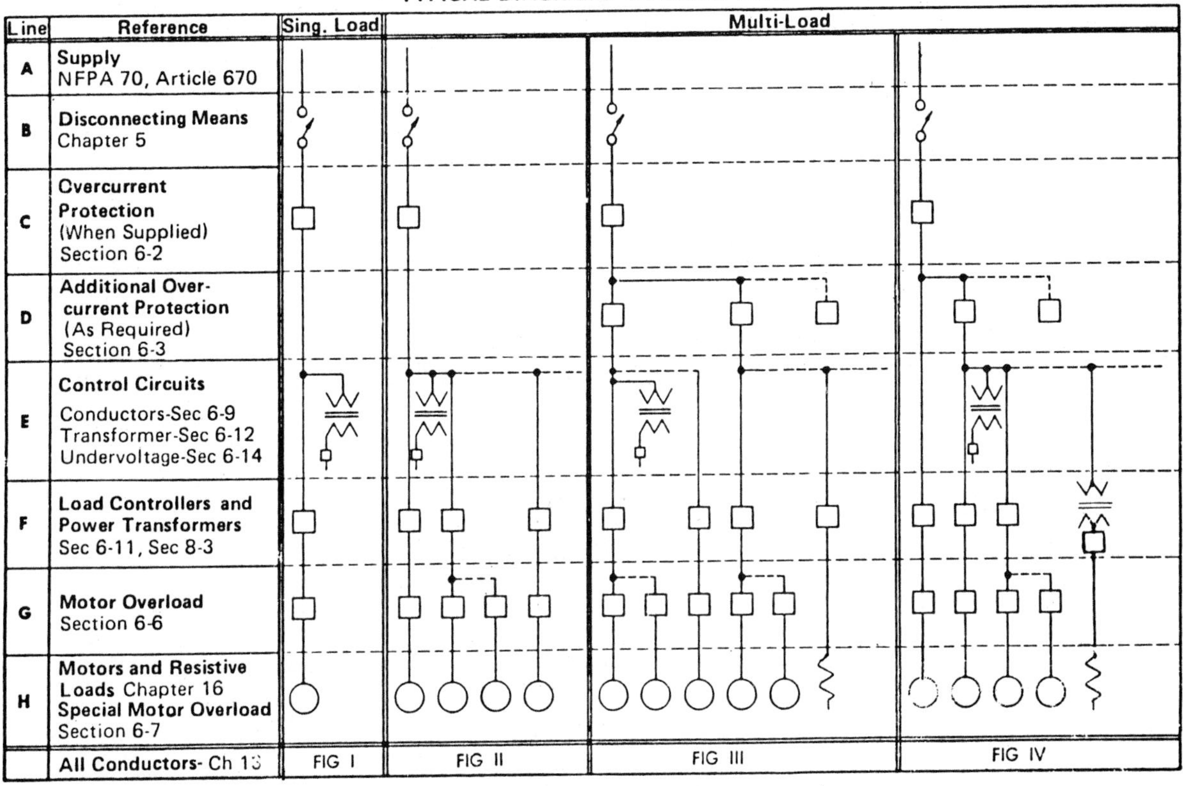

Line	Reference	Sing. Load	Multi-Load		
A	Supply NFPA 70, Article 670				
B	Disconnecting Means Chapter 5				
C	Overcurrent Protection (When Supplied) Section 6-2				
D	Additional Over-current Protection (As Required) Section 6-3				
E	Control Circuits Conductors-Sec 6-9 Transformer-Sec 6-12 Undervoltage-Sec 6-14				
F	Load Controllers and Power Transformers Sec 6-11, Sec 8-3				
G	Motor Overload Section 6-6				
H	Motors and Resistive Loads Chapter 16 Special Motor Overload Section 6-7				
	All Conductors- Ch 13	FIG I	FIG II	FIG III	FIG IV

455

Table 6-5(a)   Maximum Rating or Setting of
Motor Branch-Circuit Short-Circuit Ground-Fault
Protective Devices

Type of Motor	Percent of Full-Load Current			
	Nontime Delay Fuse	Dual-Element (Time-Delay) Fuse	Instantaneous Trip Breaker	Inverse Time Breaker
Single-phase, all types				
No code letter	300	175	700	250
All ac single-phase and polyphase squirrel-cage and synchronous motors with full-voltage, resistor or reactor starting:				
No code letter	300	175	700	250
Code letter F to V	300	175	700	250
Code letter B to E	250	175	700	200
Code letter A	150	150	700	150
All ac squirrel-cage and synchronous motors with autotransformer starting:				
Not more than 30 amps				
No code letter	250	175	700	200
More than 30 amps				
No code letter	200	175	700	200
Code letter F to V	250	175	700	200
Code letter B to E	200	175	700	200
Code letter A	150	150	700	150
High-reactance squirrel-cage				
Not more than 30 amps				
No code letter	250	175	700	250
More than 30 amps				
No code letter	200	175	700	200
Wound-rotor — No code letter	150	150	700	150
Direct current (constant voltage)				
No more than 50 hp				
No code letter	150	150	250	150
More than 50 hp				
No code letter	150	150	175	150

NOTE: Rating or Setting for Individual Motor Circuit. The motor branch-circuit short-circuit and ground-fault protective device shall be capable of carrying the starting current of the motor. The required protection shall be considered as being obtained where the protective device has a rating or setting not exceeding the values given in the above table.

An instantaneous trip circuit breaker shall be used only if adjustable, if part of a combination controller having motor-running overload and also short-circuit and ground-fault protection in each conductor, and if the combination is especially identified.

Table 6-5(b) Relationship Between Conductor Size and Maximum
Rating or Setting of Short-Circuit Protective Device for Power Circuits

Conductor Size AWG	Max. Rating Non-Time Delay Fuse or Inverse Time Circuit Breaker	Time Delay or Dual Element Fuse
14	60	30
12	80	40
10	100	50
8	150	80
6	200	100
4	250	125
3	300	150
2	350	175
1	400	200
0	500	250
2/0	600	300
3/0	700	350
4/0	800	400

Table 6-6(c)   Running Overcurrent Units

Kind of Motor	Supply System	Number and Location of Overcurrent Units (such as trip coils, relays, or thermal cutouts)
1-phase ac or dc	2-wire, 1-phase ac or dc ungrounded	1 in either conductor
1-phase ac or dc	2-wire, 1-phase ac or dc, one conductor grounded	1 in ungrounded conductor
1-phase ac or dc	3-wire, 1-phase ac or dc, grounded-neutral	1 in either ungrounded conductor
3-phase ac	Any 3-phase	*3, one in each phase

*Exception: Unless protected by other approved means.

NOTE: For 2-phase power supply systems see the *National Electrical Code*, Section 430-37.

Table 6-12 Control Transformer Overcurrent Protection
(120 Volt Secondary)

Control Transformer Size, Volt-Amperes	Maximum Rating, Amperes
50	0.5
100	1.0
150	1.6
200	2.0
250	2.5
300	3.2
500	5
750	8
1000	10
1250	12
1500	15
2000	20
3000	30
5000	50

NOTE: For transformers larger than 5000 volt-amperes, the protective device rating shall be based on 125 percent of the secondary current rating of the transformer.

Table 8-3(a)
Horsepower Ratings for Three-Phase, Single-Speed Full Voltage Magnetic Controllers for Nonplugging and Nonjogging Duty

Size of Motor Controller	Service-Limit Current Rating Amperes*	Three-Phase Horsepower at		
		200 Volts	230 Volts	460/575 Volts
00	11	1½	1½	2
0	21	3	3	5
1	32	7½	7½	10
2	52	10	15	25
3	104	25	30	50
4	156	40	50	100
5	311	75	100	200
6	621	150	200	400
7	932	—	300	600
8	1400	—	450	900
9	2590	—	800	1600

Reference ANSI/NEMA ICS-2-1978, Table 2-321-1.
*The service-limit current ratings shown in Tables 8-3(a) and 8-3(c) represent the maximum rms current in amperes which the controller may be expected to carry for protracted periods in normal service.

Table 8-3(c)
Horsepower Ratings for Three-Phase, Single-Speed
Full Voltage Magnetic Controllers for
Special Duty Applications

Size of Controller	Continuous Current Rating* Amperes	Horsepower at 60 Hertz			Service-limit Current Rating** Amperes
		200 Volts	230 Volts	460 or 575 Volts	
0	18	1½	1½	2	21
1	27	3	3	5	32
2	45	7½	10	15	52
3	90	15	20	30	104
4	135	25	30	60	156
5	270	60	75	150	311
6	540	125	150	300	621
9	2250		800	1600	2590

Reference ANSI/NEMA ICS 2-1978, Table 2-321-3.

NOTE: Refer to ANSI/NEMA ICS 2-1978 for horsepower ratings of single-phase, reduced voltage, or multispeed motor controller application.

*The continuous-current ratings shown in Tables 8-3(a) and 8-3(c) represent the maximum rms current, in amperes, which the controller may be expected to carry continuously without exceeding the temperature rises permitted by Part ICS 1-109 of NEMA Standards Publication No. ICS 1.

**The service-limit current ratings shown in Tables 8-3(a) and 8-3(c) represent the maximum rms current, in amperes, which the controller may be expected to carry for protracted periods in normal service. At service-limit current ratings, temperature rises may exceed those obtained by testing the controller at its continuous current rating. The current rating of overload relays or the trip current of other motor protective devices used shall not exceed the service-limit current rating of the controller.

Table 11-1
Color Coding for Pushbuttons, Indicator (Pilot)
Lights, and Illuminated Pushbuttons

Color	Device Type	Typical Function	Examples
RED	Pushbutton	Emergency Stop, Stop, Off	Emergency Stop button, Master Stop Button, Stop of one or more motors.
	Pilot Light	Danger or alarm, Abnormal condition requiring immediate attention	Indication that a protective device has stopped the machine e.g., overload.
	Illuminated Pushbutton		Machine stalled because of overload, etc. (use of RED illuminated pushbutton shall not be permitted for emergency stop).
YELLOW (AMBER)	Pushbutton	Return, Emergency Return, Intervention-suppress abnormal conditions	Return of machine elements to safe position, override other functions previously selected. Avoid unwanted changes.
	Pilot Light	Attention, caution/marginal condition. Change or impending change of conditions	Automatic cycle or motors running; some value (pressure, temperature) is approaching its permissible limit; Ground fault indication. Overload which is permitted for a limited time.
	Illuminated Pushbutton	Attention or caution/Start of an operation intended to avoid dangerous conditions.	Some value (pressure, temperature) is approaching its permissible limit; pressing button to override other functions previously selected.
GREEN	Pushbutton	Start-On	General or Machine Start; Start of cycle or partial sequence; Start of one or more motors; Start of auxiliary sequence; Energize control circuits.
	Pilot Light	Machine Ready; Safety	Indication of safe condition or authorization to proceed. Machine ready for operation with all conditions normal or cycle complete and machine ready to be restarted.
	Illuminated Pushbutton	Machine or Unit ready for operation/Start or On	Start or On after authorization by light; Start of one or more motors for auxiliary functions. Start or energization of machine elements.

   For illuminated pushbuttons the function(s) of the light is separated from the function(s) of the button by a virgule (/).

Table 11-1 (continued)
Color Coding for Pushbuttons, Indicator (Pilot)
Lights, and Illuminated Pushbuttons

Color	Device Type	Typical Function	Examples
BLACK	Pushbutton	No specific function assigned	Shall be permitted to be used for any function except for buttons with the sole function of Stop or Off; Inching or jogging.
WHITE or **CLEAR**	Pushbutton	Any function not covered by the above	Control of auxiliary functions not directly related to the working cycles.
	Pilot Light	Normal Condition Confirmation	Normal pressure, temperature.
	Illuminated Pushbutton	Confirmation that a circuit has been energized or function or movement of the machine has been started/Start-On, or any preselection of a function.	Energizing of auxiliary function or circuit not related to the working cycle Start or preselection of direction of feed motion or speeds.
BLUE or **GRAY**	Pushbutton, Pilot Light, or Illuminated Pushbutton	Any function not covered by the above colors.	

For illuminated pushbuttons the function(s) of the light is separated from the function(s) of the button by a virgule (/).

Table 13-1(a)
Single Conductor Construction — Type MTW

Wire Size AWG MCM	Thickness of Insulation in Mils		Minimum Stranding	
	A	B	Nonflexing	Flexing
22	30	15	7[a]	[c]
20	30	15	10[a]	10[b]
18	30	15	16[a]	16[b]
16	30	15	19[a]	26[b]
14	30	15	19[a]	41[b]
12	30	15	19[a]	65[b]
10	30	20	19[a]	104[b]
8	45	30	19[a]	[c]
6	60	30	19[a]	[c]
4-2	60	40	19[a]	[c]
1-4/0	80	50	37[a] (19[d])	[c]
250-500	95	50	61[a] (37[d])	[c]

[a] ASTM designation B-8, Class C (1977).
[b] ASTM designation B-174, Class K (1976).
[c] Nonflexing construction shall be permitted for flexing service.
[d] Shall be permitted.

Table 13-1(c) Conductor Ampacity

Conductor Size AWG	Ampacity In		Conductor Size AWG or MCM	Ampacity In*	
	Cable or Raceway	Control Enclosure		Cable or Raceway	Control Enclosure
30		0.5	00	145	225
28		0.8	000	165	260
26		1	0000	195	300
24	2	2	250	215	340
22	3	3	300	240	375
20	5	5	350	260	420
18	7	7	400	280	455
16	10	10	500	320	515
14	15	20	600	355	575
12	20	25	700	385	630
10	30	40	750	400	655
8	40	55	800	410	680
6	55	80	900	435	730
4	70	105	1000	455	780
3	80	120			
2	95	140			
1	110	165			
0	125	195			

*Sizing of conductors in wiring harnesses or wiring channels shall be based on the ampacity for cables.

Table 13-1(f)   Maximum Conductor Size for Given Motor Controller Size*

Motor Controller Size	Maximum Conductor Size, AWG or MCM
00	14
0	10
1	8
2	4
3	0
4	000
5	500

*See ANSI/NEMA ICS 2-1978, Table 2, 110-1.

Table 13-3(a) Multiconductor Cable Stranding (Constant Flexing Service)

Conductor Size AWG	No. of Strands
18	41
16	65
14	41
12	65
10	105

Table 15-3(g)
Minimum Radius of Conduit Bends

Size of Conduit (In.)	Radius of Bend Done by Hand (In.)[1]	Radius of Bend Done by Machine (In.)[2]
½	4	4
¾	5	4½
1	6	5¼
1¼	8	7¼
1½	10	8¼
2	12	9½
2½	15	10½
3	18	13
3½	21	15
4	24	16
4½	27	20
5	30	24
6	36	30

For SI units: (Radius) 1 in. = 25.4 mm
NOTE 1: For field bends done by hand, the radius is measured to the inner edge of the bend.
NOTE 2: For a single operation (one shot) bending machine designed for the purpose, the radius is measured to the center line of the conduit.

Table 17-2(c)
Size of Grounding Conductors

Column "A", Amperes	Copper Conductor Size, AWG
10	16* or 18*
15	14, 16* or 18*
20	12, 14*, 16* or 18*
30	10
40	10
60	10
100	8
200	6
300	4
400	3
500	2
600	1
800	0
1000	2/0
1200	3/0
1600	4/0

*Permitted only in multiconductor cable where connected to portable or pendent equipment.

INDEX